AF388998

Historical Ethnobotany: Interpreting the Old Records

Editors

Renata Sõukand
Raivo Kalle

Basel • Beijing • Wuhan • Barcelona • Belgrade • Novi Sad • Cluj • Manchester

Editors
Renata Sõukand
Department of Environmental
Sciences, Informatics &
Statistics, Università Ca'
Foscari Venezia
Venice, Italy

Raivo Kalle
Department of Folkloristics,
Estonian Literary Museum
Tartu, Estonia

Editorial Office
MDPI
St. Alban-Anlage 66
4052 Basel, Switzerland

This is a reprint of articles from the Special Issue published online in the open access journal *Plants* (ISSN 2223-7747) (available at: https://www.mdpi.com/journal/plants/special_issues/historical_ethnobotany).

For citation purposes, cite each article independently as indicated on the article page online and as indicated below:

Lastname, A.A.; Lastname, B.B. Article Title. *Journal Name* **Year**, *Volume Number*, Page Range.

ISBN 978-3-0365-9399-9 (Hbk)
ISBN 978-3-0365-9398-2 (PDF)
doi.org/10.3390/books978-3-0365-9398-2

Contents

Editorial

Historical Ethnobotany: Interpreting the Old Records

Raivo Kalle [1,*] **and Renata Sõukand** [2]

[1]   Estonian Literary Museum, Vanemuise 42, 51003 Tartu, Estonia
[2]   Department of Environmental Sciences, Informatics and Statistics, Ca' Foscari University of Venice,
      Via Torino 155, Mestre, 30172 Venice, Italy; renata.soukand@unive.it
*   Correspondence: raivo.kalle@mail.ee or raivo.kalle@kirmus.ee

Citation: Kalle, R.; Sõukand, R.
Historical Ethnobotany: Interpreting
the Old Records. *Plants* **2023**, *12*, 3673.
https://doi.org/10.3390/plants12213673

Received: 16 October 2023
Accepted: 20 October 2023
Published: 25 October 2023

For centuries, knowledge about the use of plants has been collected, published, or simply left in archives. Today, however, we live in a world with an abundance of information, and modern scientific standards put pressure on us, as researchers, to collect more and more data continuously. In current ethnobotany, publishing research with new data collected during fieldwork is much faster and easier. Quiet time for the qualitative analysis of previously collected data may seem unimportant; however, this Special Issue aims to emphasize that authors should utilize previously collected or stored ethnobotanical information in their research and, as such, this practice could be considered pioneering. This practice would also aid in paying tribute to our colleagues who came before us and collected and valued these data. In addition, this knowledge is essential in understanding changes in the environment and plant use as well as attitudes towards plants more broadly. It is encouraging that so many research articles focus on the cultural significance of a particular plant. In the case of medicinal plants, a growing number of scholars have begun to understand that their cultural, historical, and religious significance is also important, in addition to the active substances obtained from these plants. This Special Issue also covers the topics of ethnoveterinary medicine and dyeing with plants, which are rarely addressed in European ethnobotany. Furthermore, this collection includes more unusual approaches to studying the use of plants, for example, their representation in early folk songs and paintings. Moreover, it is indeed a pleasure that colleagues who have not dealt with and/or studied ethnobotany before have contributed to this collection.

We arranged the current introduction on a diachronic scale, starting with those who utilized the earliest available sources.

Edelman et al. [1] and Dal Cero et al. [2] mapped the most distant past and presented knowledge that is up to 2000 years old. In their comprehensive review, Edelman et al. [1] detailed the use of small freshwater plants, known as duckweeds (Lemnaceae Martinov), from ancient times to the Middle Ages. In addition, they compared the uses of this group of plants in different civilizations—Chinese, Christian, Greek, Hebrew, Hindu, Japanese, Mayan, Muslim, and Roman. They found that the use of duckweeds was already geographically widespread in antiquity and that they were integrated into classical cultures in the Americas, Europe, the Middle East, and the Far East. Dal Cero et al. [2] reviewed medicinal plant use in Central Europe from the earliest available records. They found that of the same 102 medicinal plants circulating in herbals from ancient times, more than half have retained similar uses regardless of the changes in both medicine and technology that have taken place since this period. The value of their work is that they confirm the concept of the social validation of plant uses. Thus, traditional and long-standing medicinal plants form the basis for regulatory sources of traditional herbal use.

De Vahl et al. [3] studied the period that began with the Middle Ages, revealing that the first reports of medicinal uses of *Peucedanum ostruthium* (L.) Koch (Apiaceae), naturally occurring in the mountainous regions of central and southern Europe, date back to the 13th century in the Nordic countries. This species was first cultivated in Sweden in the 17th–19th centuries and was known as the primary drug used in ethnoveterinary medicine. Today, this species has been preserved in specific locations in Sweden, in former cultural areas, and has become a part of the country's biocultural heritage. The authors emphasize that during the reconstruction of former farm gardens in open-air museums, the culturally important species of the past should also be highlighted. Pinke et al. [4] examined the time period from the late Middle Ages (1578) to the present day. They reviewed the folk uses and cultural significance of three field weeds in Hungary—*Papaver rhoeas*, *Centaurea cyanus*, and *Delphinium consolida*. They found that these species were used for medicinal and ornamental purposes, in religious celebrations, and in children's games during the time period studied. In addition, they found that they play an essential role in folk art and folk poetry. The height of the cultural importance of field weeds was in the early 20th century. Since the decline of field weeds, beginning at the end of the 20th century, their cultural importance has drastically decreased. The general conclusion of the authors is that in addition to preserving the natural species diversity of fields, it is also necessary to consider the cultural importance of plants.

Jasprica et al. [5] studied how plants were depicted in Baroque art on the eastern coast of the Adriatic Sea in the modern era. It must be mentioned that their approach is quite innovative in the field of ethnobotany. They found that 23 different plant species were portrayed in art at the time, with 71% of them considered "exotic" species. The exotic species came from the Palearctic region (Eurasia) and the American continent. *Lilium candidum*, *Acanthus mollis*, and *Chrysanthemum* cf. *morifolium* were the most represented taxa in the paintings. The researchers believe that the plants represented in the art were chosen for their decorative appearance and symbolic importance. Milani et al. [6] travel back in time with their research to 18th century Europe and conclude their research in the present day. They compared the earliest manuscript record from Valle Imagna (Bergamo, Italy) with later sources and data collected in the area during the present time period. The value of this research is that the authors worked through a vast amount of literature. Their study revealed only a few overlaps between current and 18th century plant use—only 34 species overlapped out of the 200 species mentioned in the manuscript. The most significant change occurred in this valley in the 1960s–1970s when most of the population emigrated from the region. However, the general use of medicinal plants is fading, as the use of only 42 species was identified in recent fieldwork.

Altogether, five studies focus on the 19th century. Prakofjewa et al. [7] analyzed the folk use of medicinal plants from three early sources. The examined sources were all published in the territory of today's Baltic states in 1829, 1891, and 1895, and a total of 219 species were identified in these reports. The authors also found that although the three early sources describe plant uses that overlap geographically, they were still quite different, with only 14 species overlapping in all three sources and 27 others mentioned in two of the sources. This indicates high biocultural diversity and dependence on local plant taxa in the past. Comparing these data with the book published by the Greek physician Pedanius Dioscorides (AD 40–90) revealed that as many as 46% of the plants mentioned by two or three authors overlapped. Overall, the presence of plants mentioned by Dioscorides was 26%. Köhler et al. [8] analyzed available information on plants used for dyeing in an 1883 questionnaire. It was found that 74 species are used in present-day Poland, Ukraine, and Belarus, the most popular of which was the onion. The authors state that most of the plants mentioned were widely known dyeing plants. However, plant dyeing is practically forgotten in Poland today, and this article may contribute to the re-emerging tradition of plant dyeing in the region.

Kalle et al. [9] looked at the period of 1891 to 1893. Dr Mihkel Ostrov carried out one of the first collections solely focused on ethnopharmacology, using national newspapers to distribute appeals for data collection to the population at large. With this action, Ostrov can be considered as one of the first individuals to use citizen science in ethnomedical data collection. In addition to appeals, Ostrov gave his correspondents feedback through said newspapers and provided motivation to the population to continue collecting such data. Using such a method, Ostrov obtained one of the highest quality collections of his time. Sõukand and Kalle [10] based their study on reports on herbal medicine, collected from three parishes bordering Russia between 1888 and 1996, which are stored in archives. In total, one hundred and nineteen species were identified. The authors observed a great variety of plant names and significant plant heterogeneity, especially in the earlier archival sources. Archival sources also provide a good context for understanding the use of medicinal plants in the past. The authors also emphasize that appropriate research methods must be used when identifying plants in archival sources because mistakes can easily occur in plant identification in historical sources. Fišer [11] focused on a specific analysis of plant lore in the folk songs examined by ethnologist Karel Štrekelj between 1895 and 1912. Plants were mentioned in 14% of the songs of the time. Among the 93 species mentioned, there were a surprising number of cultivated and exotic species, while only 42% were local wild taxa. Therefore, folk songs are also important for evaluating the relationship between man and nature.

Mattalia et al. [12] took the 20th century as the basis of their research, comparing their data collected in the Trentino–South Tyrol region of northern Italy (in 2022) with data previously gathered in 1989. While 75% of the species overlapped in both studies, the introduction of "new" plants has already occurred through the media. They also point out how courses teach people about local plant use in the region and that such bottom-up initiatives should be encouraged more. However, comparing the two regions showed that medicinal plants were used more in the border region (South Tyrol). Pieroni et al. [13] examined, through a case study of the Maronite community residing in the small village of Kormakitis, Northern Cyprus, how people who have lived in the same cultural space in the Mediterranean region for centuries have adapted their plant use. For many centuries, the Maronite minority living in the area have adapted their wild vegetable foraging to the Greek majority through long-standing cultural exchange. However, what was documented was mainly the memory of historical use, while currently, wild vegetables are foraged by a very limited number of people.

Funding: R.S. was supported by the European Research Council, grant number 714874, and R.K. was supported by the Estonian Research Council, grant number STP37.

Data Availability Statement: Not applicable.

Acknowledgments: The editors thank Gina Wang for her professional and kind curation of this Special Issue.

Conflicts of Interest: The author declares no conflict of interest.

References

1. Edelman, M.; Appenroth, K.-J.; Sree, K.S.; Oyama, T. Ethnobotanical History: Duckweeds in Different Civilizations. *Plants* **2022**, *11*, 2124. [CrossRef] [PubMed]
2. Dal Cero, M.; Saller, R.; Leonti, M.; Weckerle, C.S. Trends of Medicinal Plant Use over the Last 2000 Years in Central Europe. *Plants* **2023**, *12*, 135. [CrossRef] [PubMed]
3. de Vahl, E.; Mattalia, G.; Svanberg, I. "Cow Healers Use It for Both Horses and Cattle": The Rise and Fall of the Ethnoveterinary Use of *Peucedanum ostruthium* (L.) Koch (fam. Apiaceae) in Sweden. *Plants* **2023**, *12*, 116. [CrossRef] [PubMed]
4. Pinke, G.; Kapcsándi, V.; Czúcz, B. Iconic Arable Weeds: The Significance of Corn Poppy (*Papaver rhoeas*), Cornflower (*Centaurea cyanus*), and Field Larkspur (*Delphinium consolida*) in Hungarian Ethnobotanical and Cultural Heritage. *Plants* **2023**, *12*, 84. [CrossRef] [PubMed]
5. Jasprica, N.; Lupis, V.B.; Dolina, K. Botanical Analysis of the Baroque Art on the Eastern Adriatic Coast, South Croatia. *Plants* **2023**, *12*, 2080. [CrossRef] [PubMed]
6. Milani, F.; Bottoni, M.; Bardelli, L.; Colombo, L.; Colombo, P.S.; Bruschi, P.; Giuliani, C.; Fico, G. Remnants from the Past: From an 18th Century Manuscript to 21st Century Ethnobotany in Valle Imagna (Bergamo, Italy). *Plants* **2023**, *12*, 2748. [CrossRef] [PubMed]
7. Prakofjewa, J.; Anegg, M.; Kalle, R.; Simanova, A.; Prüse, B.; Pieroni, A.; Sõukand, R. Diverse in Local, Overlapping in Official Medical Botany: Critical Analysis of Medicinal Plant Records from the Historic Regions of Livonia and Courland in Northeast Europe, 1829–1895. *Plants* **2022**, *11*, 1065. [CrossRef] [PubMed]
8. Köhler, P.; Bystry, A.; Łuczaj, Ł. Plants and Other Materials Used for Dyeing in the Present Territory of Poland, Belarus and Ukraine according to Rostafiński's Questionnaire from 1883. *Plants* **2023**, *12*, 1482. [CrossRef] [PubMed]

9.   Kalle, R.; Pieroni, A.; Svanberg, I.; Sõukand, R. Early Citizen Science Action in Ethnobotany: The Case of the Folk Medicine Collection of Dr. Mihkel Ostrov in the Territory of Present-Day Estonia, 1891–1893. *Plants* **2022**, *11*, 274. [CrossRef] [PubMed]

10.   Sõukand, R.; Kalle, R. The Appeal of Ethnobotanical Folklore Records: Medicinal Plant Use in Setomaa, Räpina and Vastseliina Parishes, Estonia (1888–1996). *Plants* **2022**, *11*, 2698. [CrossRef] [PubMed]

11.   Fišer, Ž. "I Climbed a Fig Tree, on an Apple Bashing Spree, Only Pears Fell Free": Economic, Symbolic and Intrinsic Values of Plants Occurring in Slovenian Folk Songs Collected by K. Štrekelj (1895–1912). *Plants* **2022**, *11*, 458. [CrossRef] [PubMed]

12.   Mattalia, G.; Graetz, F.; Harms, M.; Segor, A.; Tomarelli, A.; Kieser, V.; Zerbe, S.; Pieroni, A. Temporal Changes in the Use of Wild Medicinal Plants in Trentino–South Tyrol, Northern Italy. *Plants* **2023**, *12*, 2372. [CrossRef] [PubMed]

13.   Pieroni, A.; Sulaiman, N.; Polesny, Z.; Sõukand, R. From *Şxex* to *Chorta*: The Adaptation of Maronite Foraging Customs to the Greek Ones in Kormakitis, Northern Cyprus. *Plants* **2022**, *11*, 2693. [CrossRef] [PubMed]

Review

Ethnobotanical History: Duckweeds in Different Civilizations

Marvin Edelman [1,*], Klaus-Juergen Appenroth [2,*], K. Sowjanya Sree [3,*] and Tokitaka Oyama [4]

[1]   Department of Plant and Environmental Sciences, Weizmann Institute of Science, Rehovot 7610001, Israel
[2]   Plant Physiology, Matthias Schleiden Institute, University of Jena, 07743 Jena, Germany
[3]   Department of Environmental Science, Central University of Kerala, Periye 671320, India
[4]   Department of Botany, Graduate School of Science, Kyoto University, Kyoto 606-8502, Japan
*   Correspondence: marvin.edelman@weizmann.ac.il (M.E.); klaus.appenroth@uni-jena.de (K.-J.A.);
    ksowsree9@cukerala.ac.in (K.S.S.)

Abstract: This presentation examines the history of duckweeds in Chinese, Christian, Greek, Hebrew, Hindu, Japanese, Maya, Muslim, and Roman cultures and details the usage of these diminutive freshwater plants from ancient times through the Middle Ages. We find that duckweeds were widely distributed geographically already in antiquity and were integrated in classical cultures in the Americas, Europe, the Near East, and the Far East 2000 years ago. In ancient medicinal sources, duckweeds are encountered in procedures, concoctions, and incantations involving the reduction of high fever. In this regard, we discuss a potential case of ethnobotanical convergence between the Chinese Han and Classical Maya cultures. Duckweeds played a part in several ancient rituals. In one, the unsuitability of its roots to serve as a wick for Sabbath oil lamps. In another reference to its early use as human food during penitence. In a third, a prominent ingredient in a medicinal incantation, and in a fourth, as a crucial element in ritual body purifications. Unexpectedly, it emerged that in several ancient cultures, the floating duckweed plant featured prominently in the vernacular and religious poetry of the day.

Keywords: duckweed; ethnobotanical convergence; Hildegard von Bingen; Paul Emile Botta; Ritual of the Bacabs; Babylonian Talmud; Kurma Purana; Ono no Komachi; Ho Ching-ming

Citation: Edelman, M.; Appenroth, K.-J.; Sree, K.S.; Oyama, T. Ethnobotanical History: Duckweeds in Different Civilizations. *Plants* **2022**, *11*, 2124. https://doi.org/10.3390/plants11162124

Academic Editor: Laura Cornara

Received: 20 May 2022
Accepted: 9 August 2022
Published: 15 August 2022

Publisher's Note: MDPI stays neutral with regard to jurisdictional claims in published maps and institutional affiliations.

1. Introduction

Duckweeds (Lemnaceae Martinov) are a globally spread family of higher plants with greatly reduced anatomies that float in slow-moving waters, such as found at river or lake edges or in still ponds and pools. Although these small plants are full-fledged monocot angiosperms, they reproduce mainly by vegetative budding at rapid rates, forming floating mats of verdant green in their natural habitat [1] (Figure 1).

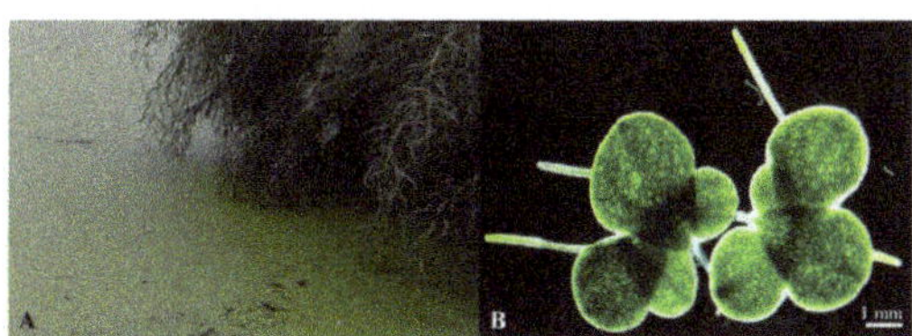

Figure 1. Duckweed in nature. (**A**) A natural population of the duckweed, *Lemna gibba* L., growing as a mat on the water surface. (**B**) Two colonies of *L. gibba* showing multiple generations of the plant vegetatively propagating. Left colony: The uppermost, large frond is the mother plant; its first daughter emerged from its left meristematic pocket and a second-generation daughter is in the process of emerging from the first daughter itself. Meanwhile, a daughter is also emerging from the mother frond's right meristematic pocket. In this species, fronds typically have one root, several of which are seen. Photographed with illumination from below to accentuate parenchymal-cell air pockets (the lighter color areas), which generate the characteristic floating property of *Lemna* plants.

The family is divided into five genera (*Spirodela* Schleid., *Lemna* L., *Landoltia* Les and Crawford, *Wolffia* Horkel ex Schleid. and *Wolffiella* Hegelm.) and has 36 species [2]. The first monograph dedicated to the duckweeds was published in 1839 [3], while biochemical studies of the family initiated around the 1950s [4]. Due to their miniature size, rapid growth rates, and ease of manipulation, interest in duckweeds both as a molecular-genetic research tool and in agrotechnology is now flourishing in the post-genomic era [5].

The paleontological record for the Lemnaceae is poor and is represented mainly by fossilized pollen grains from the Late Cretaceous period [6]. The paleolimnological record for Lemnaceae is also sparse, as duckweed fronds do not preserve well in lake and pond sediments [7]. Moreover, duckweeds rarely flower [1], resulting in a scarcity of pollen and seeds in sediment cores. Using, as an indicator, a highly significant association of *Lemna* with the epiphytic diatom *Lemnicola hungarica* (Grunow) Round et Basson (which does preserve well in sediments), it proved possible to detail the past presence of *Lemna minor* L. in a pond in England back a couple of centuries [8]. Aside from such ingenious attempts to go back in time, what is available to document duckweeds in antiquity are the surviving ancient manuscripts and texts.

One occasionally comes across indirectly documented reference to duckweeds from classical times [9]. Records from those times are mostly from religious or medicinal manuscripts, the main two repositories of writings handed down through the ages. We uncovered ethnobotanical referral to duckweeds in Chinese, Christian, Greek, Hebrew, Hindu, Japanese, Maya, Muslim and Roman cultures. Here we document these passages from the ancient literature. In some cases, we delve deeply, in others somewhat less so, but at a minimum, the exact reference and literature passage is provided along with its contextual background. The findings are grouped in sections: habitat, medicinal usage, ritual rites and poetic association. Figure 2 shows the geographical spread of the different cultures which are discussed here on a background of the global distribution of collected duckweed isolates [1].

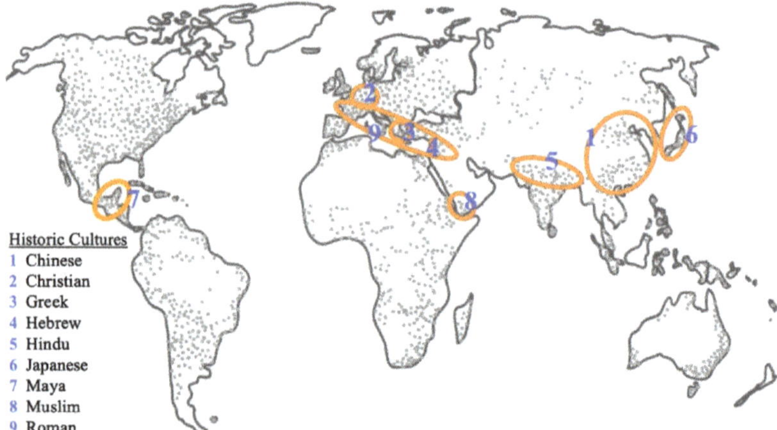

Figure 2. Map of the historic cultures investigated by us with ethnobotanical reference to duckweeds. The background map shows the multitude of locations of duckweed accessions [1], the great majority of which were collected in modern times. The circled areas show the locations of the classic cultures described here. While Theophrastus geographically pinpointed reference to duckweed to a local site in Greece [10], Dioscorides' descriptions were presented in the wider context of his travels in the Roman Empire [11]. The references to duckweed in Chinese culture are for the Later Han [12] and Ming [13] dynasties and in the Hindu religious texts to the Himalayan cedar forests [14]. The other areas encircled encompass the general regions or countries where the cultural references occurred.

2. Duckweeds in Ancient Cultures

Duckweed's Habitat

Duckweed in Ancient Greece: Theophrastus' Enquiry into Plants

It all starts with Theophrastus, the original classifier of plants. In his epic compendium *Peri phytōn historia* ("Enquiry in Plants" [10], also known as *Historia Plantarum* [15]), he used plant physiology, ecology, and agricultural methodology to arrange his treatise into ten books, nine of which are still extant. Theophrastus (371–287 BCE), born on the Greek island of Lesbos, composed his monumental work well over 2000 years ago. Rather than a completed work, the compendium is thought to represent his organized notes. He classified duckweed in terms of its aquatic habitat and coined the term *"lemna"* (water plant), which eventually became the base word for the family Lemnaceae. Mentored by Plato and then Aristotle, Theophrastus absorbed the cardinal importance of classification. His extensive botanical writings constitute a counterpart to Aristotle's zoological works. Theophrastus may have been the first botanist to systematically look for and record characteristic features which distinguish one plant from another; he is often considered the Father of Botany [10].

In 1483 CE, under the patronage of the Papacy, the nine existing books of *Peri phytōn historia* were translated into Latin (*Historia plantarum*) by Theodorus Gaza of Thessalonica (c. 1400–1475), a prominent Greek Humanist and translator of the works of Aristotle and Theophrastus for the Roman church [16,17]. While Gaza was esteemed by his contemporaries, the translation of Theophrastus' compendium was challenging, as botanical vocabulary in Latin is quite limited when compared to that in Greek. Gaza had both detractors (for his use of rare terms) and avid supporters for his translation [18].

In 1916, Sir Arthur Hort, Fellow at Trinity College, Cambridge, England, and amateur gardener and hybridist, produced an English translation of the first five books of Theophrastus' *Peri phytōn historia*, naming it "Enquiry in Plants" [10]. Theophrastus' compendium is a systematic study based on scientific observation, hence Hort's translation of *'historia'* to 'Enquiry' [19]. In his preface to the Enquiry [10], Hort pays tribute to the botanist Sir William Thiselton-Dyer, the third director of the Royal Botanical Gardens at Kew. So, too did the editors of Nature when Hort's translation came out in 1917 [20]. It was Thiselton-Dyer who provided the identification of Theophrastus' plants in the Enquiry and proofread Hort's entire translation. However, there were some who took issue with Thiselton-Dyer botanical expertise [21]. We have chosen to concentrate on Hort's translation and compare Gaza's translation to it.

Book IV deals with "trees and plants special to particular districts and positions". It starts off with a dissertation on the importance of position and climate, and continues with trees special to Egypt, shrubs special to Libya, herbs special to Asia, and the cold-seeking shrubs of Europe. The description then moves on to aquatic plants: of the Mediterranean Basin, the Atlantic Ocean, and the Persian Gulf, and continues with aquatic plants of rivers, marshes, and lakes, down to the resolution of a particular lake (Lake Copais) near Orchomenos, Greece. Book IV is completed with a discussion on the life spans of water plants (generally shorter) versus terrestrial ones (generally longer), weather-induced plant diseases, and the effects of human activities on plant life span.

Duckweed appears in Book IV, Section 10: "Plants peculiar to the lake of Orchomenos". (adapted from Hort [10]). Comments in brackets are Hort's; those in parentheses, are ours.

> "[10.2] Now in the lake near Orchomenos grow the following trees and woody plants: willow(,) goat-willow(,) water-lily reeds [both that used for making pipes and the other kind](,) galingale(,) phleos(,) bulrush; and also 'moon-flower'(,) **duckweed** and the plant called marestail: as for the plant called water-chickweed the greater part of it grows under water."

> "[10.3] Now of these most are familiar: the goat-willow(,) water-lily(,) 'moon-flower'(,) **duckweed** and marestail probably grow also elsewhere, but are called by different names. Of these we must speak. The goat-willow is of shrubby habit and like the chaste-tree: its leaf resembles that leaf in shape, but it is soft like that

of the apple, and downy. The bloom is like that of the abele, but smaller, and it bears no fruit. It grows chiefly on the floating islands; [for here too there are floating islands as in the marshes of Egypt, in Thesprotia, and in other lakes]..."

"[10.5] Of the plants of the lake they say that water-lily(,) sedge(,) and phleos bear fruit, and that of the sedge is black, and in size like that of the water-lily. The fruit of phleos is what is called the 'plume,' and it is used as a soap-powder... **Duckweed**(,) 'moon-flower' and marestail require further investigation."

The Enquiry places duckweeds among lake plants that "probably also grow elsewhere but are called by different names", acknowledging thereby, the spread of these plants geographically and awareness of them in those locations. The rareness and minute size of duckweed fruits (0.05 mm for *L. minor* [1]) may be behind Theophrastus' comment in paragraph 10.5 when discussing fruits "Duckweed...require further investigation".

In an archived copy of Gaza's *Historia Plantarum* published in 1552 [15], the relevant passage to Hort's sections [10.2] and [10.3] is a single, abbreviated account. The positionally comparable term to Hort's "duckweed" is *"icma"*, possibly one of the rare terms that were unavoidable in the translation of Theophrastus' botanically-rich Greek to Latin. Of interest is the appearance of the term *"lemna"* a few words further on in Gaza's sentence.

"... ad haec menififlora, icma, & quod ipnum appellant. Qnod enim lemna vocatur, altius mergitur in aqua." [15]

and in English,

"... menififlora, icma, and what they call ipnum. For what is called the lemna is immersed deeper in water." (translated from the Latin by Susanne Kochs).

However, when compared to Hort's translation

"... 'moon-flower'(,) duckweed and the plant called marestail: as for the plant called water-chickweed the greater part of it grows under water." [10]

it becomes evident that Theophrastus did not coin the term *"lemna"* specifically for duckweed (which, characteristically, floats on the water's surface) but rather as a general term for 'water plants' (in this case, one that "is immersed", or "grows", under water).

3. Medicinal Applications Involving Duckweeds in Ancient and Classical Sources

Duckweed, the archetype, and ever-present water plant, may have found its way into medicinal concoctions in ancient times more by simply "being there" than by strong medicinal benefit. Yet, a connecting thread may exist between sources, suggesting a remedial cooling effect on temperature-related maladies.

3.1. Duckweed in the Roman Empire: Dioscorides' Materia Medica

Pedianos Dioscorides (40–90 CE), born in Cilicia of the Roman Empire (present-day Turkey), was a physician/pharmacologist in the Roman army. He lived three centuries after Theophrastus and was a contemporary of the naturalist Pliny, the Elder. Dioscorides wrote a five-volume treatise in Greek, *Peri hules iatrikēs* (on Medical Material), commonly known in the Western world as *de Materia Medica*. As T.A. Osbaldeston, translator and editor of a modern English version, succinctly put it [11]: "Theophrastus was the scientific botanist; Pliny produced the systematic encyclopaedia of knowledge; and Dioscorides was merely a medical botanist". However, Dioscorides concentrated on the practical and pharmacopeial use of the plants he described. He was a hands-on botanist. In his dedication to his monumental work, he states, "I know many plants personally ... by questioning the local inhabitants about each type of plant, I will attempt a different classification ... I intend to assimilate things that are common knowledge and those that are somehow related so that the information will be exhaustive" [11]. As a result, Dioscorides' descriptions were often sufficient for identification, including methods of preparation, medicinal uses, and dosages. Remarkably, his *de Materia Medica* functioned as the core of Western pharmacopeia through to the 19th century.

Book 4 of *de Materia Medica* describes "other herbs and roots", among them, duckweed in Section 4.88.

Phakos epi ton telmaton (Adapted from a translation of the Greek by Susanne Kochs.)

Lens on swamps/stagnant water bodies: it is found on stagnant water; it is a moss with similarity to a lens; with its power it is cooling; against all phlegmon [diffuse inflammation], erisypelas [bacterial inflammation of the skin] and podagral [foot gout] it helps when it is applied alone or together with barley. It also seals hernias in children.

T.A. Osbaldeston identifies this entry as denoting duckweed and presents "lens" as describing *Spirodela polyrhiza* (L.) Schleid. and/or *L. minor* [11]. The phrase "a moss with similarity to a lens" may raise eyebrows; however, duckweeds have been compared to moss in other ancient texts [22], and mats of dense duckweed growth are termed "seed moss" by freshwater fishermen today in the American Lower Mississippi Alluvial Valley [23]. Dioscorides is the first to associate duckweed with the quality of "cooling" in a medicinal sense. While he then generalizes its application to "all phlegmon", in other cultures [12,24], cooling is defined more specifically in terms of alleviating a fever.

In Osbaldeston's English translation of *de Materia Medica* [11], an additional sentence is part of this section

"It is also called wild lens, or *epipteron*, the Romans call it *viperalis*, and some, *iceosmigdonos*."

This addition is lacking in the Latin translation by Janus Antonius Saracenus in 1598 [25]. As with many ancient manuscripts, the literally dozens of translations of *de Materia Medica* over the centuries doubtlessly introduced copying errors, notes, deletions and additions.

3.2. The Divine Farmer's Chinese Materia Medica Classic

The Divine Farmer's Materia Medica Classic (*Shen Nong Ben Cao Jing*) is considered the authoritative version of Chinese herbal medical literature [12]. It was first committed to writing in the later Han dynasty (c. 200 CE) by Tao Hong-jing. Tao's Classic underwent a convoluted, although not uncommon, evolution to reach its current form. The manuscript was lost but then carefully reconstructed from extant documents that had incorporated large tracts of the Classic into their own. Such behavior was common practice in the ancient and medieval world and without negative connotations. Many texts from the past would not be available to us today except for this practice [26]. Tao's Classic also underwent multiple versions and rearrangements of sections by various scribes through the ages. The English translation by Shou-zhong Yang in 1998 [12] is based on the Classic edited by Cao Yuan-yu in 1987 and published by the Shanghai Science and Technology Press, Shanghai. This version of the Classic is thought to be closer to the original than any other.

The Classic tends more toward the mystical (Daoist influence) than to the practical and systematic, often suggesting the usage of herbs to achieve immortality and supernatural abilities. There are three levels of medicinal plants in the Classic: Superior, Middle, and Inferior. Superior class plants are nontoxic and bestow longevity; they are good for a person's health. Middle-class plants modify temperament, and their prescription requires care. Inferior class plants are usually toxic and cannot be taken in large amounts or for prolonged periods without developing side effects. The plants in the Classic are also categorized as to flavors—sour, salty, sweet, bitter, or acrid. However, categorization by qualities of *qi* (literally, vapor or breath)—cold, hot, warm, and cool is a later addition to the Classic [12].

Duckweed (*Shui Ping*) is presented in the Classic as a Middle-class medicinal plant.

*"**Shui Ping** (water weed) is acrid and cold. It mainly treats fulminant heat and generalized itching, precipitates water qi, [helps] get over wine, promotes the growth of the beard and [head] hair, and quenches wasting thirst. Protracted taking may*

make the body light. Its other name is *Shui Hua* [water flower]. It grows in pools and swamps." (adapted from [12]; found just before comment 187)

Yang describes *Shui Ping / Shui Hua* as *Lemna* or *Spirodela*, adding that this medicinal is good for promoting sweating [12]. His comment emphasizes the effect of duckweed in treating explosive fever.

3.3. Duckweeds in Medieval Christian Europe

Knowledge of herbs and their use in medicine in the High Middle Ages is connected with the name St. Hildegard von Bingen (1098–1179 CE), author of *Causa et Curae* (Reasons and treatment of diseases), a medieval compilation of knowledge about plants, including the medicinal use of duckweeds. Hildegard was a nun and later an abbess [27]. she was officially promoted as Saint and Doctor of the Church on 10 May 2012 by Pope Benedict XVI.

Hildegard became a nun in the Benedictine cloister Disibodenberg in Rhineland-Palatinate, Germany already, at the age of fourteen. According to the basic rule of the Benedictines, *Ora et labora* (Latin terms for pray and work), she was responsible for the cloister garden. Following the order *Capitulare de villis* of Charles the Great from the year 812 CE, many cloisters established herb gardens and documented the knowledge of plants and their use. This way, Hildegard became familiar with plants. Later she was elected as Magistra, i.e., leader of the nun convent. She founded a new cloister, Rupertsberg, near Bingen at the river Rhine and became an abbess. This is where her name came from, Hildegard of Bingen. This new convent with the larger cloister herbal garden extended her prospects to learn more about plants and their use through her own practical experience. The second source of her knowledge was the rich cloister library, always a fixed part of each Benedictine cloister, including books about nature and medicine. It is assumed that Hildegard knew the book collection *Corpus Hippocraticum* from the fifth to second century BCE, which has part of texts from the physician Hippokrates of Kos of the physician school, Knidos. Furthermore, the oldest German medical book, *Lorscher Arzneibuch* (Engl. Lorscher Pharmacopoeia) from the 8th century, was for sure a part of the library. This book contains information about several hundred plants [27]. The third source of her knowledge was the far-reaching contacts that Hildegard could have due to her position as an abbess of a well-off Benedictine cloister. This way, she was in contact with contemporaneous scientific activities. As an example, she had intensive contact with a monastery in Salerno, south of Naples, Italy.

Popular knowledge about plants and their use in treating illnesses in medieval times was collected by women who normally did not know Latin. Therefore, their texts were not considered to be worth keeping and got lost over time. Hildegard of Bingen, however, was highly educated and wrote several books in Latin. One of her main works was called *Physica*, consisting of nine volumes, describing the scientific and medicinal properties of plants apart from giving information about animals and stones. The second main work is called *Causa et Curae* translated as Reasons and treatment of diseases [28]. In both books, the collected contemporary knowledge about plants was compiled, including the use of duckweed. Both books were originally compiled in the textbook *Liber subtilitatum diversarum rerum naturalum* (Book of diverse exact properties of all creations).

The following are a few remedies suggested by Hildegard von Bingen that make use of duckweeds. Hildegard recommended the application of an ointment against colic and described how to prepare this ointment using different plants. Her formula [28,29] is shown in Table 1.

The translator of *Causa et curae*, H. Schulz [28], assumed that the duckweed used was *L. minor*, which is most probably justified because it is the most common duckweed species in the area of the cloisters where Hildegard lived.

Table 1. Ingredients of the ointment against colic as recommended by Hildegard von Bingen.

Ingredient	Term Used by Hildegard	Plant Species	
		Latin Nomenclature	Plant Family
Feverfew	Mutterkraut	*Tanacetum parthenium* (L.) Sch.Bip.	Asteraceae
Sage	Salbei	*Salvia officinalis* L.	Lamiaceae
Zedoary	Zitwer	*Curcuma zedoaria* (Christm.) Roscoe	Zingiberaceae
Fennel	Fenchel	*Foeniculm vulgare* (L.) Mill.	Apiaceae
Duckweed	Wasserlinsen	*Lemna minor* L.	Lemnaceae
Erected cinquefoil	Tormentillwurzel	*Potentilla erecta* L. (Raeusch.)	Rosaceae
Charlock mustard	Senf	*Sinapis arvensis* L.	Brassicaceae
Burdock	Klette	*Arctium lappa* L.	Asteraceae

A second remedy, using duckweed, that Hildegard recommended was against precancerous indications and against colic, heart pain, and rheumatism for body detoxification and a weak immune system [30] (Table 2). This remedy was called "Wasserlinsenelixier" in German (duckweed elixir) and *Decoctum Lemnae cp.* in Latin. The abbreviation *"cp"*. stands most probably for *compositum*, indicating that other components apart from duckweeds are also involved. The protocol to prepare the elixir is pretty complicated [30,31].

Table 2. Ingredients of duckweed elixir as per the protocol of Hildegard von Bingen.

Ingredient (Quantity)	Term Used by Hildegard	Plant Species		Remarks
		Latin Nomenclature	Plant Family	
Component I: powder Ginger root powder (2.5 g) Cinnamon (10 g)	 Ingwerwurzel Zimtrindenpulver	 *Zingiber officinale* Roscoe *Cinnamomum verum* J.Presl	 Zingiberaceae Lauraceae	 Both powders were mixed
Component II: juice Sage juice from leaves (2 g) Fennel juice (3 g) Common tansy juice (2 g), without flowers, collected in spring	 Salbei Fenchelkrautsaft Rainfarnkrautsaft	 *Salvia officinalis* L. *Foeniculum vulgare* (L.) Mill: *Tanacetum vulgaris* L.	 Lamiaceae Apiaceae Asteraceae	All plants were homogenized, pressed out and filtered.
Component III: Honey wine 90 g Honey boiled in 1 L wine White pepper (1.2 g)	 Weisser Pfeffer	 *Piper nigrum* L.	 Piperaceae	All three components were mixed
Other components **Duckweed** (20 g) Erected cinquefoil (40 g) Charlock mustard (40 g) Cleavers (15 g)	 Wasserlinsen Blutwurz Ackersenf Labkraut	 *Lemna minor* L. *Potentilla erecta* L. (Raeusch.) *Sinapis arvensis* L. *Galium aparine* L. or *Galium verum* L.	 Lemnaceae Rosaceae Brassicaceae Rubiaceae	These other components were added to the three component mixture, mixed and filtered

This remedy is now available as "Wasserlinsenelixier" (duckweed elixir) by several producers and can be ordered via the Internet. It is recommended to activate metabolism and for detoxification of the human body. All these producers and shops stress that they strictly follow the protocols of Hildegard von Bingen.

4. Duckweeds in Ancient Religious Rituals

Rituals are essentially culture-dependent, leading to a spectrum of topics where duckweeds play a part. In one ancient text, we find duckweed (presumably its roots) in a discussion concerning its suitability as a wick for Sabbath oil lamps. In another, an ancient (the first?) reference to its use as human food in an act of asceticism. In a third, it features in a medicinal incantation, and in a fourth, in ritual purifications.

4.1. Duckweed in the Maya Civilization: Ritual of the Bacabs

The Maya civilization was, in its two periods, one of the foremost societies of Mesoamerica. The Maya Empire was centered in the tropical swamp lands of the Yucatan Peninsula (today encompassing Guatemala, Belize, and parts of Mexico, Honduras, and El Salvador). From discoveries at El Mirador [32], it appears that many of the cultural and architectural components of Maya society, such as massive stone temples and palaces built as stepped pyramids and embellished with multiple glyphs and inscriptions, were already in place in the Late Preclassic period (300 BCE to 150 CE), several centuries before the better-studied Classic period (250–900 CE). The Maya Empire famously and inexplicably collapsed twice, with most of the great stone cities abandoned at the termination of each period [33].

Ah-men (literally, he or she who knows) were the most important healers in Maya society during the Classic and Postclassic periods. They were shamans, thought of as intermediaries between the deities and the people, combining incantations to appease the gods while using their training and knowledge as herbalists to heal [34]. During the Spanish conquest of the Maya in the 16th century, almost all books and codices concerning Mayan gods, astronomy, and medicine were publicly burned by Franciscan missionaries in an effort to wipe out the local religious practices. A clandestine manuscript, discovered in the early 20th century, called the "Ritual of the Bacabs" is the only known surviving work containing the texts of the *ah-men* chants that accompanied medical treatments of the shamans. While the manuscript itself is from the Colonial Period, the chants use metaphors primarily associated with glyph inscriptions from the older Classic Period [35].

Duckweed (*Ixim ha*; literally, maize-water [35]) plays an important part in the incantation on manuscript pages 114–115 (Text 18). We considered three translations of this incantation into English [24,34,35]. Table 3 is adapted from [24], with annotations. The table's footnotes explain the ethnological meanings and render the text logical in modern terms.

Table 3. Duckweed as the symbol of a cooling plant in the Ritual of the Bicabs [a].

This is for cooling a high fever and for cooling a pox [b].
With the protecting shade of my foot, the protecting shade of my hand
      I cooled the pox.
Five [c] are my red [d] hailstones [e], my white hailstones, my black hailstones, my yellow hailstones.
      With them I cooled the pox.
Thirteen [c] are the layers of my red [d] liturgical vestment [f], my white liturgical vestment,
      my black liturgical vestment, my yellow liturgical vestment.
I seized the strength [g] of the pox.
      A black fan is my symbol when I seized the strength of the pox.
With me descends certainly my white h duckweed [i].
      I seized the strength of the pox.
With me descends my white [h] water lily.
      Then it happens that I seized the strength of the pox.
Soon I will do good with the protecting shade of my foot, the protecting shade of my hand.
Amen. [j]

---

[a] Ritual of the Bicabs, manuscript pages 114–115. Adapted and annotated from [24].
[b] Title of the incantation. The term "pox": alternatively, "eruption" [34]; "fire-pox" [35].
[c] Numbers 5 and 13 are significant in the complex, Classic-Maya 2-year calendar cycle. The first refers to a short, ominous period in the secular, agricultural, 365-day cycle; the second, to the number of 20-day months in the following sacred 260-day cycle [36].
[d] Colors are associated in the Classic Maya *materia medica* with the four cardinal directions of the world and linked to the journey of the sun deity (generator of light, time, heat, and the cardinal directions) through the sky. Red is associated with the east, where the sun rises; white with the north, from where the cooling winds of winter come; black with the west, where the sun fades and disappears; and yellow with the south, the bright broad-side of the sun [37].
[e] "Hailstones", representing coldness.
[f] "Liturgical vestment": alternatively, "dressing" [34]; "ornaments" [35].
[g] "Strength": alternatively, "*Kinam*" [35]; "force" [34].
[h] "White": representing the cooling winds of the north (see footnote "d").
[i] "Duckweed": *yxim ha* (literally, maize-water plant). It grows in the cool caves and sink holes of the Yucatan [35] and is proposed as *L. minor* or *W. brasiliensis* [24].
[j] "Amen": one of the few intrusions of Christian elements in the Ritual of the Bacabs, suggesting that Maya belief had not undergone many changes by 1779, when this Colonial period manuscript was committed to writing [24].

4.2. Duckweed in the Babylonian Talmud

The *Talmud* (from the Hebrew root 'to study') combines two ancient texts: the *Mishna*, the written version of Jewish oral law, compiled in the Land of Israel and completed around the year 200 CE, and the *Gemara*, the rabbinic interpretation of the Mishna, compiled in Babylonia (southeastern Mesopotamia between the Tigris and Euphrates rivers; present-day Iraq) and completed about 300 years later. It was edited thereafter by the *Savoraim* for an additional 40 to 187 years. The Babylonian Talmud is a monumental work consisting of over 5000 folio pages. It includes every imaginable topic, from the phases of the moon to financial investment strategy to how best to arise in the morning. Exhaustive analysis of dissenting opinions was the tool used by the Talmud for developing and then resolving issues. It operated at manifold levels, pitting multiple rabbinic sages over multiple generations and multiple locations, all against all, irrespective of time or space, to maximize discussion and arrive at a resolution. By banning time and space as limiting factors in the written discourse, the Talmud coincidentally produced what may be one of humanity's first Big Data resources, a voluminous reservoir that continues to be mined today.

Duckweed enters the Talmud in a discussion in the Mishna concerning the suitability of various materials for use as an oil lamp on the Sabbath. It is prescribed in the Talmud that the festive Sabbath evening meal be held in "well lighted quarters". The Talmud discusses, in detail, what constitutes an acceptable oil lamp for this purpose, a discussion that has since become a part of the traditional Sabbath evening services in the Synagogue:

> "With what may we light [the Sabbath lamp], and with what may we not light it? We may not light it with a wick made of cedar-bast (*lechesh*), uncombed flax (*chosen*), floss-silk (*chalach*), or with a wick of willow-fiber (*iddan*), desert weed (*petilat ha-midbar*), or **duckweed** (*yaroka on the face of the water*) [since such wicks burn unevenly]. It may not be lighted with pitch (*zefet*), liquid wax (*shaava*), castor oil *(shemen kik)*, nor with oil that must be burned and destroyed *(shemen s'raifa)*, or with tail fat *(alyah)*, nor with tallow *(chailev)*. Nahum of Media says: We may use melted tallow *(chailev mevushal)*. The sages, however, say: It is immaterial whether or not it is melted, it must not be used for the Sabbath lamp." (Mishna, Treatise Shabbat, Ch. 2; adapted from [38]).

The English translation of this text in the two leading traditional Hebrew prayer books matches closely, except for the phrase "*yaroka* on the face of the water". *Yaroka* is translated in one as "duckweed" [38] and in the other as "seaweed" [39]. The Gemara discusses *yaroka*:

> "This *yaroka*, what is its nature? If you say that it is the *yaroka* on top of the narrow channels [where water gathers and there is greenery on top], it crumbles [and a wick cannot be made from it]. However, Rav Pappa said that it is referring to the *yaroka* that accumulates on a ship [as greenery (at the water line) when a ship is stationary]." (Gemara, Tractate Shabbat, p. 20b; Translated by M.E., also the following passages. In brackets, comments by Rashi (1040–1105 CE), a leading interpreter of the Gemara).

The Gemara distinguishes between two types of *yaroka* based on location: one in seawater on the sides of ships and the other floating on the water in narrow channels. Rav Pappa in the Gemara favors seaweeds growing on the sides of stationary ships. If so, from where does the prayer-book designation of *yaroka* as "duckweed" arise? The answer is in the setting of the "narrow channels", which is interpreted by other commentators as land-based, containing freshwater algae or plants [22]. Such a locale is well supported by Tanchum of Jerusalem (1219–1291 CE), author of a noted ancient dictionary (Figure 3) explaining the terms used by Maimonides (1135–1204 CE) in his *Mishne Torah* (Code of Law extracted from the Talmud). For the entry "*Yerek*" (Greenery), Tanchum states:

> "... *Yaroka* on the surface of the water that remains on the ground without moving and is not flowing...", [40]

thus implying a stagnant, freshwater site.

Figure 3. Part of a manuscript page from Tanchum of Jerusalem's dictionary of terms in Maimonides' Code of Law. Bodleian Library MS. Huntington 621, University of Oxford. [40] Manuscript date: 1393 CE. Language: Judaeo-Arabic and Hebrew. Folio page100r includes the entry for "*Yerek*" (circled in blue; literally "Greenery"). In the description of this term, Tanchum refers in Hebrew to "*yaroka* on the surface of the water" (underlined in blue), and, in Judaeo-Arabic, continues by describing this water as: "which remains on the ground without moving and is not flowing".

From the text of the Talmud, it is clear that the fourth-century sages of the Gemara (such as Rav Pappa, c. 300–375 CE) were unfamiliar with the names of the materials for oil-lamp wicks described by the second-century sages of the Mishna. In addition to the passage of time, the two groups resided in different geographical locations. The sages of the Gemara flourished in exile on the shores of the Euphrates river (present-day Iraq), while the sages in our discussion of the Mishna lived and taught in the Land of Israel during the period and aftermath of the two large-scale revolts against the Roman Empire (70–200 CE). Early in this period, the center of rabbinic learning in Israel relocated northward from the town of Yavne on the shore of the Mediterranean Sea via several waystations to the city of Tiberias (Figure 4).

> "...and the Sanhedrin was exiled... to Yavne, and from Yavne to Usha . . . and from Usha to Shefaram, and from Shefaram to Bet She'arim, and from Bet She'arim to Tzippori, and from Tzippori to Tiberias". (Gemara, tractate Rosh Hashana, 31a)

This relocation is significant in our context, as Tiberias sat on the shore of the Sea of Galilee, a freshwater body of about 166 km^2 fed by the three sources of the River Jordan. (It is interesting to note that in Talmudic phraseology, "*Yam*", the Hebrew term for "Sea", can refer to a large body of either saline or freshwater. For example, the Mediterranean Sea was known as the "Great *Yam*" and the Sea of Galilee as the "*Yam* of Tiberias".) Thus, in terms of geographic and ecological context, the relocation to Tiberias suggests a freshwater site for the Mishnaic discussion concerning *yaroka*.

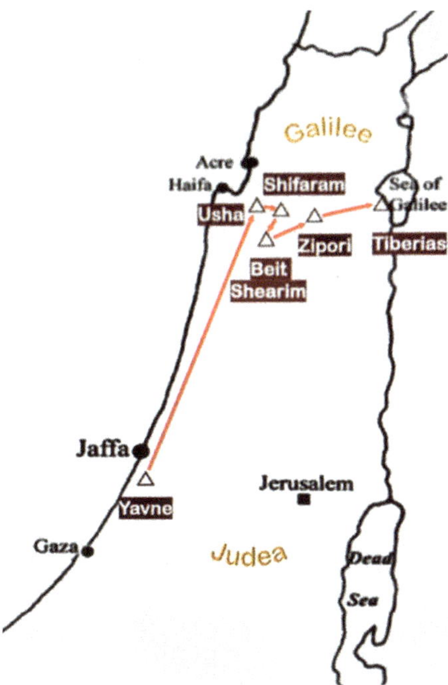

Figure 4. Exile of the Sanhedrin. To avoid restrictive edicts by the Roman Legate, the Sanhedrin (the religious and administrative leadership of the Jews at the time of the Mishna) relocated near the end of the 1st century from Yavne, a Mediterranian salt-water coastal town in Judea at the center of the country, northward to the Galilee, eventually ending up in the city of Tiberias on the shore of the freshwater Sea of Galilee by the latter part of the 2nd century.

What might be the identity of the Gemara's freshwater *yaroka*? Filamentous algae, such as *Spirogyra*, which accumulate in freshwater ponds and are buoyed to the surface by trapped bubbles of oxygen produced during active photosynthesis in the summer, might serve [22]. So too would duckweeds from the genera *Lemna* and *Landoltia*, which float on the surface of still or slow-moving freshwater bodies and possess dangling roots of several centimeters. The "narrow channels" mentioned in the Gemara fit duckweeds well, as such channels are a favored location for finding them, particularly in rain-filled narrow troughs of ancient grain mills and wine presses at archaeological sites dotting the Galilee and Golan Heights of northern Israel (personal observation, M.E.). In this regard, the designation for the entry *"Tachlav"* in Maimonides' glossary of drug names in his collected Medical Writings is quite germane. Maimonides, a 12th-century Pre-Renaissance polymath [41], served as a personal physician to the Sultan Salah ad-Din in Egypt, in addition to his well-known theological and philosophical pursuits. Maimonides defined the term *Tachlav* as *"yaroka* on top of the narrow channels", which is identified in the modern edition of his Medical Writings [42] explicitly as duckweed.

4.3. Duckweed in Medieval Hindu Literature

The *puranas* (literally, ancients) are Hindu religious texts composed in the medieval period and written in Sanskrit. The different *puranas* contain texts on several topics, including the structure of the cosmos and the course of conduct that human beings need to follow [43]. The Sanskrit scholar, Alexander Hamilton (1762–1824), collated the *purana* manuscripts at the National Library of Paris and concluded that "after the *vedas* (literally, knowledge) texts, the *puranas* are considered the most sacred of the Indian books" [44,45].

In the past century, these Hindu texts were studied by several scholars both in India and abroad.

Our focus is on the *Kurma Purana*, thought to be from the beginning of the eighth century with revisions thereafter [43]. The *Kurma purana* contains, together with other topics, teachings on acquiring knowledge by the practice of yoga and about the path to salvation. One of the stories takes place in a forest. During the course of austere penance in order to worship Lord Shiva, the sages perform several rituals, one of which is eating duckweed. Figure 5 shows an excerpt of the inscription in Sanskrit [14] referring to duckweed.

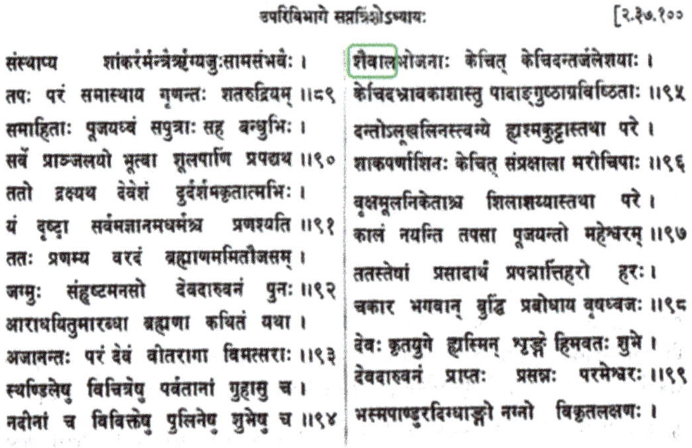

उपरिविभागे सप्तत्रिंशोऽध्यायः   [२.३७.१००

संस्थाप्य   शांकरंमन्त्रैर्ॠग्यजुःसामसंभवैः ।   शैवालभोजनाः केचित् केचिदन्तर्जलेशयाः ।
तपः परं समास्थाय गृणन्तः शतरुद्रियम् ॥८९   केचिदभ्रावकाशास्तु पादाङ्गुष्ठाग्रविष्ठिताः ॥९५
समाहिताः पूजयध्वं सपुत्राः सह बन्धुभिः ।   दन्तोऽलूखलिनस्त्वन्ये द्रुष्मकुट्टास्तथा परे ।
सर्वे प्राञ्जलयो भूत्वा शूलपाणिं प्रपद्यथ ॥९०   शाकपर्णाशिनः केचित् संप्रक्षाला मरीचिपाः ॥९६
ततो द्रक्ष्यथ देवेशं दुर्दर्शमकृतात्मभिः ।   वृक्षमूलनिकेताश्च शिलाशय्यास्तथा परे ।
यं दृष्ट्वा सर्वमज्ञानमधर्मश्च प्रणश्यति ॥९१   कालं नयन्ति तपसा पूजयन्तो महेश्वरम् ॥९७
ततः प्रणम्य वरदं ब्रह्माणममितौजसम् ।   ततस्तेषां प्रसादार्थं प्रपन्नार्त्तिहरो हरः ।
जग्मुः संहृष्टमनसो देवदारुवनं पुनः ॥९२   चकार भगवान् बुद्धिं प्रबोधाय वृषध्वजः ॥९८
आराधयितुमारब्धा ब्रह्मणा कथितं यथा ।   देवः कृतयुगे ह्यस्मिन् भृङ्गे हिमवतः शुभे ।
अजानन्तः परं देवं वीतरागा विमत्सराः ॥९३   देवदारुवनं प्राप्तः प्रसन्नः परमेश्वरः ॥९९
स्थण्डिलेषु विचित्रेषु पर्वतानां गुहासु च ।   भस्मपाण्डुरदिग्धाङ्गो नग्नो विकृतलक्षणः ।
नदीनां च विविक्तेषु पुलिनेषु शुभेषु च ॥९४

Figure 5. Page 507 of the *Kurma purana* [14]. The term *shaivaal* (alternatively, *saivala*) in Sanskrit is enclosed in the green box. It is described figuratively in [14] as "a kind of green grass-like plant growing in pools" and directly translated in [43] as duckweed.

The following translation by R.H. Davis of a section of Figure 5 is adapted from [43].

"They bowed to the beneficent Brahma, unlimited in his power, and returned to the Pine (Himalayan cedar) Forest, their hearts rejoicing. They began to worship just as Brahma had advised them. Still not knowing the highest god, but without desire and without jealousy, some worshiped him on multicolored ritual platforms, some in mountain caves, and some on empty, auspicious riverbanks. Some ate **duck-weed** for food, some lay in water, and some stood on the tips of their toes, abiding amid the clouds. Others ate unground grain, or ground it with a stone. Some ate vegetable leaves, and some purified themselves by subsisting on moonbeams (rays of light [14]). Some dwelled at the foot of trees, and others made their beds upon rocks. In these ways they passed their time performing austerities and worshiping Siva."

The translation of the Sanskrit term as per the context has been appropriately performed by Davis as duckweed [43]. This is the first historical mention of duckweed being eaten as food by human beings in the Hindu texts. Today, in-depth research has shown that these aquatic plants are a source of nutritious food for humans. Duckweeds contain high-quality protein and fatty acids [46,47] together with other phytonutrients such as phytosterols, vitamins, and minerals [48], and they do not show any adverse effects on the human system [49,50].

4.4. Duckweed and Ritual Purification by Yemeni Muslims

Duckweeds have a high capacity for water purification resulting from their facile ability to take up minerals and nutrients from the medium in which they grow [51]. An ancient exploitation of this property was brought to our attention several years ago by Pierre Goloubinoff, known, among other things, for his adventurous travels in Yemen [52].

Goloubinoff had read of a curious Yemini Muslim custom reported by the naturalist Paul-Emile Botta, who had been commissioned by the Natural History Museum of Paris in 1836 to explore the local flora of Yemen and collect specimens for the museum. Botta, a guest of the provincial governor, was staying at a chateau in 1837 on the flanks of mount Maammara in the province of Taiz in the southwest of the country. Nearby were some stone huts which served as housing for the families of soldiers and servants living in the governor's castle at the mountain top. It is in this setting that Botta described a cistern used by the denizens of the stone huts [53].

> … For their use and that of travelers, a large cemented cistern had been dug which received the rainwater, and in which they (the residents of the huts) not only drew their drink, but also bathed and performed their ritual purifications. … It is permitted, according to them (the residents of the huts), to wash and bathe in water which is not flowing and therefore does not renew itself, provided that it is abundant enough; while a Sunni after immersion in this way would consider himself impure, religiously speaking, as before. … the Yemenites claim, and perhaps believe, that the **duckweed** which covers the surface of stagnant waters, including their cisterns, is able to purify them (the waters), and they (the Yeminites) would not want to use standing water for purification where they would not see some (duckweed) floating. I must point out that this sect to which the inhabitants of Yemen belong is Zaydism. (translated by KSS).

The duckweed species growing in the Taiz region is *L. minor*. While Botta was skeptical of the water cleaning powers of *Lemna*, experimental results [54,55] are on the side of the Zaydis. The driving force for the water-purifying phenomenon is the unusually rapid growth of duckweeds in nutrient-rich water (biomass doubling in about two days). Feeding ensues from the entire underside of the floating, leaf-like frond, and growth proceeds essentially exponentially, since mother fronds and then successive daughter fronds bud multiple times before aging or until crowding sets in as the water surface is covered [1]. The result is extensive depletion of many dissolved substances in the pool [54,55], which are taken up and either metabolized or stored by the duckweed, hence the clearing of the pool water. Moreover, the floating mat or moss-like cover afforded by the plant (see Figures 1A and 6) retards the growth of contamination by light-seeking algae and bacteria. With careful cultivation by a local caretaker, the result is stagnant pool water that does not change its taste, color, or smell for an extended period. Botta's singular report of stagnant pools being permitted for ablutions by Zaydi sectarians needs confirmation from additional sources and references in fiqh or fatwa.

Figure 6. Duckweed mat covering the water surface of the cistern at Jabal éIbrahim, Himlan. From [56] with permission.

An additional reason suggested for why the Yemeni maintain the presence of duckweed in their water sources comes from an ethnographic study of rainwater-harvesting cisterns *ad locum* in the Governorate of Hajja, Yemen [56]. Figure 6 shows a thick carpet of duckweed left in place in the cistern. As opposed to water hyacinths that increase evapotranspiration at the water surface, duckweed reduces evaporation [57]. As the natives expressed it to E. Hovden [56]: "it prevents the wind from taking the water".

5. Duckweed in Ancient Secular and Religious Poetry

Duckweeds were widely spread globally already in ancient times. This gave impetus to their use as conventional imagery in poetry circles. While in Japanese culture, duckweed took on a stylized figurative meaning at the popular level, in China, its cultural imagery was habitat and biologically oriented, and so, too, its poetic context in biblical Israel.

5.1. Poetic Duckweed in Japanese Culture

All the references to duckweed in antiquity brought so far are from medicinal-related or religious manuscripts. Not so the history of duckweed in Japanese culture. It first appeared in the written record early in the Heian period (794–1185 CE) in collections of informal poetry and prose sponsored by the imperial court. Poetry served both political and social roles in Heian culture and became the main means of an intimate dialogue between the sexes. Much of this vernacular literature was of high quality and composed by women of the court. Remarkably, formalized poetry has remained a major means of written expression in Japanese up until modern times [58].

The Japanese islands have a mild, humid climate, and since ancient times, large areas of land have been used for rice production in paddy fields [59]. The presence of duckweed in flooded rice fields [60] was a common phenomenon in Heian-era Japan. As a result, duckweeds were familiar to the general population. *Ukikusa* (literally, floating weeds) is the term for duckweed in Japanese and is frequently found in poetry from the 9th and 10th centuries. The tiny plant became a symbol of the transience of life and mind due to its floating nature, or it was used in a rhetorical manner to evoke the feeling of melancholy or woefulness [61]. Thus, duckweeds were useful as an intermediary in the Heian poetry world and remain so in Japan today.

Ono no Komachi (born c. 850 CE) was a prominent female Heian poet. Her poems in the Kokin-waka-shu (the first and most prestigious of the imperial thirty-one-syllable waka anthologies) are mostly love poems. In Heian aristocratic society, it was impossible to function, in either public or private, without the ability to compose waka. Komachi's poetry makes use of pivot words that have more than one meaning, allowing compression of multiple connotations within the prescribed length of the poem [58]. The best-known example involving *ukikusa* is Poem 938 (Table 4).

Table 4. A flirtatious poem by Ono no Komachi (adapted and annotated from a translation by A. Commons in [58]).

When Fun'ya no Yasuhide became the third-ranked official of Mikawa Province and invited me to come sightseeing in the provinces, this was my reply: [a]	
wabinureba	Lonely and forlorn
mi o ukikusa no	as a **duckweed** {*uki* [b] *kusa*}:
ne o taete [c]	should flowing waters [d]
sasou mizu araba	beckon
inamu to zo omou	I think I'd follow.

[a] The headnote identifies this poem as a response of Ono no Komachi to a poem sent to her by another prominent poet of the Heian period, Fun'ya no Yasuhide.

[b] The "uki" of *ukikusa* is a pivot word meaning both "floating" (of the duckweed) and "miserable" (the poet). *Lemna aoukikusa* T.Beppu and Murata was, for a time, a synonym name (now retired) for *Lemna aequinoctialis* Welw. [62], which is a prevalent duckweed throughout most of Japan [1].

[c] *ne o taete* means "without a root" and often appears alongside of *ukikusa*; thus, embedding an additional subtle reference to duckweed.

[d] The "flowing waters" represent the message from Yasuhide. This is guided by the name of the province where Yasuhide officiated, "Mikawa"; its literal meaning is "three rivers".

5.2. Duckweed in Classical Chinese Poetry of the Ming Dynasty

Alongside the Japanese culture of vernacular poetry, classical poetry among well-educated Chinese constituted a dominant form of social interaction up until recent times. During the Ming dynasty (1368–1644 CE), it became a conventional skill. In his excellent monograph ("The Great Recreation" [13]) on Chinese poetry of the Ming dynasty, D. Bryant focused on the life of a mid-level civil servant, Ho Ching-ming (1483–1521 CE), an Archaist poet in search of a return to an imagined ideal Chinese society of yester years. Ho passed his examination degrees, without which it was almost impossible to reach a position of influence in the civil administration [63], and spent a good part of his career years in Peking with his literary friends trying to avoid the pitfalls inherent in the politically corrupt atmosphere of the civil service. However, in 1508 and near retirement [13], Ho, together with over a hundred other officials of high character who opposed the ongoing corruption, was sacked from his post by the all-powerful and corrupt palace eunuch Liu Chin [64]. Ho then retired to his provincial village and home. This is the contextual background for Ho's poem "Spring meditation", translated and edited by Bryant [13], with its interesting reference to duckweed.

"The east wind comes, and in a moment the end of spring is here;

Day after day, on the clear river, I sorrow for white **duckweed**.

Toward the north, the cloudy sky is lacking any road;

From the west, over heaven and earth, haze and dust are seen.

Having known high station and low, I see how they are related,

When things come up, in safely or peril, remember the men of old … "

Here we understand "white duckweed" as faithfully describing aged duckweed fronds, in which chlorophyll is catabolized and the green color lost. Such naturally-aged, dead plants remain intact and visible in floating patches of live, green duckweed plants for quite some time before disintegrating (Figure 7). This is in contrast to the meaning of "white duckweed" in the Classic Maya, Ritual of the Bicabs, where the color "white" takes on a ritual and cultural meaning signifying coolness (see Table 3, footnote "h") rather than a natural physiological meaning as here.

Figure 7. Naturally-aged white duckweed with younger green ones in a patch.

Ho composed several aquatic poems with duckweed imagery. For example, "On the Pond" begins with two couplets featuring duckweed, as brought by Bryant [13]

"Reeds grow at the mouth of a wintry pond;

Daily mated to the floating **duckweed**.

Breeze and ripples rock their stems;

They eddy and drift like a traveler's roaming."

The duckweed species with "stems" that provided poetic inspiration to Ho in this poem was most likely of the genus *Lemna*, members of which display a dangling root, with *L. aequinoctialis* and *Lemna japonica* Landolt naturally populating the Beijing area today [1].

5.3. Duckweed in the Hebrew Scriptures, Book of Psalms

The Book of Psalms is a collection of individual religious hymns in the Hebrew Scriptures, composed from the ninth to fifth century BCE. The hymns appear in poetic and song formation in the traditional parchment scrolls (two columns versus one for biblical prose). Their authorship is popularly attributed to King David (1040–970 BCE), who is mentioned in the titles of about half of the 150 individual Psalms [65]. However, it is quite possible that in addition to "written by David", the meaning in Hebrew of *"l'david"* ("by", "of," or "to" David) in a psalm title could indicate "dedicated to", "sung by", "played by" David. Or, maybe all of these, as David is portrayed more than once in the Hebrew Bible as an accomplished poet–musician, a harpist, and a musical conductor.

The term *"yawvein"* is connected to watery sediments. It appears twice in the Book of Psalms but nowhere else in the Hebrew Scriptures. At the beginning of Psalm 40, *yawvein* appears following the title verse. God is praised for deliverance from some previous misfortune of the psalmist. The remaining text (not shown) of the 18-verse psalm then precedes with supplications regarding the psalmist's present problems. (translations by M.E. based on [65,66]).

For the choir master; a Psalm, by David.

I fervently hoped for the Lord,
and He turned to me and heard my cry.

And He lifted me, from a turbulent watery dungeon,
from the mud of the **yawvein**.
And He set my feet upon a rock,
directing my steps . . .

The second appearance of duckweed is at the beginning of Psalm 69, where the psalmist calls out to God in a similar fashion:
For the conductor; upon shoshanim (a musical instrument), [a Psalm] of David.

Save me O God,
for the waters have reached my neck.

I have sunk,
in the depths of **yawvein** there is no foothold.

I have entered deep waters,
the current is sweeping me . . .

Most commentators have understood *yawvein* to refer to sticky mud, a mire, or a swamp. However, the important sage, Saadia Gaon (Saadia ben Joseph Al-Fayyumi; 882–942 CE), who wrote in Judaeo-Arabic and pioneered a form of rational biblical criticism based on deep knowledge of the language of the text, translated it in Hebrew as *tachlav* [67]. As mentioned in Section 3.2, in Maimonides' authoritative 12th-century Medical Writings [42], the term *tachlav* refers to duckweed (although filamentous algae, found in fresh and salt-water bodies, were sometimes also referred to as such [22]). In the context of Saadia Gaon's interpretation of the unique term *yawvein*, the psalmist may have been picturing duckweed plants covering water pools or swamps so densely that they appeared as moss or carpet of grass (Figures 1A and 6) and one who stepped on the "carpet" unexpectedly sank to the bottom of the waters [68].

6. Local Names for Duckweed in Antiquity and in the Middle Ages

The ancient local names for duckweed in the various cultures studied are listed in Table 5.

Table 5. Local names of duckweed in ancient cultures.

Culture	Period of History	Local Name	Literal Meaning	Ref.
	Han Dynasty, c.200 CE	*Shui Ping*	water weed	[12]
Chinese		*Shui Hua*	water flower	[12]
	Ming Dynasty, c.1500 CE	*Fu Ping*	floating duckweed	[13]
Christian	Hildegard von Bingen c.1150 CE	*Lemna*	Duckweed	[28]
Greek	Theophrastus c.330 BCE	*Lemna* [a]	water plant	[10]
	Book of Psalms c.1000 BCE	*Yawvein*	duckweed	[65,66]
Hebrew		*Tachlav* [b]	duckweed	[42,67]
	Talmud (Mishna) c.200 CE	*Yaroka*	greenery on water	[38]
Hindu	Kurma purana c.700 CE	*Shaivaal*	weed on water [c]	[69,70]
Japanese	Ono no Komachi c.850 CR	*Ukikusa*	floating weeds	[58]
		Ne-nashi-k(g)usa	weeds without a root	[58]
Maya	Ritual of the Bicabs c.250 CE	*Ixim ha*	maize-water plant	[35]
Roman	Dioscorides c.70 CE	*Lens*	lentil-shaped	[11]
Yemini	Zaydism c.1000 CE [d]	*Simsim* [e]	Sesame-seed [e]	

[a] Theophrastus coined the Greek term "lémna", which became the base word of the family Lemnaceae.
[b] Judaeo-Arabic, from the Pre-Islamic Arabic for duckweed, *al-ṭuḥlubū* [69].
[c] Shaivaal translated as "weed on water" by K.S.S. based on [70].
[d] Zaidism reached Yemen in the 11th century [71].
[e] Dialect term for *L. minor* in the Al-Shu'ayb District, Al-Ḍāli' Governorate (Daniel Varisco, personal communication). The same term can refer to different plants and the same plant can have different names depending on the locality, even between villages. Simsim as a moniker for duckweed is likely based on sesame seed's size and shape.

In naming duckweeds, Theophrastus concentrated on the plant's aquatic habitat [10], Dioscorides on the characteristic lens shape of the Spirodela and Lemna fronds local to Turkey (his country of birth) and the Greek-speaking eastern Mediterranean where he was stationed [11]. The Rabbis of the Mishna focused on duckweed's green color and freshwater location [38], the Japanese poets, its floating nature [58], the Maya [35] possibly on its ubiquitous presence as was the maize plant in their society, and the Zaydi Yemenis on its water decontamination properties.

7. Discussion

A Case of Ethnobotanical Convergence?

It is of interest to note that two ancient cultures, the later Han dynasty (c. 200 CE) in eastern China and the Maya of the Classical Period (250–900 CE) in Mesoamerica, each unaware of the other, used duckweeds as a major component in their concoctions for relieving a high fever [12,24,34,35]. This can be understood in the context of natural water bodies being associated with the quality of coolness and floating duckweeds as the visible example par excellence of an aquatic plant. Yet, the match in several details (Table 6) raises the possibility that we have uncovered a putative case of ethnobotanical convergence, defined as independent origins by at least two cultures, of a given plant or family's specific usage [72].

Table 6. Ethnobotanical convergence suggested in ancient duckweed medicinal usage.

Divine Farmer's *Materia Medica* [12]	Ritual of the Bacabs [24,34,35]
Later Han Dynasty, eastern China (c. 200 CE)	Maya Classical Period, the Americas (250–900 CE)
Lemnaceae: *Lemna, Spirodela*	**Lemnaceae:** *L. minor, W. brasiliensis*
"treats fulminant heat"	**"cooling a high fever"**
Daoist influence	shaman incantation
"precipitates water *qi*"	"seized the *kinam* (strength) of the pox"

The Divine Farmer of the Chinese Han Dynasty, with its strong Daoistic influence, promoted duckweeds (presented as *Lemna* or *Spirodela*) as a medicinal for cooling "fulmanent heat" (eruptive fever) and "precipitating" (initiating) "water qi" (cooling strength of water) [12]. The Bicabs, with glyphs of the Maya Classical period in Mesoamerica, under the influence of the am men (shaman) healers, promoted duckweeds (presented as Lemnaceae species *L. minor* or *W. brasiliensis*) for "cooling a high fever" that "descends" (cools down) as "I seized the kinam" [35] (strength) of the "pox" (the eruptive heat) [24]. The putative case here for ethnobotanical convergence lies in the shared alleviation of a sudden or high fever by members of the family Lemnaceae and, therein, the genus *Lemna*. These points are shown in bold in Table 6. The spiritual comparisons of the Daoistic and shaman healers and their tools of trade provide some depth to the case in that the spiritual practitioners of the late Han Dynasty, like those of the Classical Maya culture, were among the most schooled healers of the period, lending an added modicum of credence to their medicinal diagnoses.

8. Conclusions

Our by no means exhaustive quest into the ethnobotany of duckweed in ancient cultures revealed a number of expected and unexpected references to the tiny, floating plant recorded in ancient texts, manuscripts, and glyphs. We look forward to the current presentation motivating researchers from cultures not represented here to build on the present studies. Some general points emerged. Duckweeds were widely distributed geographically already in antiquity and were integrated into classical cultures in the Americas, Europe, the Near East, and the Far East 2000 years ago. Another point that emerged is that duckweed plants infrequently served alone as a primary medicine or drug. Apparently, the plant's strategy is to asexually outgrow the competition (mainly algae) rather than produce toxins as protectants. Yet, duckweeds appear to be of identifiable medicinal value. We described the Classic Maya and Chinese Han cultures, separated geographically and one unaware of the other, each promoting duckweed as a significant component in alleviating a high fever; possibly a novel instance of ethnobotanical convergence which needs further study. Unexpectedly, we also found that duckweeds played a role in ancient secular and religious poetry. While plant inflorescence is clearly poetically evocative, that is not the situation here. The peculiarities of tiny, floating duckweeds apparently evoked a poetic intimacy of the classical cultures with the plant itself.

Author Contributions: Conceptualization, M.E.; writing, review and editing, M.E., K.-J.A. and K.S.S.; providing data about duckweed in Japanese culture, T.O. All authors have read and agreed to the published version of the manuscript.

Funding: Internal institutional support to M.E.

Data Availability Statement: All the data are available in the manuscript.

Acknowledgments: We thank Susanne Kochs, University of Jena, Faculty of Theology, for help in understanding sections of the original Greek text of Theophrastus' *Peri phytōn historia* and the Latin text of T. Gaza's *Historia plantarum*. Likewise, we thank Daniel Varisco, President, American Institute for Yemini Studies, for discussions and assistance in navigating Paul-Emile Botta's report on his voyage to Yemen. We express our gratitude to Muhammad Gerhoum and Abdul Wali al Khulaidi for details on duckweeds in Yemen and Eirik Hovden for discussions and direction.

Conflicts of Interest: The authors declare that there is no conflict of interest.

References

1. Landolt, E. *The Family of Lemnaceae—A Monographic Study*; Veröffentlichungen des Geobotanischen Institutes der ETH, Stiftung Ruebel: Zurich, Switzerland, 1986; Volume 1.
2. Bog, M.; Appenroth, K.J.; Sree, K.S. Duckweed (Lemnaceae): Its Molecular Taxonomy. *Front. Sustain. Food Syst.* **2019**, *3*, 117. [CrossRef]
3. Schleiden, M.J. Prodromus Monographiae Lemnacearum oder Conspectus generum atque specierum. *Linnaea* **1839**, *13*, 385–392.
4. Hillman, W.S. Nonphotosynthetic light requirement in *Lemna minor* and its partial satisfaction by kinetin. *Science* **1957**, *126*, 165–166. [CrossRef] [PubMed]
5. Acosta, K.; Appenroth, K.J.; Borisjuk, L.; Edelman, M.; Heinig, U.; Jansen, M.A.K.; Oyama, T.; Pasaribu, B.; Schubert, I.; Sorrels, S.; et al. Return of the Lemnaceae: Duckweed as a model plant system in the genomics and postgenomics era. *Plant Cell* **2021**, *33*, 3207–3234. [CrossRef] [PubMed]
6. Kvaček, Z. *Limnobiophyllum krassilov*—A fossil link between Araceae and Lemnaceae. *Aquat. Bot.* **1995**, *50*, 49–61. [CrossRef]

7. Gallego, J.; Gandolfo, M.A.; Cúneo, N.R.; Zamaloa, M.C. Fossil Araceae from the Upper Cretaceous of Patagonia, Argentina, with implications on the origin of free-floating aquatic aroids. *Rev. Palaeobot. Palynol.* **2014**, *211*, 78–86. [CrossRef]
8. Emson, D. Ecology and Palaeoecology of Diatom—Duckweed Relationships. Ph.D. Thesis, Department of Geography University College London and Department of Botany Natural History Museum, University College London, London, UK, 2015. Available online: https://discovery.ucl.ac.uk/id/eprint/1462713/1/Dave_Emson_PhD.pdf (accessed on 8 August 2022).
9. Petrova-Tacheva, V.; Alekova, S.; Ivanov, V. *Lemna minor* L. and folk medicine. *Rheumatism* **2019**, *9*, 19–22.
10. Theophrastus. *Enquiry into Plants, Books 1–5*; Hort, A., Translator; Loeb Classical Library; Harvard University Press: Cambridge, MA, USA; London, UK, 1916. [CrossRef]
11. *Dioscorides. De Materia Medica—The Herbal by Dioscorides the Greek. A New Indexed Version in Modern English*; Osbaldeston, T.A. (Ed.) Osbaldeston, T.A., Translator; IBIDIS Press: Johannesburg, South Africa, 2000. Available online: http://www.cancerlynx.com/dioscorides.html (accessed on 8 August 2022).
12. *The Divine Farmer's Materia Medica (Shen Nong Ben Cao Jing)*; Shou-Zhong, Y., Translator; Blue Poppy Press, Inc.: Boulder, CO, USA, 1998. Available online: https://archive.org/stream/ShenNongBenCaoLingTheDivineFarmersMateriaMedica/ShenNong%20Ben%20Cao%20ling%28The%20Divine%20Farmers%20Materia%20Medica%29_djvu.txt (accessed on 8 August 2022).
13. Bryant, D. *The Great Recreation: Ho Ching-Ming (1483–1521) and His World*; Brill. Leiden: Boston, MA, USA, 2008.
14. Gupta, A.S. *The Kurma Purana*; All-India Kashiraj Trust: Varanasi, India, 1972; p. 507. Available online: https://archive.org/details/kurmapuranaTRkashirajtrust1972/page/n559/mode/2up (accessed on 8 August 2022).
15. Theophrastus. *De Historia Plantarum*; Gaza, T., Translator; Apud Gulielmum Gazeium; Free eBook from the Internet Archive; Nicolaus Bacquenoius: Lyon, France, 1552. Available online: https://archive.org/details/mobot31753000811833 (accessed on 8 August 2022).
16. Chisholm, H. (Ed.) Gaza, Theodorus. In *Encyclopædia Britannica*, 11th ed.; Cambridge University Press: Cambridge, UK, 1911; Volume 11, pp. 543–544.
17. Herbermann, C. (Ed.) Theodore of Gaza. In *Catholic Encyclopedia*; Robert Appleton Company: New York, NY, USA, 1913.
18. Theophrastus. Gaza, Theodorus: De Historia Plantarum, in: Noscemus Wiki. 1913. Available online: http://wiki.uibk.ac.at/noscemus/De_historia_plantarum (accessed on 17 May 2021).
19. Papadopoulos, J.K. Greek protohistories. *World Archaeol.* **2018**, *50*, 690–705. [CrossRef]
20. Theophrastus. Enquiry into plants, and minor works on odours and weather signs. *Nature* **1917**, *99*, 282. [CrossRef]
21. Scarborough, J. *A Review of Paul Millett's, Theophrastus and His World*; Proc. Camb. Philol Soc, (Suppl.) 2007, 33; Bryn Mawr Classical Review: Bryn Mawr, PA, USA, 2009. Available online: https://bmcr.brynmawr.edu/2009/2009.10.55/ (accessed on 8 August 2022).
22. Raanan, M. Velo Beyeroka Shehalpnei Hamayim (Hebrew). Available online: https://daf-yomi.com/DYItemDetails.aspx?itemId=55044 (accessed on 8 August 2022).
23. *Wetland Management for Waterfowl Handbook. Common Moist-Soil Plants Identification Guide 70–127*; Nelms, K.D., Ballinger, B., Boyles, A., Eds.; Natural Resources Conservation Service: Mississippi, USA, 2007. Available online: https://www.nrcs.usda.gov/Internet/FSE_DOCUMENTS/nrcs142p2_016986.pdf (accessed on 8 August 2022).
24. Bolles, D. A Translation of the Edited Text of Ritual of the Bacabs. 2003. Available online: http://davidsbooks.org/www/Bacabs.pdf (accessed on 8 August 2022).
25. De Materia Medica: Libri Veiusdem de Venenis Libri Duo. Translated by Janus Antonius Saracenus. 1598. Available online: https://www.digitale-sammlungen.de/en/view/bsb10994207?page=488 (accessed on 8 August 2022).
26. Russell, M. *Arthur and the Kings of Britain: The Historical Truth behind the Myths*; Amberley Publishing: Stroud, UK, 2017; p. 320.
27. Feuerstein-Prasser, K.; Kanbay, F.; Koethe, R. *Die Bewährte Heilkunde der Hildegard von Bingen (The Approved Physic of Hildegard of Bingen)*; FR text edition; Reader's Digest: Hamburg, Germany, 2021.
28. Von Bingen, H. *Ursachen und Behandlung der Krankheiten (Causa et Curae). (Reasons and Treatment of Diseases)*; Schulz, H., Translator; Verlag der Aeztlichen Rundschau Otto Gmelin: Muenchen, Germany, 1933.
29. Von Bingen, H. *Ursachen und Behandlung der Krankheiten (Causa et Curae). (Reasons and Treatment of Diseases)*; Schulz, H.; Karl, F., Translators; Haug Verlag: Heidelberg, Germany, 1992.
30. Hertzka, G.; Strehlow, W. *Große Hildegard-Apotheke. (Great Hildegard Apothecary)*; Christiana-Verlag: Kisslegg-Immenried, Germany, 2020.
31. Von Bingen, H. *Heilwissen (Knowledge of Healing)*; Kaiser, P., Translator; FV Editions; Shambhala Pubs: Berkeley, CA, USA, 2021.
32. Brown, C. El Mirador, the Lost City of the Maya. 2011. Available online: https://www.smithsonianmag.com/history/el-mirador-the-lost-city-of-the-maya-1741461/ (accessed on 7 December 2021).
33. Freidel, D.A.; Schele, L. Kingship in the late preclassic Maya lowlands: The instruments of power and places of ritual power. *Am. Anthropol.* **1988**, *90*, 547–567. [CrossRef]
34. Doemel, K. Mayan Medicine: Rituals and Plant Use by Mayan Ah-Men. Ph.D. Thesis, University of Wisconsin, La Crosse, WI, USA, 2013. Available online: https://minds.wisconsin.edu/bitstream/handle/1793/66629/Doemel_Thesis.pdf?sequence=1 (accessed on 8 August 2022).
35. Knowlton, T.W. Flame, icons and healing: A colonial Maya ontology. *Colon. Latin Am. Rev.* **2018**, *27*, 392–412. [CrossRef]
36. Mills, D.L. The Classic Maya Calendar and Day Numbering System. 1995. Available online: https://www.eecis.udel.edu/~{}mills/maya.html (accessed on 8 August 2022).

37. Ferrier, J.; Pesek, T.; Zinck, N.; Curtis, S.; Wanyerka, P.; Cal, V.; Balick, M.; Arnason, J.T. A classic Maya mystery of a medicinal plant and Maya hieroglyphs. *Heritage* **2020**, *3*, 275–282. [CrossRef]
38. *Daily Prayer book, Ha-Sidur Ha-Shalem*; Birnbaum, P. (Ed.) Hebrew Publ. Co.: New York, NY, USA, 1949; p. 252.
39. *The Complete Artscroll Siddur*; Scherman, N. (Ed.) Mesorah Publications, Ltd.: New York, NY, USA, 1984; p. 323.
40. Bodleian Library MS. Huntington 621. Rabbi Tanhum Hayerushalmi's Murfshid Al-Kafi: Dictionary of Terms in Maimonides' Mishneh Torah (in Judaeo-Arabic, Hebrew). Folio 100r. Available online: https://digital.bodleian.ox.ac.uk/objects/06abea16-07ae-4bd6-80cc-be7773bb33ae/surfaces/cefe511e-365f-41d7-ba2e-3535273a5946/ (accessed on 7 December 2021).
41. Kraemer, J.L. *Maimonides: The Life and World of One of Civilization's Greatest Minds*; Doubleday Religion: Cordoba, Spain, 2010; p. 640.
42. *Medical Works of Maimonides (in Hebrew)*; Montner, S. (Ed.) Mosad Harav Kook: Jerusalem, Israel, 1959; Volume 4.
43. Davis, R.H. The origin of Linga worship. In *Religions of India in Practice*; Lopez, D.S., Jr., Ed.; Princeton University Press: Princeton, NJ, USA, 1995; pp. 637–648.
44. Rocher, R. *Alexander Hamilton (1762–1824). A Chapter in the Early History of Sanskrit Philology*; American Oriental Society: New Haven, CO, USA, 1968; p. 41.
45. Rocher, L. The Puranas. In *A History of Indian Literature*; Gonda, J., Ed.; Otto Harrassowitz: Wiesbaden, Germany, 1986; Volume II, p. Fasc. 3.
46. Edelman, M.; Colt, M. Nutrient value of leaf vs. seed. *Front. Chem.* **2016**, *4*, 32. [CrossRef] [PubMed]
47. Appenroth, K.J.; Sree, K.S.; Böhm, V.; Hammann, S.; Vetter, W.; Leiterer, M.; Jahreis, G. Nutritional value of duckweeds (Lemnaceae) as human food. *Food Chem.* **2017**, *217*, 266–273. [CrossRef]
48. Appenroth, K.J.; Sree, K.S.; Bog, M.; Ecker, J.; Seeliger, C.; Böhm, V.; Lorkowski, S.; Sommer, K.; Vetter, W.; Tolzin-Banasch, K.; et al. Nutritional value of the duckweed species of the genus *Wolffia* (Lemnaceae) as human food. *Front. Chem.* **2018**, *6*, 483. [CrossRef] [PubMed]
49. Sree, K.S.; Dahse, H.M.; Chandran, J.N.; Schneider, B.; Jahreis, G.; Appenroth, K.J. Duckweed for human nutrition: No cytotoxic and no anti-proliferative effects on human cell lines. *Plant Foods Hum. Nutr.* **2019**, *74*, 223–224. [CrossRef]
50. Mes, J.J.; Esser, D.; Somhorst, D.; Oosterink, E.; van der Haar, S.; Ummels, M.; Siebelink, E.; van der Meer, I.M. Daily intake of *Lemna minor* or spinach as vegetable does not show significant difference on health parameters and taste preference. *Plant Foods Hum. Nutr.* **2022**, *77*, 121–127. [CrossRef] [PubMed]
51. Landolt, E.; Kandeler, R. *The Family of Lemnaceae—A Mono-Graphic Study*; Veroeffentlichungen des Geobotanischen Instutes der ETH; Stiftung Ruebel: Zurich, Switzerland, 1987; Volume 2.
52. Tawil, H.; Miodoenik, S.; Goloubinoff, P. *Operation Esther: Opening the Door for the Last Jews of Yemen*; Belkis Press: New York, NY, USA, 1998.
53. Botta, P.E. *Relations d'un Voyage Dans l'Yémen, Entrepris en 1837 Pour le Museum D'Histoire Naturelle de Paris*; B. Duprat: Paris, France, 1841; p. 148. Available online: https://gallica.bnf.fr/ark:/12148/bpt6k58013253/f21.item.texteImage.zoom (accessed on 8 August 2022).
54. Ozengin, N.; Elmaci, A. Performance of Duckweed (*Lemna minor* L.) on different types of wastewater treatment. *J. Environ. Biol.* **2007**, *28*, 307–314.
55. Ziegler, P.; Sree, K.S.; Appenroth, K.J. Duckweeds for water remediation and toxicity testing. *Toxicol. Environ. Chem.* **2016**, *98*, 1127–1154. [CrossRef]
56. Hovden, E. Rainwater Harvesting Cisterns and Local Water Management. Master's Thesis, University of Bergen, Bergen, Norway, 2006. Available online: https://bora.uib.no/bora-xmlui/bitstream/handle/1956/2001/Masteroppgave_Hovden.pdf?sequence=1&isAllowed=y (accessed on 8 August 2022).
57. DeBusk, T.A.; Ryther, J.H.; Williams, L.D. Evapotranspiration of *Eichhornia crassipes* (Mart.) Solms and *Lemna minor* L. in central Florida: Relation to canopy structure and season. *Aquat. Bot.* **1983**, *16*, 31–39. [CrossRef]
58. Shirane, H. *Traditional Japanese Literature: An Anthology, Beginnings to 1600, Abridged*; Columbia University Press: Boston, MA, USA, 2012.
59. Sato, Y.-I. History and culture fostered by rice. *Highlighting Jpn.* **2020**, *150*, 6–7. Available online: https://www.gov-online.go.jp/pdf/hlj/20201101/06-07.pdf (accessed on 8 August 2022).
60. Li, H.; Liang, X.Q.; Lian, Y.F.; Xu, L.; Chen, Y.X. Reduction of ammonia volatilization from urea by a floating duckweed in flooded rice fields. *Soil Sci. Soc. Am. J.* **2009**, *73*, 1890–1895. [CrossRef]
61. Oyama, T. Background. Duckweed 2015 Kyoto. Available online: http://www.duckweed2015.cosmos.bot.kyoto-u.ac.jp/background.html (accessed on 8 August 2022).
62. Borisjuk, N.; Chu, P.; Gutierrez, R.; Zhang, H.; Acosta, K.; Friesen, N.; Sree, K.S.; Garcia, C.; Appenroth, K.J.; Lam, E. Assessment, validation and deployment strategy of a two barcode protocols for facile genotyping of duckweed species. *Plant Biol.* **2015**, *17*, 42–49. [CrossRef] [PubMed]
63. Hucker, C.O. *The Ming Dynasty, Its Origins and Evolving Institutions*; E-book; University of Michigan Center for Chinese Studies: Ann Arbor, MI, USA, 1978. [CrossRef]
64. Eberhard, W. A History of China. 1950. (The Project Gutenberg EBook). Available online: https://www.gutenberg.org/cache/epub/11367/pg11367.html (accessed on 8 August 2022).
65. *Book of Psalms, Sefer Tehillim Daat Mikrah (in Hebrew)*; Israel's Leading Publishers: Jerusalem, Israel, 1990; Volume 1.
66. *Tehilim, a New Translation with a Commentary Anthologized from Talmudic, Midrashic and Rabbinic Sources*; Feuer, A.C., Translator; Mesorah Publ. Ltd.: Rahway, NJ, USA, 1985.

67. Book of Psalms. Tehillim, with Commentary by Rabbi Saaadia Gaon (in Judaeo-Arabic, Translated into Hebrew by Y. Chacham. 1966). Available online: https://tablet.otzar.org/book/book.php?book=8066&pagenum=1 (accessed on 8 August 2022).
68. Amar, Z. *Plants of the Bible (in Hebrew)*; Rubin Maas Inc.: Jerusalem, Israel, 2009; p. 247. Available online: https://kotar.cet.ac.il/KotarApp/Viewer.aspx?nBookID=102357889#3.118.6.default (accessed on 8 August 2022).
69. Miller, N.A. Tribal Poetics in Early Arabic Culture: The Case of Ash'ār al-Hudhaliyyīn. Ph.D. Thesis, The Faculty of the Division of the Humanities, Department of Near Eastern Languages and Civilizations, Chicago University Press, University of Chicago, Chicago, IL, USA, 2016.
70. Hinkhoj Dictionary. Available online: https://dict.hinkhoj.com/shaivaal-meaning-in-english.words (accessed on 8 August 2022).
71. Salmoni, B.A.; Loidolt, B.; Wells, M. *Regime and Periphery in Northern Yemen. The Huthi Phenomenon. Appendix B. Zaydism: Overview and Comparison to Other Versions of Shi'ism*; RAND Corp., National Defence Research Institute: California, CA, USA, 2010.
72. Hawkins, J.A.; Teixidor-Toneu, I. Defining 'Ethnobotanical Convergence'. *Trends Plant Sci.* **2017**, *22*, 639–640. [CrossRef] [PubMed]

Article

Trends of Medicinal Plant Use over the Last 2000 Years in Central Europe

Maja Dal Cero [1], Reinhard Saller [2], Marco Leonti [3] and Caroline S. Weckerle [1,*]

1　Department of Systematic and Evolutionary Botany, University of Zurich, Zollikerstrasse 107, 8008 Zurich, Switzerland
2　Albisstrasse 20, 8038 Zurich, Switzerland
3　Department of Biomedical Sciences, University of Cagliari, Cittadella Universitaria, 09042 Monserrato, Italy
*　Correspondence: caroline.weckerle@systbot.uzh.ch

Abstract: Medicinal plant knowledge in Central Europe can be traced back from the present to antiquity, through written sources. Approximately 100 medicinal plant taxa have a history of continuous use. In this paper, we focus on use patterns over time and the link between historical and traditional uses with the current scientific evidence. We discuss our findings against the backdrop of changing eras and medicinal concepts. Based on use-records from totally 16 historical, popular and scientific herbals, we analyze how use categories of 102 medicinal plant taxa developed over time. Overall, 56 of the 102 taxa maintained continuous use throughout all time periods. For approximately 30% of the continuous uses, scientific evidence supporting their use exists, compared to 11% for recently added uses and 6% for discontinuous uses. Dermatology and gastroenterology are use categories that are relevant across all time periods. They are associated with a high diversity of medicinal taxa and continuously used medicinal species with scientific evidence. Antidotes, apotropaic (protective) magic, and humoral detoxification were important use categories in the past. New applications reflecting biomedical progress and epidemiological challenges are cardiovascular and tonic uses. Changes in medicinal concepts are mirrored in plant use and specifically in changes in the importance of use categories. Our finding supports the concept of social validation of plant uses, i.e., the assumption that longstanding use practice and tradition may suggest efficacy and safety.

Keywords: historical ethnobotany; medicinal plants; Central Europe; traditional use; historical ethnopharmacology

Citation: Dal Cero, M.; Saller, R.; Leonti, M.; Weckerle, C.S. Trends of Medicinal Plant Use over the Last 2000 Years in Central Europe. *Plants* **2022**, *12*, 135. https://doi.org/10.3390/plants12010135

Academic Editors: Renata Sõukand and Raivo Kalle

Received: 18 November 2022
Revised: 20 December 2022
Accepted: 21 December 2022
Published: 27 December 2022

1. Introduction

Different types of historical studies on medicinal plant use exist. Historical ethnobotanical studies in Europe have been interested in the mechanisms of knowledge transmission, e.g., by Dioscorides and Galen [1,2], or the influence of ancient herbals on recent medicinal plant use, e.g., Tabernaemontanus 16th century [3], Hildegard von Bingen 12th century [4,5], Iatrosophia texts in Cyprus [6,7], Corpus Hippocraticum 5th century BC [8,9], Nordic countries [10], Northeastern Europe 19th century [11], Celtic Provenance Medieval Wales [12] and several Western pharmacopeias [13]. Ancient herbals were also used for extracting information that appears to be relevant for drug discovery programs (e.g., [14–16]).

In this paper, we are interested in patterns in historical and traditional medicinal plant uses and their links with current scientific evidence. We discuss our findings in the context of social validation of medicinal plant uses, which is relevant for assessing the efficacy and safety of traditional herbal remedies in Europe [17,18]. Use patterns are also discussed against the backdrop of changing eras over the last 2000 years. These include epidemiological factors, alterations in philosophical, scientific and medicinal theories, and key medical discoveries and major historical events (Figure 1). For our investigation, we focused on around 100 medicinal plant species which were uninterruptedly used for therapeutic purposes in Central Europe over the last two millennia [19].

The following research questions are addressed: (1) Which are the general trends of medicinal plant use patterns over time? For example, which uses are restricted to specific periods and which are practiced across time? (2) What percentage of continuously used medicinal plants show a link between historical and traditional uses and current scientific evidence?

Historical Context of Medicinal Plant Use in Central Europe

The history of medicine has been well documented since Antiquity and era-specific changes in the prevailing medical philosophy can easily be traced [20–22]. Over almost two millennia, the prevailing medical theory was based on the idea of an analogy between microcosm and macrocosm. This idea originated in ancient Greek philosophy at around the 5th century BC [23]. The theory of humoral pathology arose from this concept and provided a framework for the systematic analysis of complex relationships between humans and their environment. Through Galen's (ca. 131–201 AD) writings, humoral pathology became the prevailing medical theory until the early 18th century [21].

During medieval times, written knowledge of ancient medicine was retained in Christian monasteries. Old codices were newly compiled, and the Mediterranean *materia medica* was substituted with local species [24]. Ancient predilections and slogans such as "diet over drugs" [8,9,21] are reflected in monastic medicine, e.g., in Hildegard von Bingen's (1098–1179 AD) *Physica*, where she describes healthy qualities of food plants [25]. Additionally, Christian ethos and charity brought new aspects to medicine and became the drivers for the development of hospitals in Central Europe [21].

During the Renaissance, the ancient sources of medical knowledge were revisited, with an attempt to delete Arabic influences from the texts [21,26]. At the same time, detachment from ancient medical authorities and Christian religion began. The enlightenment movement (18th century) stands for the beginning of modern times and was paralleled by a scientific revolution, resulting in new ideas and theories replacing ancient concepts with an increasingly mechanistic worldview. The reliance on medicinal herbs as the principal resource for multi-target drugs decreased and was largely replaced by the application of mono-substance remedies [27,28].

Thus, since antiquity, the medical landscape of the Old World has been diverse and changeful. Written and institutionalized medicine existed along various forms of oral traditions, which finally resulted in todays' Central European medical pluralism [26,29–31]. In parallel to the scientific revolution leading to biomedicine, naturopathy, as a (health-) political countermovement, arose in the late 18th century [32,33]. This laid the foundation for today's complementary and alternative medicine, which still considers ancient ideas of bodily humors as so-called 'constitutional factors' and the idea that a body in balance prevents sickness. Additionally, 'blood cleansing' and detoxifying strategies are still commonly used in popular medicine [30,34].

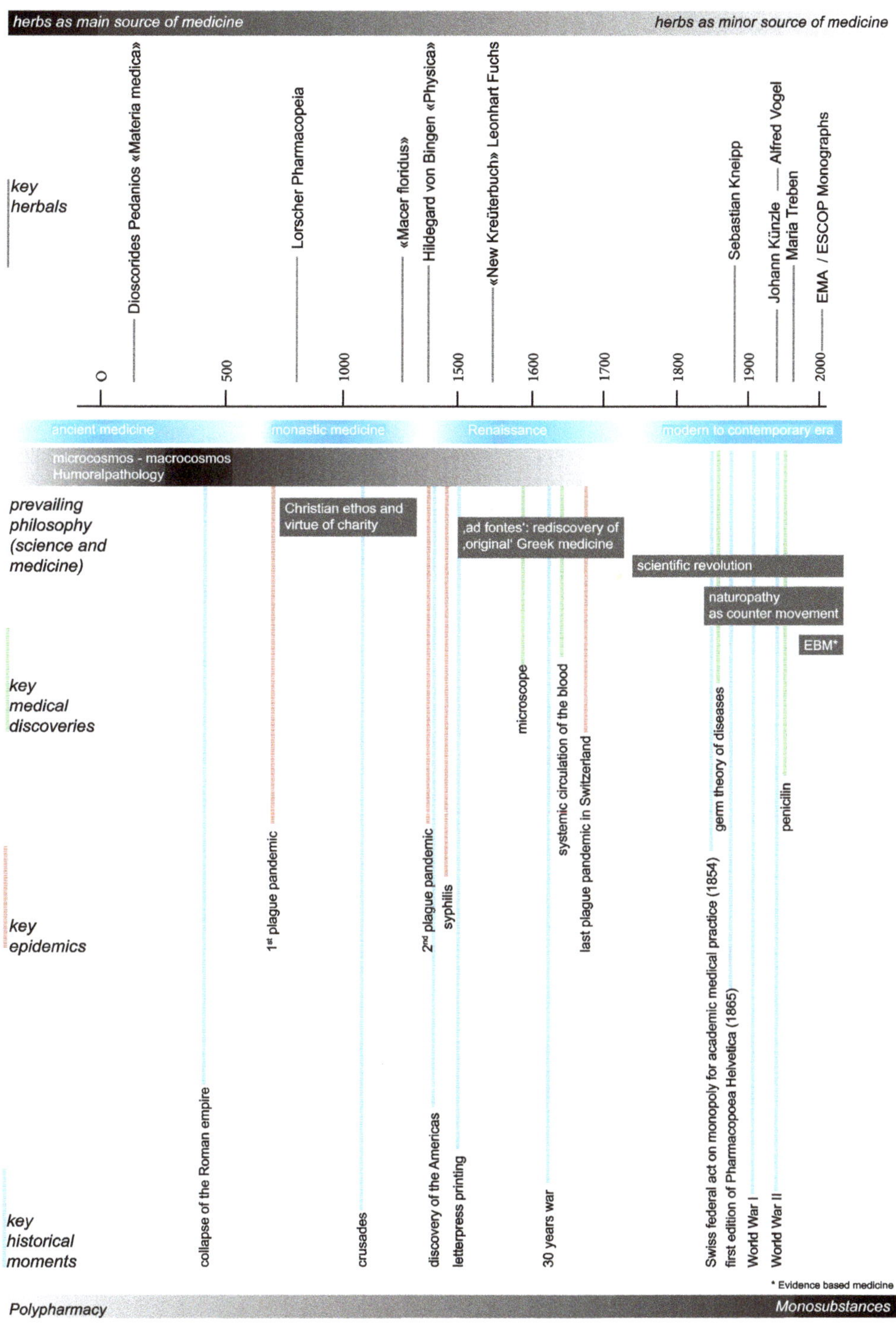

Figure 1. Synchronoptic view of key drivers influencing medicinal plant use.

2. Methods

2.1. Written Sources

For the present analysis, we reassessed the selection of 24 written documents used in Dal Cero et al. [19], which covered the most important Central European herbals from classical antiquity to Renaissance [35]. The selection of books was based as far as possible on medical texts compiled by doctors and not on recipe collections. It can, therefore, be assumed that there was a practical review and critical appreciation of the texts corresponding to the time of the authors. The distinguished time periods are Antiquity, monastic medicine, Renaissance, and the modern to contemporary era (Table 1). In total, 14 documents were selected, which provided detailed information about the medical uses of 102 taxa that were uninterruptedly used for therapeutic purposes through all time periods. We omitted herbals which did not add new uses [36–40] or did not provide detailed information about the medical use of single species (index of *Capitulare de villis* [41] and index of Lonicero [42]).

For modern and contemporary herbals, we differentiated between (1) popular herbals based on folk medicinal practices and personal experience, and (2) scientific herbals with evidence of efficacy and safety [43]. The choice of modern herbals was largely based on interviews with 61 herbalists, who were asked about the medicinal plant books they use [34]. We did not consider homeopathy [44], anthroposophic medicine [45] and Bach flowers [46]. In addition to the scientific herbals, we used ESCOP [47,48] and EMA Monographs [49] (accessed 2022) to check for scientific evidence of efficacy. Table 1 shows the written sources on which our analysis was based.

Primary data are provided as Supplementary Material.

Table 1. Books used for the analysis of use categories.

Time Period	Book Title «Short Title»	Author	First Edition/ Edition Used	Abbreviation
Antiquity 1st century CE	*De Materia Medica*	Dioscorides Pedanios from Anazarbos	1st century CE/ Berendes (1902) [50]	DIOS [1)]
Monastic medicine 8th–12th century	Lorscher Pharmacopoeia	Anonymus	8th century/ Stoll (1992) [24]	LO
	«Macer floridus»	Odo Magdunensis	ca. 1100/ Mayer and Goehl (2001) [51]	MF
	«Physica»	Hildegard von Bingen	ca. 1151 Portmann (1991) [25]	HvB
Renaissance 16th–17th century	«New Kreüterbuch»	Leonhart Fuchs	1543/ Dobat and Dressendörfer (2001) [52]	LF
	«Neuw Kreuterbuch»	Tabernaemontanus; Jacob Theodor	1588/ Edition anno 1625 [53]	TAB [2)]
	Popular herbals [3)]			
	So sollt ihr leben	Sebastian Kneipp	1889/ Kneipp (2010) [54]	KN
Modern to contemporary era since 19th century	Das grosse Kräuterheilbuch	Johann Künzle	1945/ Künzle (1945) [55]	JK
	Der kleine Doktor	Alfred Vogel	1952/Vogel (1952) [56]	AV
	Phytothérapie: Traitement des Maladies par les Plantes	Jean Valnet	1983/Valnet (1992) [57]	VAL

Table 1. *Cont.*

Time Period	Book Title «Short Title»	Author	First Edition/ Edition Used	Abbreviation
	Gesundheit aus der Apotheke Gottes	Maria Treben	1980/Treben (2011) [58]	MT
	Natürlich gesund mit Heilpflanzen	Bruno Vonarburg	1988/Vonarburg (1988) [59]	BVA
	Praxis-Lehrbuch der modernen Heilpflanzenkunde	Ursel Bühring	2005/Bühring (2005) [60]	UB
Scentific herbals				
	Teedrogen und Phytopharmaka	Max Wichtl (ed.)	1984/Wichtl (2008) [43]	WI
	ESCOP Monographs and supplement	European Scientific Cooperative of Phytotherapy	2003 and 2009 [47,48]	ESCOP
	EMA Monographs	Committee on Herbal Medicinal Products (HMPC)	Webpages 1995 –2022/accessed Oct. 2022 [49]	EMA

[1] We used the modern translation of Dioscorides' *De Materia Medica* from Berendes (1902) [50] as a surrogate for earlier Dioscorides translations. We crosschecked for ethnotaxa with Matthioli (1568) [61] as one of the most widespread Renaissance translations of Dioscorides' *De Materia Medica* [2]. [2] For those species not included in Leonhart Fuchs (*Acorus calamus, Malus sylvestris,* and *Pyrus communis*) we consulted the herbal of Tabernaemontanus (1625) [53]. [3] These popular herbals are the sources of information for herbalists in Switzerland at present (cf. [34]).

2.2. Use-Records and Use Categories

For the analysis, we recorded each documentation of a specific taxon for a specific use as one use-record. All uses were grouped into 18 use categories related to organs, symptoms and route of administration (Table 2). The categories follow [2] and [34]. To match historical uses with modern use categories, we consulted Hoefler (1899) [62].

Table 2. Use categories related to organs and symptoms.

Abbreviation	Organ/Symptom	Notes
ANT	Antidote	bites and stings of poisonous and mad animals, intoxication
APH	Aphrodisiac	and anaphrodisiac
APO	Apotropaic	against 'bad influence' and ailments [no internal use], charms
CAR	Cardiovascular	blood circulation, heart diseases, systemic applications for hemorrhoids and veins
DER	Dermatological	skin, wounds, ulcers, topic applications for hemorrhoids and veins
EAR	Ear	ear infections, deafness
EYE	Ophthalmic	eye infections, blindness
FEV	Fever	including malaria
GAS	Gastrointestinal	digestion, stomachache, diarrhea, icterus
GYN	Gynecological	menstrual problems, perinatal
HUM [1]	Humoral detoxification	general indication for purification and detoxification
NER	Nerves	sleeplessness, nervousness, general analgesics
RES	Respiratory	cough, lungs

Table 2. *Cont.*

Abbreviation	Organ/Symptom	Notes
SKE	Skeletomuscular	musculoskeletal pain and disability, rheumatism, injuries
TEE	Teeth	toothache
TON [2]	Tonic	general strengthening, immunomodulatory, roborants, anemia
URO	Urological	bladder, kidney disease
VAR	Varia	including anti-inflammatory, blood, cancer, diabetes, diet, metabolic disorders, parasites, spleen

[1] 'humoral detoxification' is used only for general detoxifying indications without a link to diuretic (->URO) or laxative effects (->GAS), mainly for 'removing of bad humors' (blood, cholera, phlegm), in the sense of the ancient theory of the four humors and humoral pathology. [2] 'tonic' is used in a strict sense and all indications with a link to appetite and digestion (e.g., orexygenic) are allocated in GAS; indications with a link to fatigue or nervous exhaustion are allocated in NER.

We considered a use category with scientific evidence when the use category appeared to be validated for a specific taxon in either Wichtl (2008) [43], ESCOP [47,48] or EMA [49].

2.3. Medicinal Plant Taxa

For species identification, we relied on recent editions of ancient, monastic and Renaissance herbals, which include Latin names of the plants (Table 1). For a few species, which were not mentioned in the recent edition of Fuchs' «New Kreüterbuch», we relied on the original plant list of Tabernaemontanus (1588) [53]. All of these taxa were easily identifiable, such as, *Acorus calamus*.

Taxonomically, this study was based on the 'Flora indicativa' [63] which covers plants of the Swiss flora and the Alps. For several species, we used species complexes (aggregates, agg.) [63]. These aggregates comprise closely related Swiss and Alpine species and tend to reflect so-called ethnotaxa, i.e., species with identical or similar local names and uses. The following adjustments were made with respect to Dal Cero et al. (2014) [19]: we added *Helleborus* spp., *Peucedanum* spp., *Teucrium* spp., *Salvia* spp. and *Urtica* spp. as ethnotaxa. Different species of these genera, also as local substitutes for Mediterranean species, have been used since Antiquity. In addition, we merged the following species into ethnotaxa as they have been used interchangeably in one or several time periods: *Abies alba* and *Larix decidua* (*Abies* spp.), *Lepidium officinale* and *Nasturtium officinale* (*Lepidium* spp.), *Matricaria chamomilla* and *Anthemis* spp. (*Matricaria chamomilla*), *Mercurialis annua* and *M. perennis* (*Mercurialis* spp.), *Prunus avium*, *P. domestica* and *P. spinosa* (*Prunus* spp.), *Sambucus nigra* and *Sambucus ebulus* (*Sambucus* spp.), *Sinapis alba* and *Brassica nigra* (*Sinapis* spp.), as well as *Solanum nigrum* and *Solanum dulcamara* (*Solanum* spp.). In total, we analysed 102 taxa (species, aggregates and ethnotaxa). Accordingly, we used the term 'plant taxa' instead of 'plant species'. Nomenclature follows Plants of the World Online [64], and the APG system [65].

2.4. Analysis of Data and Diachronic Patterns

Diachronic patterns were analyzed from the perspective of (1) use categories, i.e., diversity of medicinal taxa over time per use categories; and (2) medicinal taxa, i.e., diversity of use categories over time per taxon. In addition, typical diachronic patterns were highlighted with the example of a few medicinal taxa.

Descriptive statistics (mean ± standard deviation) was used to describe changes in taxa per use category and use categories per taxa.

2.5. Abbreviations

UCat	Use category
UCat$_{const}$	Use category constant since Antiquity
UCat$_{recent}$	Use category added in contemporary period
UR	Use-record

3. Results

3.1. Use-Records Per Time Period

In total, 3993 use-records were found for the102 medicinal plant taxa: Antiquity 891 use-records, monastic medicine 677 UR, Renaissance 1036 UR, modern to contemporary era 1154 UR from popular herbals, and 235 UR for 53 taxa from scientific herbals.

3.2. Diachronic Changes: The Use Category Perspective

The plant taxa used for specific use categories change over time. The share of taxa per use-category utilized uninterruptedly across all time periods ranges between 0–29% (Figure 2, pie charts). The highest numbers of taxa constantly used across all time periods were found for categories GAS (33 taxa), DER (28), GYN (16), RES (12), and URO (11). For the categories GAS, DER and RES, the highest percentage was found for constantly used taxa with scientific evidence (GAS: constantly used 33 taxa [34%], 16 taxa with scientific evidence; DER: constantly used 28 taxa [27.5%], 12 taxa with scientific evidence; RES constantly used 12 taxa [15.5%], 4 taxa with scientific evidence).

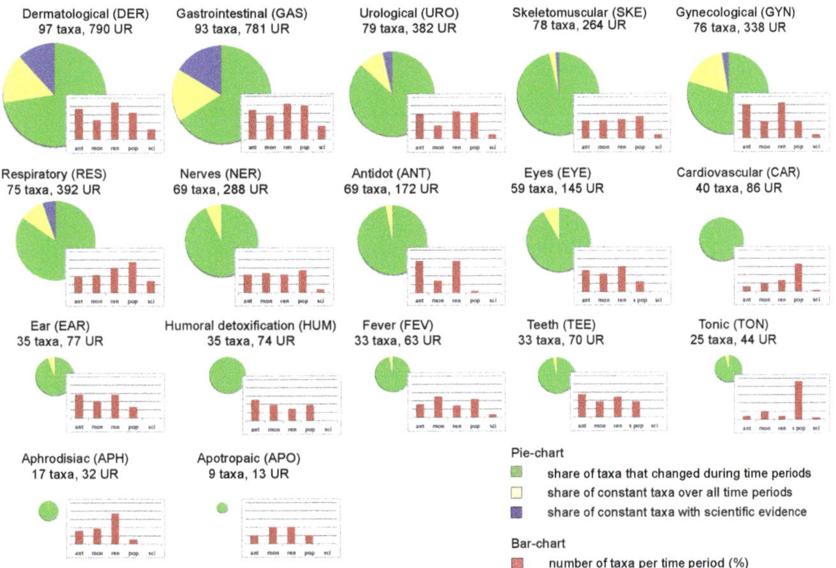

Figure 2. Pie charts show the total number of taxa used for a specific use category. Bar charts show the percentage of total taxa used in different eras.

A steady increase is observable over time in the number of taxa used for the categories RES, CAR and TON (Figure 2, bar charts). Taxa used for cardiovascular problems increased from five in Antiquity, to eight in monastic medicine, 11 in the Renaissance period to 28 in modern and contemporary herbals, but not one single taxon was used through all time periods. TON is associated with a higher number of taxa in contemporary popular herbals (22 taxa, e.g., *Avena sativa*, *Origanum vulgare*, *Thymus vulgaris*, *Urtica dioica*), whereas in Antiquity only *Artemisia absinthium* and *Ficus carica* were considered as general tonics. The concept of antidots was important until the Renaissance, with 14 documented taxa since

Antiquity, whereas in contemporary herbals, this indication is only documented for *Allium sativum* and *Ruta graveolens*.

3.3. Diachronic Changes: The Medicinal Taxon Perspective

Table 3 provides an overview of use categories per taxa over time. In 129 cases (12.6% of all possible cases, i.e., all use categories across all taxa), use categories for a specific taxon remained constant since Antiquity (Table 3, black fields; 1283 UR; found among 56 taxa). For 31.8% of these constant use categories, scientific evidence exists (Table 3, black field with white x; 41 cases, 93 UR). This includes, e.g., *Achillea millefolium* for DER, *Allium sativum* for RES, *Artemisia absinthium* for GAS, *Foeniculum vulgare* for GAS and GYN, *Urtica dioica* for SKE. Taxa with high numbers of constant use categories since Antiquity are: *Urtica dioica* (6 UCat$_{const}$ with 102 UR), *Ruta graveolens* (8 UCat$_{const}$, 88 UR), *Artemisia absinthium* (6 UCat$_{const}$, 67 UR), *Allium sativum* (5 UCat$_{const}$, 61 UR), *Rosa* spp. (5 UCat$_{conat}$, 55 UR), *Thymus* spp. (5 UCat$_{const}$, 54 UR), and *Matricaria chamomilla* (5 UCat$_{const}$, 51 UR).

Table 3. Use categories per taxa documented over the last two millennia. black field: category occurs through all time periods; grey field: category occurs in contemporary era only; light grey: category occurs in several time periods; X: scientific evidence.

Taxon	Use Categories per Taxon per Era [Mean ± Sd]	Min per Era	Max per Era	ANT (Antidote)	APH (Aphrodisiac)	APO (Apotropaic)	CAR (Cardiovascular)	DER (Dermatological)	EAR (Ear)	EYE (Ophthalmic)	FEV (Fever)	GAS (Gastrointestinal)	GYN (Gynecological)	HUM (Humoral detoxification)	NER (Nerves)	RES (Respiratory)	SKE (Skeletomuscular)	TEE (Teeth)	TON (Tonic)	URO (Urological)	VAR (Varia)	Total Use-Records
Abies spp.	6.0 ± 2.5	4	9																			38
Achillea millefolium agg.	3.8 ± 3.6	1	9					X				X	X									67
Acorus calamus L.	4.3 ± 2.5	1	7					X				X										30
Adiantum capillus-veneris L.	4.3 ± 3.2	1	7																			21
Agrimonia eupatoria L.	4.5 ± 3.2	3	9					X				X				X						30
Allium cepa L.	9.8 ± 2.5	7	13																			72
Allium sativum L.	10.5 ± 2.5	8	14				X									X	X					104
Althaea officinalis L.	7.0 ± 1.8	5	9					X				X	X			X						63
Anagallis arvensis agg.	5.3 ± 3.0	1	6																			22
Anethum graveolens L.	5.5 ± 1.7	4	7																			30
Arctium lappa agg.	3.0 ± 1.0	2	4				X														X	21
Artemisia abrotanum L.	7.5 ± 2.5	4	10																			36
Artemisia absinthium L.	11.0 ± 2.2	9	14									X										96
Arum maculatum agg.	3.0 ± 1.2	2	4																			22
Asarum europaeum agg.	5.0 ± 3.4	1	9																			25
Avena sativa agg.	4.8 ± 4.2	2	11					X														34
Beta vulgaris L.	5.0 ± 2.7	1	7																			23
Cannabis sativa L.	2.3 ± 1.9	1	5																			12
Capsella bursa-pastoris agg.	3.5 ± 2.5	1	7					X					X									33
Carum carvi L.	3.3 ± 2.6	1	7									X										28
Chelidonium majus L.	4.5 ± 2.4	3	8									X										27
Cichorium intybus L.	5.0 ± 2.2	2	7									X										34

Table 3. *Cont.*

	Use Categories per Taxon per Era [Mean ± Sd]	Min per Era	Max per Era	ANT (Antidote)	APH (Aphrodisiac)	APO (Apotropaic)	CAR (Cardiovascular)	DER (Dermatological)	EAR (Ear)	EYE (Ophthalmic)	FEV (Fever)	GAS (Gastrointestinal)	GYN (Gynecological)	HUM (Humoral detoxification)	NER (Nerves)	RES (Respiratory)	SKE (Skeletomuscular)	TEE (Teeth)	TON (Tonic)	URO (Urological)	VAR (Varia)	Total Use-Records
Clematis vitalba L.	2.3 ± 1.3	1	4																			9
Colchicum autumnale agg.	1.0 ± 0.0	1	1																			5
Conium maculatum L.	2.3 ± 1.9	1	5																			11
Convallaria spp.	2.5 ± 1.7	1	5																			13
Coriandrum sativum L.	2.8 ± 1.5	1	4									X										18
Corylus avellana L.	3.0 ± 1.4	2	5																			14
Crocus sativus L.	4.5 ± 4.0	1	8																			23
Cucurbita pepo L.	3.8 ± 2.2	1	6																	X		30
Cydonia oblonga Mill.	4.2 ± 2.2	2	5																			29
Daucus carota L.	4.8 ± 3.0	1	8																			35
Eryngium campestre L.	4.0 ± 2.9	1	7																			19
Euphorbia esula agg.	1.8 ± 1.0	1	3																			12
Ficus carica L.	7.3 ± 4.5	2	11																			66
Foeniculum vulgare agg.	7.0 ± 2.5	5	10									X	X			X						76
Fumaria officinalis agg.	2.8 ± 1.3	1	4									X										22
Gentiana lutea agg.	4.0 ± 2.3	2	6									X							X			28
Hedera helix L.	6.0 ± 4.1	1	9													X						34
Helleborus spp.	6.0 ± 4.2	1	10																			32
Heracleum sphondylium agg.	5.3 ± 3.1	1	8																			25
Hordeum vulgare agg.	3.8 ± 2.5	1	7																			32
Hyoscyamus niger L.	8.0 ± 1.8	6	10																			41
Hypericum perforatum agg.	3.6 ± 3.1	1	8					X				X		X								52
Inula helenium L.	5.8 ± 1.6	4	7																			30
Iris germanica agg.	7.5 ± 1.5	1	11																			55
Juglans regia L.	5.3 ± 1.9	4	8					X														48
Juniperus communis agg.	6.3 ± 2.2	3	8									X				X						70
Juniperus sabina L.	2.8 ± 1.0	2	4																			23
Lepidium spp.	7.5 ± 3.0	5	11									X				X						51
Levisticum officinale L.	5.8 ± 1.7	4	8									X								X		38
Linum usitatissimum agg.	4.0 ± 2.0	1	5					X				X										38
Malus sylvestris agg.	6.2 ± 2.5	1	7																			23
Malva sylvestris agg.	8.0 ± 2.0	7	11									X				X						68
Marrubium vulgare L.	6.8 ± 1.0	6	8									X				X						57
Matricaria chamomilla L.	7.8 ± 2.5	5	11					X				X	X				X					73
Melissa officinalis L.	7.8 ± 0.5	7	8					X				X				X						65
Mentha pulegium L.	8.0 ± 1.0	2	11																			48
Mentha spicata agg.	8.3 ± 0.5	8	9					X				X				X						86
Mercurialis spp.	2.5 ± 2.1	1	5																			11
Meum athamanticum Jacq.	3.3 ± 1.0	1	5																			18
Morus nigra L.	3.5 ± 1.0	2	5																			23
Ocimum basilicum L.	5.5 ± 3.5	2	8																			29
Onopordum acanthium L.	2.3 ± 2.9	1	6																			11

Table 3. *Cont.*

Taxon	Use Categories per Taxon per Era [Mean ± Sd]	Min per Era	Max per Era	ANT (Antidote)	APH (Aphrodisiac)	APO (Apotropaic)	CAR (Cardiovascular)	DER (Dermatological)	EAR (Ear)	EYE (Ophthalmic)	FEV (Fever)	GAS (Gastrointestinal)	GYN (Gynecological)	HUM (Humoral detoxification)	NER (Nerves)	RES (Respiratory)	SKE (Skeletomuscular)	TEE (Teeth)	TON (Tonic)	URO (Urological)	VAR (Varia)	Total Use-Records
Origanum vulgare agg.	7.3 ± 5.3	1	11																			45
Papaver somniferum L.	7.0 ± 0.6	3	9																			49
Petasites hybridus (L.) P. Gaertn.	4.5 ± 4.1	1	9									X				X				X		33
Petroselinum crispum (Mill.) Fuss	4.3 ± 1.0	4	5									X								X	X	30
Peucedanum spp.	8.3 ± 1.5	6	10																			52
Pimpinella saxifraga agg.	3.0 ± 2.3	1	5													X						20
Polygonum aviculare agg.	5.0 ± 3.6	1	8																			34
Polypodium vulgare L.	2.8 ± 1.3	3	9									X				X						17
Potentilla spp.	5.8 ± 2.8	3	9							X		X										43
Prunus spp.	4.5 ± 1.5	2	8																			37
Pyrus communis agg.	2.0 ± 1.2	1	3																			10
Quercus robur agg.	4.0 ± 2.2	1	6							X		X										36
Raphanus sativus L.	5.3 ± 3.3	2	9																			36
Rosa spp.	9.3 ± 2.6	7	13					X								X	X				X	91
Rubia tinctorum L.	4.3 ± 3.2	1	7																			20
Rubus idaeus L.	3.3 ± 2.1	1	5									X	X									23
Rumx spp.	6.3 ± 2.2	4	7																			41
Ruta graveolens L.	11.5 ± 1.3	10	13																			114
Salix alba agg.	4.8 ± 2.5	1	6							X					X		X					41
Salvia officinalis agg.	8.0 ± 2.9	4	11				X					X	X									78
Sambucus nigra L.	6.8 ± 2.9	3	10									X				X				X		62
Saponaria officinalis L.	5.8 ± 2.6	3	8													X						34
Secale cereale	1.3 ± 0.6	1	2																			6
Sinapis spp.	7.5 ± 4.2	2	12													X					X	49
Solanum spp.	4.5 ± 1.0	4	6				X															24
Symphytum officinale agg.	3.3 ± 2.2	1	6				X										X					46
Teucrium spp.	5.0 ± 4.2	1	9																			31
Thymus spp.	8.8 ± 1.3	7	9				X									X	X					87
Triticum aestivum agg.	4.0 ± 2.0	1	7																			36
Tussilago farfara L.	2.6 ± 2.2	1	6													X						26
Urtica dioica L.	9.3 ± 1.3	8	11				X										X			X		127
Valeriana officinalis agg.	4.8 ± 3.3	1	9												X							44
Veratrum album agg.	4.8 ± 3.8	2	10																			24
Verbascum thapsus agg.	5.5 ± 2.5	2	8													X						40
Verbena officinalis L.	6.8 ± 2.1	6	11																			60
Vinca minor L.	3.0 ± 1.7	1	4																			18
Viola hirta agg.	6.8 ± 2.5	4	10																			56
Vitis vinifera agg.	5.8 ± 3.3	1	8				X															39

In 159 cases, specific use categories occurred for the first time in contemporary popular herbals (Table 3, grey boxes; 14.1% of total cases; 301 UR; 31 taxa). For 11.3% of these recent use categories, scientific evidence exists (Table 3, gray field with white x; 13 cases, 28 UR). The following species show relatively high numbers of recent use categories: *Achillea millefolium* (7 UCat$_{recent}$, 33 UR), *Sambucus nigra* (5 UCat$_{recent}$, 18 UR), *Valeriana officinalis* (4 UCat$_{recent}$, 21 UR), *Salix alba* (3 UCat$_{recent}$, 13 UR) and *Hypericum perforatum* (3 UCat$_{recent}$, 11 UR).

In 745 cases (73.3% of all cases; 2174 UR), use categories of a specific taxon were documented in one or several time periods, but without continuity (Table 3, light grey fields). For 6.5% of these categories, scientific evidence exists (Table 3, light grey fields with white x; 48 cases; 104 UR).

The average number of total use categories per taxon over all time periods is 10.1 ± 2.9. Per time period, the average number of use categories per taxon is 5.2 ± 2.2. This varies from 11.3 ± 1.3 (*Ruta graveolens*), with a total of 14 use categories over all eras, to 1.0 ± 0.0 (*Colchicum autumnale*), with a total of 2 use categories over all eras. Other species with many use categories are, for example: *Artemisia absinthium* (11.0 ± 2.2, total 16 UCat), *Allium sativum* (10.5 ± 2.5, total 17 UCat), *Allium cepa* (9.8 ± 2.5, total 15 UCat), *Urtica dioica* (9.3 ± 1.3, total 11 UCat) and *Rosa* spp. (9.3 ± 2.6, total 16 UCat).

Few use categories were found for, e.g., *Clematis vitalba* (2.3 ± 1.3, total 7 UCat over all eras), *Cannabis sativa* (2.3 ± 1.9, total 6 UCat), *Conium maculatum* (2.3 ± 1.9, total 7 UCat), *Onopordum acanthium* (2.3 ± 2.9, total 8 UCat), *Pyrus communis* (2.0 ± 1.2, total 6 UCat), *Euphorbia esula* (1.8 ± 1.0, total 5 UCat), and *Secale cereale* (1.3 ± 0.6, total 3 UCat).

3.4. Diachronic Patterns at the Example of Specific Taxa

Relatively few use categories were documented through time, but many new categories in modern and contemporary era were found for, e.g., *Achillea millefolium* (1 $\mathrm{UCat_{const}}$, 26 UR; 7 $\mathrm{UCat_{recent}}$, 33 UR; Table 3 and Figure 3). *Achillea millefolium* was broadly used, with a total of 11 use categories over all time periods. Only dermatological uses have been stable since antiquity and are also documented in the EMA Monograph (2020) [62,63]. In the contemporary era, seven use categories, CAR, GAS, RES, SKE, TEE, TON, and URO, were added. For application in GAS, scientific evidence exists. GYN was documented since the Renaissance and is backed by scientific evidence (EMA Monograph 2020) [66,67].

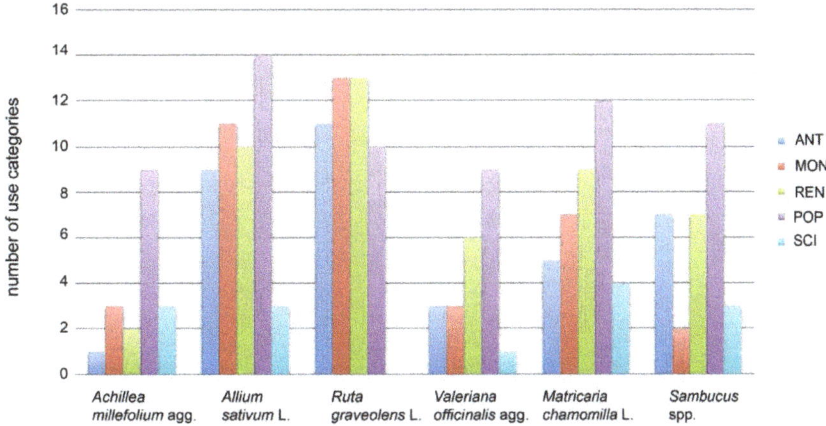

Figure 3. Number of use categories for 6 species during different eras, showing different diachronic trends.

High numbers of use categories through all or several time periods were observed for, e.g., *Allium sativum* (17 UCat, 107 UR); FEV is the only category in which *Allium sativum* was never documented. ANT, DER, GAS, RES remained stable during all time periods and, for RES, scientific evidence exists.

Use categories remained stable over all or several time periods and no additional use categories occurred during the modern and contemporary era for *Ruta graveolens* (14 UCat, 107 UR). *Ruta graveolens* is an example for a species with a very constant use over time. Eight out of 14 categories remained stable, including ANT, DER, EAR, EYE, GAS, GYN, NER, RES (88 UR of 115 UR). However, there is no scientific evidence for any of the uses.

No stable use category existed over time, but a use category added in the modern and contemporary era, backed by scientific evidence, was found for *Valeriana officinalis* (11 UCat, 50 UR; 0 UCat$_{const}$; 3 UCat$_{recent}$, 26 UR). *Valeriana officinalis* was broadly used but without any constant use over time. Contemporary popular herbals added four categories, CAR, GAS, NER and RES, where NER has scientific evidence.

A high number of stable use categories over all time periods plus scientific evidence can be observed for, e.g., *Matricaria chamomilla* (5 UCat$_{const}$, 74 UR); DER, GAS and GYN (58 UR of 100 UR) remained stable over time and are backed by scientific evidence. RES was added in the contemporary era and is also sustained by scientific evidence (EMA-monograph 2015) [68].

4. Discussion

4.1. Medicinal Plant Use Patterns over Time

While some medicinal plants were constantly used for the same reason over the last two millennia, others have a changing use history. A general diversification or decrease in uses over time does not exist; instead, different use trends occur for different species.

More than half of the analyzed taxa (56 out of 102) show specific use categories that were continuously recommended through all time periods. This adds to 12.6% of use categories across all taxa and stands for a body of medicinal plant knowledge and uses continuously practiced in Central Europe over the last two millennia [9].

Changes in medicinal concepts are mirrored in plant use, and specifically in the changing importance of use categories. While some categories are heavily bound to specific medicinal concepts, others remain stable, independently of changing eras and worldviews. For example, dermatology (DER) and gastroenterology (GAS) are use categories that were relevant across all periods, with high species diversity, and a high share of constantly used species sustained by scientific evidence. These use categories also figure prominently in neighboring Mediterranean medicinal floras [2,6], as well as medicinal floras from all over the world, e.g., [69–71]. Obviously, the universal need for effective GAS and DER treatments is largely independent of medicinal concepts and time periods.

Other categories were more susceptible to change. For example, antidotes (ANT), apotropaic magic (APO) and humoral detoxification (HUM) were important use categories in the past, but rarely play a role in contemporary herbals. Instead, new applications reflecting scientific progress and epidemiological challenges arose, such as cardiovascular (CAR) and tonic (TON) uses. The anatomic understanding of blood circulation in the 17th century fueled uses for cardiovascular disorders. At present, they are prominently found in popular herbals, as cardiovascular diseases are among the most common causes of death in Central Europe [72]. Some of the plants used for cardiovascular applications do not directly influence heart activity, but rather have a relaxant and stress reducing effect (e.g., *Melissa officinalis*, *Rosa* spp.). It is thus little surprising that they were formerly used for nerves (NER) and only recently became important for cardiovascular problems.

Furthermore, the humoral (HUM) applications of the past seem to be replaced by tonic (TON) applications in more recent times. Interestingly, the general purpose of a 'tonic', namely, to restore and maintain physiological functioning of an organ system, largely corresponds with the circumscription of humoral detoxification according to the theory of four humors [21,73]. Plants that are used for both categories, HUM and TON, such as, e.g., *Artemisia absinthium* and *Urtica dioica*, usually support digestion and/or have a diuretic effect [74,75].

4.2. Link between Historical and Traditional Uses of Taxa and Scientific Evidence

Approximately 30% of the continuous uses have scientific evidence, compared to 11% among recently added uses and 6% among the discontinuous uses.

This finding seems to support the concept of the social validation of specific plant uses, i.e., the assumption that longstanding use practice suggests efficacy and safety [17,76]. In many European countries, it is possible to register traditionally used medicinal plants as

Traditional Herbal Medicinal Products (Directive 2004/24/EC) [77]. As a proof of traditional use, an uninterrupted use of the product for at least 30 years, 15 of which in the European Union, is required. From an ethnological and historical perspective, this time period does not adequately represent the multifaceted concept of tradition [17]. In particular, products that have been "forgotten" cannot be reintroduced under the concept of tradition. The present data may be used as a resource for traditional herbal medicinal products.

All plants in Table 3 were scientifically investigated to different degrees, but not necessarily tested for specific use categories. Some are considered toxic and, therefore, no longer recommended, such as, e.g., *Tussilago farfara* (pyrrolizidine alkaloids) or *Arum maculatum* (oxalate needles, saponins). For half of the plants, a monograph of Commission E (predecessor of HMPC and EMA) exists. In the case of *Iris germanica* agg., according to Commission E, clinical efficacy was not proven. Since the EMA monographs are prepared in a regulatory context for simplified approvals based on traditional or well-established use, the listed areas of applications are often very narrow. Therefore, many use categories have not been investigated, and the abovementioned 30% cases of continuous use with scientific evidence can be seen as a conservative estimate.

4.3. Diversity of Diachronic Use Patterns Exemplified by Specific Taxa

Allium sativum has been used for all categories throughout time but fever (17 out of 18 UCat). Since antiquity, *Allium sativum* was seen as both a medicine and food [8]. This might be one of the reasons for its very broad use. Its blood-thinning properties have been documented during Renaissance [52] and, since the early 20th century, its popularity increased as pharmacological and clinical studies showed cardiotonic and anti-atherosclerotic effects [78,79].

A feedback loop and mutual impact of scientific discoveries and local popular knowledge can be assumed for *Valeriana officinalis* and its prominent contemporary use as 'nervinum' (neurotonic) [80,81]. The common use of *Valeriana officinalis* as a sedative at present has been known since the late middle-ages [24]. However, broader acceptance only came with pharmaceutical studies in the late 19th century [82].

The use history of *Ruta graveolens* in the Mediterranean is impressive. Gynecological and respiratory uses have been documented in the Hippocratic corpus but dermatological uses and the uses for swollen spleen are also very old [83]. These ancient uses are still practiced in the Mediterranean [2,6]. *Ruta graveolens* is also described in Central European popular herbals. However, at least for Switzerland, there is little evidence of its current use, although the plant is cultivated in gardens [34,84–88]. It seems that *Ruta graveolens* never fully arrived in Central European medicinal practice but is instead a Mediterranean relict.

Both *Achillea millefolium* and *Matricaria chamomilla* are very popular in modern and contemporary times. *Matricaria chamomilla* is by far the most-used medicinal plant among laypeople and experts [34,89]. Abundant phytopharmacological and clinical studies show scientific evidence for use categories documented since antiquity, such as dermatological- (DER), gastroenterological- (GAS), gynecological- (GYN) and respiratory (RES) applications [90,91]. In contrast, *Achillea millefolium* shows a broad expansion of uses in modern and contemporary popular herbals. An expansion of uses is visible for many of the 102 taxa used over the last two millennia and probably reflects an intensive exchange among different cultures and schools of knowledge related to the medicinal landscape.

5. Conclusions

Diachronic insight into medicinal plant use over two millennia highlights changes in specific use categories, which are in line with changes in medicinal concepts, pharmaceutical technologies and new needs. Many medicinal plants show a general extension of uses over time. However, a constant body of specific uses over time for a number of taxa was also identified. These medicinal plants are used in the same way as in Antiquity, monastic medicine and the Renaissance, regardless of basic changes in medicinal concepts and technological development. Overall, they show the highest share of scientific evidence, which supports the concept of social validation, stressing that longstanding use practice may suggest efficacy and safety. With our results, we present a historically based dataset that can be used as source of traditional plant use in a regulatory context. A more detailed look into use patterns through the consideration of herbal drugs and their mode of preparation would deepen our understanding of the linkage between traditional uses, scientific evidence, and the concept of social validation.

Supplementary Materials: The following supporting information can be downloaded at: https://www.mdpi.com/article/10.3390/plants12010135/s1.

Author Contributions: Conceptualization, M.D.C. and C.S.W.; Data curation, M.D.C. and R.S.; Formal analysis, M.D.C.; Methodology, M.D.C., R.S. and C.S.W.; Supervision, C.S.W.; Writing–original draft, M.D.C.; Writing–review and editing, M.L. and C.S.W. All authors have read and agreed to the published version of the manuscript.

Funding: *Stiftung für Komplementärmedizin.*

Institutional Review Board Statement: Not applicable.

Informed Consent Statement: Not applicable.

Data Availability Statement: Not applicable.

Acknowledgments: We thank Franz Huber and Gary Stafford for their kind support with the statistics and figures, Beat Meier and Sandra Claire for their helpful discussions and comments on pharmaceutical issues and medical history, and the *Stiftung für Komplementärmedizin* for their financial support.

Conflicts of Interest: The authors declare no conflict of interest.

References

1. Leonti, M.; Staub, P.O.; Cabras, S.; Castellanos, M.E.; Casu, L. From cumulative cultural transmission to evidence-based medicine: Evolution of medicinal plant knowledge in Southern Italy. *Front. Pharmacol.* **2015**, *6*, 207–2015. [CrossRef] [PubMed]
2. Leonti, M.; Cabras, S.; Weckerle, C.S.; Solinas, M.N.; Casu, L. The causal dependence of present plant knowledge on herbals–Contemporary medicinal plant use in Campania (Italy) compared to Matthioli (1568). *J. Ethnopharmacol.* **2010**, *130*, 379–391. [CrossRef] [PubMed]
3. Clair, S. Die Kräutermedizin des Renaissance-Arztes Tabernaemontanus (16. Jh.) und Phytotherapie heute—Was ist geblieben, was hat sich verändert? *Schweiz. Z. Ganzheitsmed.* **2011**, *23*, 46–52. [CrossRef]
4. Uehleke, B.; Hopfenmueller, W.; Stange, R.; Saller, R. Are the Correct Herbal Claims by Hildegard von Bingen Only Lucky Strikes? A New Statistical Approach. *Complement. Med. Res.* **2012**, *19*, 187–190. [CrossRef] [PubMed]
5. Mayer-Nicolai, C. *Vergleich der Durch die Historischen Autoren Hildegard von Bingen und Leonhart Fuchs Pflanzlichen Arzneimitteln Zugeschriebenen mit Aktuell Anerkannten Indikationen*; Universität Würzburg: Würzburg, Germany, 2008. Available online: https://nbn-resolving.org/urn:nbn:de:bvb:20-opus-33967 (accessed on 22 October 2022).
6. Lardos, A.; Prieto-Garcia, J.; Heinrich, M. Resins and gums in historical *iatrosophia* texts from Cyprus–a botanical and medico-pharmacological approach. *Front. Pharmacol.* **2011**, *2*, 32. [CrossRef] [PubMed]
7. Lardos, A.; Heinrich, M. Continuity and change in medicinal plant use: The example of monasteries on Cyprus and historical *iatrosophia* texts. *J. Ethnopharmacol.* **2013**, *150*, 202–214. [CrossRef] [PubMed]
8. Totelin, L. When foods become remedies in ancient Greece: The curious case of garlic and other substances. *J. Ethnopharmacol.* **2015**, *167*, 30–37. [CrossRef]
9. Touwaide, A.; Appetiti, E. Food and medicines in the Mediterranean tradition. A systematic analysis of the earliest extant body of textual evidence. *J. Ethnopharmacol.* **2015**, *167*, 11–29. [CrossRef]
10. Teixidor-Toneu, I.; Kool, A.; Greenhill, S.J.; Kjesrud, K.; Sandstedt, J.J.; Manzanilla, V.; Jordan, F.M. Historical, archaeological and linguistic evidence test the phylogenetic inference of Viking-Age plant use. *Philos. Trans. R. Soc. Lond. B Biol. Sci.* **2021**, *376*, 20200086, Erratum in *Philos. Trans. R. Soc. Lond. B Biol. Sci.* **2022**, *377*, 20220158. [CrossRef]
11. Prakofjewa, J.; Anegg, M.; Kalle, R.; Simanova, A.; Prūse, B.; Pieroni, A.; Sõukand, R. Diverse in Local, Overlapping in Official Medical Botany: Critical Analysis of Medicinal Plant Records from the Historic Regions of Livonia and Courland in Northeast Europe, 1829–1895. *Plants* **2022**, *11*, 1065. [CrossRef]
12. Wagner, C.; De Gezelle, J.; Komarnytsky, S. Celtic Provenance in Traditional Herbal Medicine of Medieval Wales and Classical Antiquity. *Front. Pharmacol.* **2020**, *11*, 105. [CrossRef]
13. Vos, P. European materia medica in historical texts: Longevity of a tradition and implications for future use. *J. Ethnopharmacol.* **2010**, *132*, 28–47. [CrossRef] [PubMed]

14. Kong, L.Y.; Tan, R.X. Artemisinin, a miracle of traditional Chinese medicine. *Nat. Prod. Rep.* **2015**, *32*, 1617–1621. [CrossRef] [PubMed]
15. Adams, M.; Berset, C.; Kessler, M. Medicinal herbs for the treatment of rheumatic disorders—A survey of European herbals from the 16th and 17th century. *J. Ethnopharmacol.* **2009**, *121*, 343–359. [CrossRef] [PubMed]
16. Adams, M.; Gmünder, F.; Hamburger, M. Plants traditionally used in age related brain disorders—A survey of ethnobotanical literature. *J. Ethnopharmacol.* **2007**, *113*, 363–381. [CrossRef]
17. Jütte, R.; Heinrich, M.; Helmstädter, A.; Langhorst, J.; Menge, G.; Niebling, W.; Pommerening, T.; Tramisch, H.J. Herbal medicinal products—Evidence and tradition from a historical perspective. *J. Ethnopharmacol.* **2017**, *207*, 220–225. [CrossRef]
18. Helmstädter, A.; Staiger, C. Traditional use of medicinal agents: A valid source of evidence. *Drug Discov. Today* **2014**, *19*, 4–7. [CrossRef]
19. Dal Cero, M.; Saller, R.; Weckerle, C.S. The use of the local flora in Switzerland: A comparison of past and recent medicinal plant knowledge. *J. Ethnopharmacol.* **2014**, *151*, 253–264. [CrossRef]
20. Ackerknecht, E.H. *Therapie von den Primitiven bis zum 20 Jahrhundert*; Enke: Stuttgart, Germany, 1970.
21. Porter, R. *The Greatest Benefit to Mankind: A Medical History of Humanity*; W.W. Norton & Company: New York, NY, USA; London, UK, 1997.
22. Eckart, W.U. *Geschichte der Medizin*, 4th ed.; Springer: Berlin, Germany, 2000.
23. Rothschuh, K.E. Idee und Methode in ihrer Bedeutung für die geschichtliche Entwicklung der Physiologie. *Sudhoffs Arch. Z. Für Wiss.* **1962**, *46*, 97–119.
24. Stoll, U. Das "Lorscher Arzneibuch"; Ein medizinisches Kompendium des 8. Jahrhunderts (Codex Bambergensis Medicinalis 1). *Sudhoffs Arch. Z. Für Wiss.* **1992**, *28*, 1–542.
25. Portmann, L. *Hildegard von Bingen; Heilkraft der Natur: "Physica"*; Herder: Freiburg, Germany, 1991.
26. Bruchhausen, W.; Schott, H. *Geschichte, Theorie und Ethik der Medizin*; Vandenhoeck & Ruprecht: Göttingen, Germany, 2008.
27. Gertsch, J. Botanical Drug, Synergy, and Network Pharmacology: Forth and Back to Intelligent Mixtures. *Planta Med.* **2011**, *77*, 1086–1098. [CrossRef] [PubMed]
28. Ledermann, F. Von Nephentes zur modernen Pharmakognosie; ein kurzer Abriss der Geschichte der Phytotherapie. *Schweiz. Med. Z. Phytother.* **2001**, *3*, 32–35.
29. Jenny, M.; Sharma, R. *Heilerinnen und Heiler in der Deutschschweiz: Magnetopathen, Gebetsheiler, Einrenker*; Favre: Lausanne, Switzerland, 2009.
30. Glaser, C. *Das Sachranger Rezeptbuch*; S. Hirzel: Stuttgart, Germany, 2006; Volume 1.
31. Hoefert, H.W.; Uehleke, B. *Komplementäre Heilverfahren im Gesundheitswesen. Analyse und Bewertung*; Huber: Bern, Switzerland, 2009.
32. Wolff, E. Volksmedizin. *In* Historisches Lexikon der Schweiz. 2014. Available online: https://hls-dhs-dss.ch/de/articles/025626/2014-12-27/ (accessed on 22 October 2022).
33. Roth, S. *Im Streit um Heilwissen, Zürcher Naturheilvereine Anfangs des 20. Jahrhunderts*; Chronos: Zürich, Germany, 1991.
34. Dal Cero, M.; Saller, R.; Weckerle, C.S. Herbalists of Today's Switzerland and Their Plant Knowledge. A Preliminary Analysis from an Ethnobotanical Perspective. *Forsch. Komplementärmed.* **2015**, *22*, 238–245. [CrossRef]
35. Heilmann, K.E. *Kräuterbücher in Bild und Geschichte*, 2nd ed.; Konrad Kölbl: München-Allach, Germany, 1973.
36. Wurzer, W. *Die grosse Enzyklop.die der Heilpflanzen; ihre Anwendung und ihre Natürliche Heilkraft*, 2nd ed.; Neuer Kaiser Verlag: Klagenfurt, Austria, 2000.
37. Saller, R.; Reichling, J.; Hellenbrecht, D. *Phytotherapie; Klinische, Pharmakologische und Pharmazeutische Grundlagen*; Haug: Heidelberg, Germany, 1995.
38. Flück, H. *Unsere Heilpflanzen*; Ott: Thun, Switzerland, 1941.
39. Leclerc, H. *Précis de Phytothérapie; Thérapeutique par les Plantes Francaises*; Masson: Paris, France, 1976.
40. Weiss, R.F. *Lehrbuch der Phytotherapie*, 6th ed.; Hippokrates: Stuttgart, Germany, 1985.
41. Strank, K.J.; Meurers-Balke, J. (Eds.) *Obst, Gemüse und Kräuter Karls des Grossen*; Philip von Zabern: Mainz am Rhein, Germany, 2008.
42. Lonicero, A. *Kreuterbuch*; Druckts und Verlegts Matthäus Wagner: o.O., 1679.
43. Wichtl, M. (Ed.) *Teedrogen und Phytopharmaka: Ein Handbuch für die Praxis auf Wissenschaftlicher Grundlage*; Wissenschaftliche Verlagsgesellschaft: Stuttgart, Germany, 2008.
44. Madaus, G. *Lehrbuch der Biologischen Heilmittel*; G. Thieme: Leipzig, Germany, 1938.
45. Schramm, H. *Heilmittel- Fibel zur Anthroposophischen Medizin*; Oratio: Schaffhausen, Switzerland, 1997.
46. Scheffer, M. *Die Original-Bachblüten-Therapie: Das Gesamte Theoretische und Praktische Bachblüten-Wissen*; Südwest: München, Germany, 2011.
47. ESCOP. (Ed.) *ESCOP Monographs: The scientific Foundation for Herbal Medicinal Products: Supplement*, 2nd ed.; Thieme: Stuttgart, Germany, 2009.
48. ESCOP (Ed.). *ESCOP Monographs: The Scientific Foundation for Herbal Medicinal Products*, 2nd ed.; Thieme: Stuttgart, Germany, 2003.
49. EMA (European Medicines Agency). 2014. Available online: http://www.ema.europa.eu/ema/index.jsp?curl=/pages/medicines/landing/herbal_search (accessed on 22 October 2022).
50. Berendes, J. *Des Pedanios Dioskurides aus Anazarbos Arzneimittellehre in fünf Büchern*; Enke: Stuttgart, Germany, 1902.

51. Mayer, J.G.; Goehl, K. (Eds.) *Höhepunkte der Klostermedizin; Der «Macer floridus» und das Herbarium des Vitus Auslasser*; Reprint-Verlag: Leipzig, Germany, 2001.

52. Dobat, K.; Dressendörfer, W. (Eds.) *Leonhart Fuchs. The New Herbal of 1543*; Taschen: Köln, Germany, 2001.

53. Tabernaemontanus, J.T.N. *Kräuter-Buch, Jetzt Widerumb Auffs Newe Übersehen vnd Anderm Vermehret durch Hieronymus Bauhin*, 4th ed.; König: Basel, Switzerland, 1664.

54. Kneipp, S. *Meine Wasserkur; So Sollt ihr Leben. Die Weltberühmten Ratgeber in Einem Band*, 8th ed.; Trias: Stuttgart, Germany, 2010.

55. Künzle, J. *Das grosse Kräuterheilbuch; Ratgeber für Gesunde und Kranke Tage*; Walter: Olten, Switzerland, 1945.

56. Vogel, A. *Der kleine Doktor*; Vogel: Teufen, Teufen, 1952.

57. Valnet, J. *Phytotherapie: Traitement des Maladies par les Plantes*; Maloine: Paris, France, 1992.

58. Treben, M. *Gesundheit aus der Apotheke Gottes*, 91st ed.; Ennsthaler: Steyr, Austria, 2011.

59. Vonarburg, B. *Natürlich Gesund mit Heilpflanzen*; AT Verlag: Aarau, Switzerland, 1988.

60. Bühring, U. *Praxis-Lehrbuch der Modernen Pflanzenheilkunde*; Sonntag: Stuttgart, Germany, 2005.

61. Matthioli, P.A. *I Discorsi di M. Pietro Andrea Matthioli. Sanese, Medico Cesareo, et del Serenissimo Principe Ferdinando Archiduca d'Austria & c*; Nelli sei libri di pedacio dioscoride anazarbeo della Materia Medicale, Vincenzo Valgrisi: Venezia, Italy, 1568; Reprinted in Facsimile, Stabilimento Tipografico Julia: Rome, Italy, 1967–1970.

62. Höfler, M. *Deutsches Krankheitsnamen-Buch*; Piloty & Loehle: München, Germany, 1899.

63. Landolt, E. *Flora indicativa; Ökologische Zeigerwerte und Biologische Kennzeichen zur Flora der Schweiz und der Alpen*; Haupt: Bern, Switzerland, 2010.

64. KEWScience, Plants of the World online. Available online: https://powo.science.kew.org/ (accessed on 22 October 2022).

65. Stevens, P.F. Angiosperm Phylogeny Website; 2001. Available online: http://mobot.org/MOBOT/research/APWeb/ (accessed on 22 October 2022).

66. Committee on Herbal Medicinal Products (HMPC) (Ed.). Community Herbal Monograph on *Achillea millefolium* L., Flos Final. EMA/HMPC/143949/2010. 2011. Available online: https://www.ema.europa.eu/en/medicines/herbal/millefolii-flos (accessed on 22 October 2022).

67. Committee on Herbal Medicinal Products (HMPC) (Ed.). European Union Herbal Monograph on *Achillea millefolium* L., herba Final—Revision 1. EMA/HMPC/376416/2019. 2020. Available online: https://www.ema.europa.eu/en/medicines/herbal/millefolii-herba (accessed on 22 October 2022).

68. Committee on Herbal Medicinal Products (HMPC) (Ed.). European Union Herbal Monograph on *Matricaria recutita* L., flos Final. EMA/HMPC/55843/2011. 2015. Available online: https://www.ema.europa.eu/en/medicines/herbal/matricariae-flos (accessed on 22 October 2022).

69. Weckerle, C.S.; Huber, F.; Yongping, Y.; Weibang, S. Plant Knowledge of the Shuhi in the Hengduan Mountains, Southwest China. *Econ. Bot.* **2006**, *60*, 3–23. [CrossRef]

70. Bradacs, G.; Heilmann, J.; Weckerle, C.S. Medicinal plant use in Vanuatu: A comparative ethnobotanical study of three islands. *J. Ethnopharmacol.* **2011**, *137*, 434–448. [CrossRef] [PubMed]

71. Monigatti, M.; Bussmann, R.; Weckerle, C.S. Medicinal plant use in two Andean communities located at different altitudes in the Bolìvar Province, Peru. *J. Ethnopharmacol.* **2012**, *145*, 450–464. [CrossRef] [PubMed]

72. B.F.S. Bundesamt für Statistik. 2015. Available online: http://www.bfs.admin.ch/bfs/portal/de/index/themen/14/02/04/key/01.html#parsys_60885 (accessed on 22 October 2022).

73. Götti, R.P.; Melzer, J.; Saller, R. An Approach to the Concept of Tonic: Suggested Definitions and Historical Aspects. *Forsch. Komplementärmed.* **2014**, *21*, 413–417. [CrossRef]

74. Joshi, B.C.; Mukhija, M.; Kalia, A.N. Pharmacognostical review of Urtica dioica L. *Int. J. Green Pharm.* **2014**, *8*, 4.

75. Szopa, A.; Pajor, J.; Klin, P.; Rzepiela, A.; Elansary, H.O.; Al-Mana, F.A.; Mattar, M.A.; Ekiert, H. Artemisia absinthium L.-Importance in the History of Medicine, the Latest Advances in Phytochemistry and Therapeutical, Cosmetological and Culinary Uses. *Plants* **2020**, *9*, 1063. [CrossRef]

76. Crellin, J.K. Social validation: An historian's look at complementary/alternative medicine. *Pharm. Hist.* **2001**, *31*, 43–51.

77. Directive 2004/24/EC. Available online: https://health.ec.europa.eu/medicinal-products/herbal-medicinal-products_en (accessed on 14 November 2022).

78. Suleria, H.A.R.; Butt, M.S.; Khalid, N.; Sultan, S.; Raza, A.; Aleem, M.; Abbas, M. Garlic (*Allium sativum*): Diet based therapy of 21st century—A review. *Asian Pac. J. Trop. Dis.* **2015**, *5*, 271–278. [CrossRef]

79. Chan, W.-J.J.; McLachlan, A.J.; Luca, E.J.; Harnett, J.E. Garlic (*Allium sativum* L.) in the management of hypertension and dyslipidemia—A systematic review. *J. Herb. Med.* **2020**, *19*, 100292. [CrossRef]

80. Abascal, K.; Yarnell, E. Nervine Herbs for Treating Anxiety. *Altern. Complement. Ther.* **2004**, *10*, 309–315. [CrossRef]

81. Fischer, B.; Hartwich, C. *Hagers Handbuch der Pharmazeutischen Praxis*; Springer: Berlin, Germany, 1903; Volume 1–2.

82. Anagnostou, S. Johanniskraut, Baldrian und Passionsblume; Die Geschwister der Seele. Pharmazeutische Zeitung Online Ausgabe. Volume 48. 2011. Available online: http://www.pharmazeutische-zeitung.de/index.php?id=40172 (accessed on 22 October 2022).

83. Pollio, A.; Natale, A.; Appetiti, E. Continuity and change in the Mediterranean medical tradition: *Ruta* spp. (Rutaceae) in Hippocratic medicine and present practices. *J. Ethnopharmacol.* **2008**, *116*, 469–482. [CrossRef] [PubMed]

84. Wegmann, U. Ethnobotanik im Prättigau. Medizinalpflanzen—Nutzung und Wissen. Master's Thesis, University of Zürich, Zürich, Switzerland, 2013.
85. Poncet, A. Pflanzen und Menschen im Emmental; Eine Ethnobotanische Studie über den Kräuterhandel einer Bauernfamilie des Voralpengebiets. Master's Thesis, Université Neuchâtel, Neuchâtel, Switzerland, 2005.
86. Broquet, C. Chasseral, à la Rencontre de L'homme et du Vegetal: Enquêtes Ethnobotaniques sur L'utilisation des Plantes dans une Région de la Chaîne Jurassienne. Master's Thesis, Université Neuchâtel, Neuchâtel, Switzerland, 2006.
87. Brühschweiler, S. *Plantes et Savoirs des Alpes: L'exemple du Val d'Anniviers*; Monographic SA: Sierre, Switzerland, 2008.
88. Poretti, G. *La Malva Tücc i maa i a Calma: Inventario Etnobotanico delle Piante Medicinali del Cantone Ticino*; Centro di dialettologia e di Etnografia: Bellinzona, Switzerland, 2011.
89. Kummer, G. *Schaffhauser Volksbotanik; 1. Die Wildwachsenden Pflanzen*; Neujahrsblatt, Naturforschende Gesellschaft: Schaffhausen, Switzerland, 1953.
90. Gardiner, P. Complementary, Holistic, and Integrative Medicine: Chamomile. *Pediatr. Rev.* **2007**, *28*, 16–18. [CrossRef]
91. McKay, D.L.; Blumberg, J.B. A Review of the Bioactivity and Potential Health Benefits of Chamomile Tea (*Matricaria recutita* L.). *Phytother. Res.* **2006**, *20*, 519–530. [CrossRef]

Article

"Cow Healers Use It for Both Horses and Cattle": The Rise and Fall of the Ethnoveterinary Use of *Peucedanum ostruthium* (L.) Koch (fam. Apiaceae) in Sweden

Erik de Vahl [1], Giulia Mattalia [2,3] and Ingvar Svanberg [2,*]

[1] Department of Landscape Architecture, Planning and Management, Swedish Agricultural University, POM, SE-23422 Lomma, Sweden
[2] Institute for Russian and Eurasian Studies, Uppsala University, SE-75120 Uppsala, Sweden
[3] Department of Environmental Sciences, Informatics and Statistics, Università Ca' Foscari Venezia, 30123 Venice, Italy
* Correspondence: ingvar.svanberg@ires.uu.se

Abstract: Masterwort, *Peucedanum ostruthium* (L.) Koch, is an Apiaceae species originally native to the mountain areas of central and southern Europe. Written sources show that it was used in northern Europe. This study explores the cultivation history of masterwort and its past use in Sweden. Although only few details are known about the history of this taxon, it represents a cultural relict plant of an intentionally introduced species known in Sweden as early as the Middle Ages. In Sweden, the masterwort was mainly used as an ethnoveterinary herbal remedy from the seventeenth to nineteenth centuries. However, medicinal manuals, pharmacopoeias and some ethnographical records indicate that it was once also used in remedies for humans. Today, this species remains as a living biocultural heritage in rural areas, especially on the surviving shielings, which were once used as mountain pastures in Dalecarlia, and at former crofts that were inhabited by cattle owners in the forest areas of southern Sweden.

Keywords: cultural relict plants; herbal remedies; historical ethnobotany; living biocultural heritage; silvopastoral system

Citation: de Vahl, E.; Mattalia, G.; Svanberg, I. "Cow Healers Use It for Both Horses and Cattle": The Rise and Fall of the Ethnoveterinary Use of *Peucedanum ostruthium* (L.) Koch (fam. Apiaceae) in Sweden. *Plants* **2022**, *12*, 116. https://doi.org/10.3390/plants12010116

Academic Editor: Ain Raal

Received: 20 November 2022
Revised: 20 December 2022
Accepted: 21 December 2022
Published: 26 December 2022

1. Introduction

1.1. Background

Between 2007 and 2017, a nationwide inventory organised by the Programme for Diversity of Cultivated Plants (POM) collected data and living material for cultural heritage species in Sweden. The plant material was documented together with information regarding its cultivation history, use and traditions [1]. Many interesting landraces and old cultivars of herbs, vegetables and ornamental plants were found [2–4]. In the POM inventory, a specimen of masterwort, *Peucedanum ostruthium* (L.) Koch, synonym *Imperatoria ostruthium* L. (fam. Apiaceae), was found in Leksand parish in the province of Dalecarlia, central Sweden. This taxon has survived as a cultural relict plant for centuries, especially in Dalecarlia, but also in some other areas of Sweden [5].

When the young Carl Linnaeus and his travel companions passed by Nås parish in western Dalecarlia in the rainy summer of 1734, he noted in his diary on 11 August that *P. ostruthium* was cultivated there [6]. Later, Linnaeus also mentioned its presence in the mountainous areas of Lima parish (i.e., in the Transtrand area) in the same province [7]. Linnaeus's earlier observations on the cultivation of masterwort at the location were confirmed when topographer and writer Abraham Hülphers described Lima parish in 1762 [8]. Hülphers wrote about the importance of cultivating medicinal plants in such a remote and isolated area, where the vicar also served as the local physician. *P. ostruthium* was reported to be used with hard liquor (Swedish: *brännvin*) against colic, and as a smoked substance for calming nosebleeds, ear pain and toothache [8]. Currently, the use of *P. ostruthium* is reported as a medicinal plant for human and veterinary purposes by several ethnobotanical studies in the Alpine region (i.e., in Austria [9,10], Italy [11] and Switzerland [12]) (Figure 1).

Figure 1. *Peucedanum ostruthium* in Orsa parish, Dalecarlia (Photo Arne Holmer).

1.2. Biology

Masterwort is a hemicryptophyte and stem plant. It is a perennial herb with large rhizomes and a round stalk that grows 30 to 100 cm high. The stalk is erect, hollow, round, leafy and slightly branched. Lower leaves are on long stalks, twice ternate. The upper leaves are less compound and on shorter stalks, with a sheeting, membranous dilatation at the base. The flowers are white and slightly reddish, and sit in a pair of large, somewhat flat umbels. The plant does not bloom every year, but propagation occurs vegetatively through underground runners [13].

The rhizomes contain between 0.18 and 0.78 percent essential oils, especially sabinene. The plant is a source of coumarins, including 1.3 percent oxypeucedanin ($C_{13}H_{12}O_2$), 0.3 percent ostruthol ($C_{24}H_{24}O_8$), imperatorin, 0.1 percent osthole ($C_{12}H_{18}O_2$), isoimperatorin and 0.5 percent ostruthin ($C_{18}H_{20}O_8$) [10,14,15].

1.3. Distribution and Ecology

P. ostruthium grows wild at altitudes between 800 m and 2800 m in the mountains of central and southern Europe, in meadows and along streams. Its native distribution includes the Carpathians, the Alps, the northern Apennines, the Massif Central, the Jura Mountains and scattered occurrences in the Iberian Peninsula. In Germany, it is found in the Harz area, the Thuringia and the Ore Mountains (Erzgebirge). Since it is widely introduced and cultivated as a medicinal plant, its native range is not entirely clear [9,13,16]. In the British Isles (northern England, mid to northern Scotland, Isle of Man and northwest Ireland), it is considered introduced and naturalised in moist meadows and riverbanks [17]. In the Scandinavian countries, it has been introduced in Denmark, Norway and Sweden. It reached Scandinavia in the late medieval times [18–20].

From modern provincial floras, it appears to be found as a cultural relict plant in forest areas from Skåne up to Dalecarlia, where it can survive as a reminder of older times [21]. In Dalecarlia, it was reported from 140 recording sites in 32 parishes until 1960, especially at shielings. However, it has also been found in abandoned places of residence in the lowland areas of Dalecarlia [22]. SLU Swedish Species Information Centre reports known observations (until 2022) from the provinces of Skåne (20), Blekinge (9), Småland (422), Halland (68), Bohuslän (86), Gotland (8), Öland (3), Västergötland (361), Östergötland (191), Dalsland (68), Södermanland (24), Värmland (25), Närke (56), Västmanland (60), Dalecarlia (302), Gästrikland (47), Hälsingland (22), Härjedalen (2), Jämtland (4), Åsele lappmark (1), Pite lappmark (14), Lycksele lappmark (1) and Västerbotten (7) (Figure 2) [23].

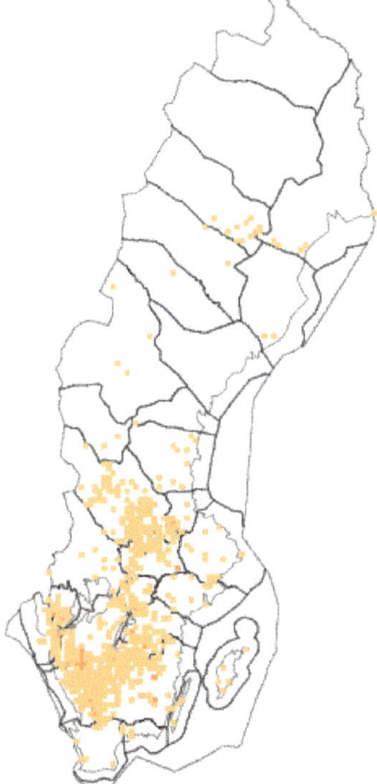

Figure 2. Contemporary distribution of *Peucedanum ostruthium* in Sweden. Source: SLU Swedish Species Information Centre. Light squares < 10 observations, dark squares 10–500 observations.

It can survive for a long time in old yards if there is not too much competition with other plants or from forests. *P. ostruthium* does not flower very often, and the flowers are rarely fully formed. It is highly uncertain whether the plant can form viable seeds in Sweden [21].

Today, the plant is rarely cultivated in Sweden, other than at botanical gardens (e.g., Uppsala Botanical Garden, the Friends of the Garden Society garden in Vadstena, Floras Garden in Sollentuna, Vallby open air-museum in Västerås and Svanå Garden in Boden, according to our knowledge). It is also preserved at the Swedish National Gene Bank for Vegetatively Propagated Horticultural Crops at the Swedish Agricultural University in Alnarp.

1.4. Aim of the Article

The main aim of this study is to bring together historical and contemporary data about *P. ostruthium* and to describe its earlier importance as a medicinal plant in Sweden. Specifically, we document its introduction and cultivation history in Sweden and its folk botanical importance, especially within ethnoveterinary medicine, among the pre-industrial cattle-breeding peasantry.

2. Material and Methods

Data Collection and Analysis

Older kitchen herbs, vegetables and medicinal plants were searched for when producing POM's nationwide inventories of cultivated plants [2–4]. Only four notifications regarding *P. ostruthium* were received, all of which were collected for cultivation trials and conservation [2]. In addition to POM's plant inventories, botanical handbooks, ethnographic archive materials and local historical works have also been used as sources [24]. A diachronic perspective was chosen in order to outline and analyse the regression and changes in the use of masterwort, during the course of which we took into account the social, ecological and chemical aspects of this plant's usage [4,24].

3. Results

3.1. Ethnographic Context

Several sources agree on linking masterwort to silvopastoral activities. For instance, in the eighteenth and nineteenth centuries, travellers and botanists in Dalecarlia noticed that *P. ostruthium* was growing at the shielings (Swedish: *fäbodar*) used by peasants as mountain pastures for grazing their livestock in the summertime [25,26]. The female herders cultivated the plant in order to use it for ethnoveterinary purposes [27]. However, it is also known to had been grown by cattle-keeping crofters in forest areas of central and southern parts of Sweden [21,26].

One specific biocultural context for the species is the transhumance pasturing system used in the province of Dalecarlia. From June until the end of September, the young peasant women would spend the summer in the shielings up in the forest-covered mountains, tending the animals and making various dairy products, while the men spent their days further away, fishing and haymaking in the wet meadows [27,28]. Although most of these shielings were abandoned in the mid-twentieth century or earlier, a few remain still active. The remaining shielings have now been converted into summerhouses. The lifestyle and the human–animal relationship associated with the shielings included many noteworthy cultural traits of great interest for ethnobiologists, for instance the special kind of song form (*kulning*) used by the women as a way to communicate not only with each other, but also with the cattle [29]. The material culture adapted to the herding way of life is also of great interest. While spending their days in the forests, the girls and women also gathered local resources which they could sell after returning to the villages on Michaelmas (29 September) [30]. These products included bundles of *Equisetum hyemale* L., which were used to scour or clean wooden milk vessels, and lumps of resin from *Picea abies* (L.) H. Karst, used as a kind of 'chewing gum' by the peasantry during church services [27]. Few plants were cultivated at the shielings, although some herbaceous taxa used as veterinary herbs were grown, for instance lovage (*Levisticum officinale* L.), tansy (*Tanacetum vulgare* (L.) Bernh.) and the previously mentioned masterwort [31]. The species has also been cultivated by livestock farmers in southern and central Sweden [29,32,33]. In some areas, horseheal (*Inula helenium* L.) has also been grown as a livestock medicine [34,35].

The relationship between humans and *P. ostruthium* dates back to medieval times in Sweden. However, this interaction has changed over time. Some other plants, like *L. officinale* and *T. vulgare*, were also planted in order to control diseases among livestock. Today, these taxa remain as cultural relics on the surviving shielings [25]. These plants have been cultivated for centuries, probably as early as the medieval times in Sweden [20]. *L. officinale* is still used as a vegetable and for seasoning, and is therefore cultivated in many Swedish gardens [36], while the use of *P. ostruthium* has mostly fallen into oblivion. Remaining *T. vulgare* is regarded as an ornamental plant at best [37]. *I. helenium* might be naturalised in some areas [35].

3.2. Cultivation History

The medieval Latin name *Imperatoria* (from *imperatoris*, meaning 'ruler' or 'master') was translated into French as *impératoire*, and rendered in English as *masterwort* (first recorded in 1653), in German as *Meisterwurz* (recorded since 1480), in Danish as *mesterrod* (recorded since 1678) and in Swedish as *mästerrot* (recorded since 1632), alluding to its reputation as a plant with superior healing properties [38]. 'Master' was once a title for a physician, and the plant was known by these names because it was regarded as a divine medicine [21,32,38].

Very little is known about its earlier history as a cultivated plant. However, it seems that in ancient times it was primarily cultivated for human ailments. Archaeological finds show that it was introduced into the British Isles by the tenth century. Seeds dated around 850 to 950 CE have been found in Antrim, Northern Ireland [39]. Swiss natural historian Conrad Gessner describes it as a cultivated plant in 1560 [38].

3.3. Medicinal Plant for Curing Humans

This taxon seems to have entered medicine as recently as the Middle Ages, probably in Germanic territory. Evidence from antiquity is lacking. The use of *P. ostruthium* in human remedies was first recorded in medieval times. A review of herbal books and medicinal manuals from the late Middle Ages shows that it was known as *Ostruhium, Astrantia and Magistrantia* [38]. In a Danish manuscript by the thirteenth-century Roskilde Cathedral canon Henrik Harpestraeng (who died in 1244), it is described as a kind of panacea used in remedies for liver diseases, jaundice, gallstones and cough. It is also mentioned as being grown in Danish physic gardens in the 1530s [40]. Italian physician Pietro Andrea Mattioli propagated it for its usefulness in making remedies in sixteenth century. So did German botanist Jakob Tabernaemontanus in the same century. English herbalist John Gerard wrote in his herbal 1597 that "the rotes and leaves stumped, doth dissolve and all pestilential carbunchles and botches, and such other apostemetions and swellings" [38] (Figure 3).

The canon Christiern Pedersen in Lund wrote in 1533 that it was grown in physic gardens, and recommended it for treating lower back pain [41]. Henrick Smid, who practiced medicine in Malmö, confirms in his herbal its use in human medicine [42].

In Swedish medicinal handbooks from the fifteenth and early sixteenth centuries, it is known as *Astrice*. The plant was recommended by Laurentius Gothus Paulinus in 1623 as a remedy against pestilence [43]. Nyström's review of plants grown in apothecary gardens in Sweden during the eighteenth century confirms that the masterwort was among the species appearing in all gardens studied [44]. Apothecary gardens emerged due to the professionalisation of the art of medicine that developed during the century, and the species in cultivation did not always correspond to the plants used in folk medicine. *P. ostruthium* was included in the Swedish pharmacopoeia from 1698 (as *Radix imperatoriae*), but was withdrawn in 1869 [45,46]. However, it had fallen out of use in academic medicine much earlier.

It did, however, continue to have some use in folk medicine [38]. In 1806, Retzius mentioned that the peasantry dried rhizomes for use against colic and mental illness in women, and it was given to small children twice a day in powder form to prevent intestinal worms [47]. We have some fragmentary data regarding its continued use for treating livestock (Upper Dalecarlia and Norrbotten) in the late nineteenth century [26,27]. Nowadays, it is forgotten in folk medicine. It is known among some current practitioners of alternative medicine, but we have no indication of whether it is actually used. Its supposed medicinal properties have also been claimed by mountain dwellers elsewhere in Europe, even up to the present day. In the Alpine regions of Europe, its rhizomes have been used for a great number of ailments, including gastrointestinal complaints, wounds, skin problems and toothache [48–50]. Among the Saami people in northern Sweden, the sharp tasting roots of a related wild species, milk parsley, *Peucedanum palustre* L., was earlier harvested and used against various ailments [51].

Figure 3. Masterwort in a colourised illustration from Pietro Andrea Mattioli's *De plantis epitome utilissima*, published by Joachim Camerarius in 1611.

3.4. Ethnoveterinary Medicine

Despite having largely ceased to be used for treating humans in academic and folk medicine, it began to be cultivated in the eighteenth century for ethnoveterinary purposes by the peasantry in southern Sweden, as far north as the province of Dalecarlia [52,53]. It was also occasionally grown for ethnoveterinary purposes in the northern part of Sweden [33]. In POM's nationwide surveys, the four accessions of masterwort described were given cultivar names relating to either the person who had cultivated them or how they were used [2]. Two of these, labelled 'Brita-Sofia' and 'Skogen', were described as surviving relict plants from old private gardens that could be dated through the living memories of the donors. However, they had no connection with any known medicinal use. The other two, known as 'Kampkur' ('horse cure') and 'Kobota' ('cow healer'), bear witness to their use in folk veterinary medicine [2] (Figure 4).

'Kampkur' was collected from Rällsjögården in Bjursås parish, Dalecarlia, where the farmer Rällsjö Anders Jansson (1863–1951) was active as a folk healer at the turn of the twentieth century. Inspired by the book *Pharmaca Composita*, published in 1896 [54], Rällsjö Anders used the medicinal plants that had long been grown at the family's shieling in Axmor for home remedies. The masterwort, called 'Kampkur', was included in a medicine for horses that also included rhubarb (*Rheum* sp.), mezereum (*Daphne mezereum* L.) and garden angelica (*Angelica archangelica* L.), and was preserved at the summer pasture by his daughter Rällsjö Brita (1901–2006), who had realised the value of maintaining both plants and memories from an older garden culture [55] (Figure 5).

Figure 4. *Peucedanum ostruthium* 'Kobota'. One of the samples of masterwort collected, evaluated and described in Swedish nationwide inventories of cultivated plants (Photo: Erik de Vahl).

Figure 5. Rällsjö Anders Jansson (1863–1951) used masterwort in a remedy for horses called 'Kamp-kur' (horse remedy). Many plants from his garden in Dalecarlia are preserved in the Swedish National Gene Bank (Photo from private collection).

'Kobota' was collected from Rosa Backman's garden in Orust in Bohuslän, western Sweden. It bears the name Mrs Backman gave it to describe how it was traditionally used in the area. In the spring, when the cows were weak, leaves from *P. ostruthium* were gathered, rolled in rye flour and then stuffed down the animals' throats. Mrs Backman recalls that the animals got up after a number of treatments, and she explained the effect of the plant with its high vitamin content.

No chemical analyses of the four collections have been carried out to date, but sensory tests, made by E.D.V. at Alnarp, have shown that the chemical content of the roots differs greatly in terms of acidity, fruitiness, bitterness and aroma. They cannot be morphologically separated, but differences in vitality and plant vigour indicate a genetic variation within the collected and preserved plant material.

These ethnoveterinary uses of *P. ostruthium* are also confirmed by the cultural historical and ethnographic data. It was known and used against animal diseases in the early seventeenth century. Masterwort is mentioned in Mårten Behms's 1648 medicinal handbook for treating horse diseases [56]. Åke Claesson Rålamb also mentioned it as remedy plant for horses, cattle and other domestic animals in his medicinal handbook dated 1690 [57]. It has been widely used against livestock diseases in particular, and was used as a diuretic and laxative for horses and against swine fever and rinderpest [58]. Still in the nineteenth century, sick swines were washed with decoction of masterwort in Dalecarlia [59].

Its use became popular in ethnoveterinary medicine in the eighteenth century. It was mainly the rhizomes that were used: fresh or dried, cut into pieces, taken as a decoction or chewed. The leaves were also used as treatments. Retzius summarised in 1806 how cattle healers 'used it for both horses and cattle', for example when treating loss of appetite [47]. Its use is recorded at many shielings [60,61]. The same use is also mentioned in other parts of Sweden [25,33]. Detailed data from Vilhelmina in Lapland describes how a local healer named Hans Persson in Latikberg used *P. ostruthium* to cure horses and cows [62].

Masterwort was also used as cattle medicine in other Northern European countries, for instance at the shielings in Norway. Høeg recorded that it was planted for use as a cow medicine, and that the herding women came from other shielings in the vicinity to harvest leaves to be used as medication [63]. Its use for ethnoveterinary purposes is recorded in Germany, Austria, Switzerland and northern Italy [64]. A record from Danish West Jutland states that masterwort was a common home remedy against "destruction by evil forces" in both animal and human folk medicine [40]. The addition of chopped masterwort and flax seed to beer, or sprinkled over the patient's head, was claimed to protect against the devil [65].

4. Discussion

Our knowledge of plants used in ethnoveterinary medicine by the peasantry in pre-industrial Sweden is still limited [66]. POM's inventories of cultural plants all over the country show that *P. ostruthium* was an important plant in folk veterinary medicine, and this has been a factor for finding and evaluating old plant material for long-term conservation. However, we can also learn that medicinal plants that have lost their function and place in modern gardens were sometimes neglected, and were only collected from donors with knowledge of old traditions.

Sweden's Virtual Herbarium includes records of masterwort from the areas around Lima [67]. All of these are from the early twentieth century. Reported localities in the Swedish Species Observation System (Artportalen) database show that the species was still growing at shielings in mountainous areas in the late 1980s. This indicates that the species has been growing continuously in the area [68]. Nevertheless, available sources do not mention how the plant was propagated between different shielings, while this information is reported for many other plant species in the inventory. We cannot tell why this plant was common among Dalecarlian pastoralists, but not in the regions north of Dalecarlia where shieling culture was also common.

POM's inventories of heirloom plants have been based on information from the public and trained volunteers working in the field, and the method of selection was based on

opportunities to link knowledge of the cultivation history of a plant to living memories or documentation of older horticultural culture [4]. This method has probably excluded species that have lost their importance, and in those cases where masterwort has been found, the donors' knowledge of the plant's historical use has been decisive. A different method based on information from the literature, such as that describing masterwort grown in Lima, would probably give a partially different result, but would also bring benefits in terms of securing the provenience of the plant material. Older herbarium records are a source material that can be important in such cases for establishing links between relict plants and historical cultivation.

The preservation value of masterwort plant material might not be as obvious as for other previously important medical plants that now have new functions in horticulture. For example, the social values of southernwood (*Artemisia abrotanum* L.) are linked to practices and memories of human interaction [3]. Multifunctional species such as tansy, *T. vulgare*, have survived in horticulture as ornamental plants, while lovage, *L. officinale*, is now used as a herb. Other former medical plants, such as rhubarb (*Rheum* sp.), have instead achieved new functions as vegetables, while knowledge of the species' medicinal qualities has now been forgotten. Masterwort has lost its functions in Swedish gardens and practices, first as a pharmaceutical drug and then also in ethnoveterinary medicine due to traditional livestock breeding being abandoned and the modernisation of veterinary medicine [69]. This is evident in the more recent (and sometimes current) use of this plant for dietary, medicinal and veterinary purposes in other sociocultural and geographical contexts, such as Alpine pastures, where pastoralism is still practised. Indeed, *P. ostruthium* prefers nitrogen rich soil which is a characteristic of grazed pastures, especially those where cattle and herds stop for a while (e.g., for the night), or in the immediate proximity of shielings or mountain huts. Thus, the conservation of this species (and biodiversity more generally) may be fostered by pastoral activities [70], and it may also be used as a medicinal resource (Figure 6).

Figure 6. Pasture at the shieling Rismyren, Lima parish, Dalecarlia, in the early twentieth century (Photo Georg Renström, Courtesy: The Nordic Museum, Stockholm, NMA.0038914).

Masterwort has yet not gained popularity as an ornamental plant and is seldom included in reconstructed herb gardens. However, preserving the genetic spread of relict plant material, recording their use and documenting ethnobotanical knowledge are important tasks of the Swedish Programme for Diversity of Cultivated Plants in relation to the *Global Plan of Action for the conservation and sustainable utilization of plant genetic resources for food and agriculture, developed by FAO*. Collecting additional plant material from northern localities in Sweden, such as Lima, might be important in terms of contributing to the gene pool of the species from a global perspective [71].

5. Conclusion

A deeper understanding of traditional practices in connection with species commonly labelled as medicinal herbs might be important when reconstructing historical landscapes and gardens. The reconstruction of an eighteenth-century physic garden differs from the cultivation at shielings connected to silvopastoral systems, even though the same species are sometimes likely to be included. Both the preserved plant material and knowledge of folk medicine and ethnoveterinary uses are crucial in order to understand and preserve biocultural values linked to different cultural heritage elements.

Author Contributions: Conceptualisation: I.S.; writing, review and editing: I.S., E.d.V. and G.M. All authors have read and agreed to the published version of the manuscript.

Funding: This research received no external funding.

Data Availability Statement: All data generated or analysed during this study are available in this article.

Acknowledgments: We are grateful to Arne Holmer for providing Figure 1 and to the Nordic Museum, Stockholm, for Figure 6.

Conflicts of Interest: The authors declare no conflict of interest.

References

1. Weibull, J.; Jansson, E.; Wedelsbäck Bladh, K. Swedes revisited: A landrace inventory in Sweden. In *European Landraces: On Farm Conservation, Management and Use*; Veteläinen, M., Negri, V., Maxted, N., Eds.; Bioversity Technical Bulletin No. 15; Biodiversity International: Roma, Italy, 2009; pp. 155–160.
2. Strese, E.-M.; de Vahl, E. *Kulturarvsväxter för Framtidens Mångfald: Köksväxter i Nationella Genbanken*; SLU: Alnarp, Sweden, 2018.
3. Svanberg, I.; de Vahl, E. «It may also have prevented churchgoers from falling asleep»: Southernwood, *Artemisia abrotanum* L. (fam. Asteraceae), in the church. *J. Ethnobiol. Ethnomed.* **2020**, *16*, 49. [CrossRef] [PubMed]
4. de Vahl, E.; Svanberg, I. Traditional uses and practices of edible cultivated *A.llium* species (fam. Amaryllidaceae) in Sweden. *J. Ethnobiol. Ethnomed.* **2022**, *18*, 14. [CrossRef] [PubMed]
5. Almquist, E.; Björkman, G. Tillägg till Dalarnas flora. *Sv. Bot. Tidskr.* **1960**, *54*, 1–68.
6. Linnaeus, C. *Linnés Dalaresa Iter Dalecarlicum, Jämte Utlandsresan Iter ad Exteros och Bergslagsresan Iter ad fodinas. Med Utförlig Kommentar*; Hugo Gebers Förlag: Stockholm, Sweden, 1953.
7. Linnaeus, C. *Flora Svecica*; Laurentius Salvius: Stockholm, Sweden, 1755.
8. Hülphers, A.A. *Dagbok Öfwer en Resa Igenom de, under Stora Kopparbergs Höfdingedöme Lydande Lähn och Dalarne år 1757*; Joh. Laur. Horn: Wästerås, Sweden, 1762.
9. Pickl-Herk, W. *Volksmedizinische Anwendung von Arzneipflanzen im Norden Südtirols*; Unpublished Diploma Thesis; Naturwissenschaftliche Fakultät d. Universität Wien: Wien, Austria, 1995.
10. Vogl, S.; Picker, P.; Mihaly-Bison, J.; Fakhrudin, N.; Atanasov, A.G.; Heiss, E.H.; Kopp, B. Ethnopharmacological in vitro studies on Austria's folk medicine. An unexplored lore in vitro anti-inflammatory activities of 71 Austrian traditional herbal drugs. *J. Ethnopharmacol.* **2013**, *149*, 750–771. [CrossRef]
11. Vitalini, S.; Puricelli, C.; Mikerezi, I.; Iriti, M. Plants, people and traditions: Ethnobotanical survey in the Lombard Stelvio National Park and neighbouring areas (Central Alps, Italy). *J. Ethnopharmacol.* **2015**, *173*, 435–458. [CrossRef] [PubMed]
12. Abbet, C.; Mayor, R.; Roguet, D.; Spichiger, R.; Hamburger, M.; Potterat, O. Ethnobotanical survey on wild alpine food plants in Lower and Central Valais (Switzerland). *J. Ethnopharmacol.* **2014**, *151*, 624–634. [CrossRef]
13. Hegi, G. *Illustrierte Flora von Mitteleuropa*; 5, 2. Teil.; J. F. Lehmanns Verlag: München, Germany, 1926.
14. Joa, H.; Vogl, S.; Atanasov, A.G.; Zehl, M.; Nakel, T.; Fakhrudin, N.; Heiss, E.H.; Picker, P.; Urban, E.; Wawrosch, C.; et al. Identification of ostruthin from *Peucedanum ostruthium* rhizomes as an inhibitor of vascular smooth muscle cell proliferation. *J. Nat. Prod.* **2011**, *74*, 1513–1516. [CrossRef]

15. Sarkhail, P. Traditional uses, phytochemistry and pharmacological properties of the genus *Peucedanum*: A review. *J. Ethnopharmacol.* **2014**, *156*, 235–270. [CrossRef]

16. Kopecký, K. Die Beziehung zwischen Siedlungsgeschichte und Verbreitung von *Imperatoria ostruthium* im Gebirge Orlické hory (Adlergebirge, Nordostböhmen). *Folia Geobot. Phytotax.* **1973**, *8*, 241–248. [CrossRef]

17. Godwin, H. *History of the British Flora: A Factual Basis for Phytogeography*; Cambridge University Press: Cambridge, UK, 1984.

18. Hultén, E. *Atlas of North European Vascular Plants North of the Tropic of Cancer*; Koeltz Scientific Books: Königstein, Germany, 1986; Volume 3.

19. Fremstad, E. Mesterrot *Peucedanum ostruthium* i Midt-Norge. *Blyttia* **2004**, *62*, 82–90.

20. Lange, J. *Kulturplanternes Indførselshistorie i Danmark: Indtil Midten af 1900-Talle*; DSR Forlag: Fredriksberg, Sweden, 1999.

21. Svedjemyr, O. Mästerroten, *Peucedanum ostruthium*—Läkeväxt från medeltid. *Sv. Bot. Tidskr.* **1989**, *83*, 149–155.

22. Almquist, E.; Björkman, G. Nya *tillägg* till *Dalarnas kärlväxtflora. Sv. Bot. Tidskr.* **1970**, *64*, 335–379.

23. SLU Swedish Species Information Centre: Mästerrot, Peucedanum ostruthium. Available online: https://artfakta.se/naturvard/taxon/Peucedanum%20ostruthium-219701 (accessed on 18 October 2022).

24. Myrdal, J. Source pluralism as a method of historical research. In *Historical Knowledge: In Quest of Theory, Method and Evidence*; Hellman, S., Rahikainen, M., Eds.; Cambridge Scholars Publishing: Newcastle upon Tyne, UK, 2012; pp. 155–189.

25. Almquist, E. *Dalarnes Flora. Förteckning över Kärlväxterna, Grundad på Professor G. Samuelssons Samlade Material och Nyare Tillägg, Jämte Några Växtgeografiska Synpunkter på Dalafloran Samt dess Utforskning*; Nordiska Bokhandeln: Stockholm, Sweden, 1949.

26. Svanberg, I. *Folklig Botanik*; Dialogos: Stockholm, Sweden, 2011.

27. Svanberg, I. Boskapsskötsel och jordbruk. In *Lima och Transtrand: Ur Två Socknars Historia*; Björklund, S., Pettersson, T.J.-E., Eds.; Malungs Kommun: Malung, Sweden, 1987; Volume 2, pp. 209–256.

28. Frödin, J. *Siljansområdets Fäbodbygd*; Gleerup: Lund, Sweden, 1925.

29. Johnson, A. Sången i Skogen: Studier Kring den Svenska Fäbodmusiken. Ph.D. Thesis, Department of Musicology, Uppsala University, Uppsala, Sweden, 1986.

30. Dahllöf, T. *Mikaeli: Sommarens Slut och Vinterns Smygande Ankomst*; Maral Production: Uppsala, Sweden, 2003.

31. Ljung, T. *Vårt Levande Arv. Minnen och Spår i Landskapet*; Dalarnas Fornminnes-Och Hembygdsförbund: Falun, Sweden, 2017.

32. Nilsson, E. Mästerroten—Hemkär men nyckfull. *Sv. Bot. Tidskr.* **1982**, *76*, 83–89.

33. Julin, E. En förekomst av *Peucedanum ostruthium* i Norrbotten. *Bot. Not.* **1961**, *114*, 322–338.

34. Svanberg, I. Odlarmöda och trädgårdsnöje. In *Signums Svenska Kulturhistoria: Frihetstiden*; Christensson, J., Ed.; Signum: Lund, Sweden, 2006; pp. 185–217.

35. Malmgren, U. *Västmanlands Flora*; Naturvetenskapliga Forskningsrådet: Stockholm, Sweden, 1982.

36. Svanberg, I. Lovage, *Levisticum officinale*. In *Encyclopedia of Cultivated Plants*; Cuno, C., Ed.; ABC-CLIO: Santa Barbara, CA, USA, 2013; Volume 2, pp. 603–606.

37. Svanberg, I. Tansy, *Tanacetum vulgare*. In *Encyclopedia of Cultivated Plants*; Cuno, C., Ed.; ABC-CLIO: Santa Barbara, CA, USA, 2013; Volume 3, pp. 1033–1035.

38. Svanberg, I. Masterwort, *Peucedanum ostruthium*. In *Encyclopedia of Cultivated Plants*; Cuno, C., Ed.; ABC-CLIO: Santa Barbara, CA, USA, 2013; Volume 2, pp. 637–640.

39. Allen, D.E.; Hatfield, G. *Medicinal Plants in Folk Tradition: An Ethnobotany of Britain and Ireland*; Timber Press: Portland, OR, USA, 2004.

40. Brøndegaard, V.J. Mesterrod som Veterinærplante. In *Dansk Veterinærhistorisk Årbog 22*; Carl Fr. Mortensens Forlag: København, Denmark, 1962; pp. 122–132.

41. Pedersen, C. *En Nøttelig Lege Bog Faar Fattige och Rige Unge och Gamle om Mange Atskillige Siwgdomme som Menniskene Hender til ath Kome paa Deris Legeme i Mage Honde Maade oc om Gode Raad oc Legedomme til Dem ath de Som Lese Kunde Mwe Hielpe dem Selffue oc Andre Flere met Urter som her Voxe i Riget*; Hoochstraten: Malmø, Sweden, 1533.

42. Smid, H. *Een Skøn Loestig. Ny Vrtegaardt. Prydet Med Mange Atskillige Vrter, som Tiene til Menniskens Legemmes Sundheds Opholdelse, Desligest Huorledis Electuaria, Syruper, Conserva, oc Olier, Skulle ret Konstelige Gøres, oc Beredes, aff Denne Vrtegaards Vrter, Deres Røder oc Blomster, oc Andre Saadane Subtilige Nyttelige Ting, Aldrig Tilforn Seet paa Vort Danske Tungemaal*; Oluff Vlricksen: Malmø, Sweden, 1546.

43. Paulinus, G.L. *Loimoscopia, Eller Pestilentz Speghel: Thet är Andeligh och Naturligh Vnderwijsning, om Pestilenzies Beskriffwelse, Orsaker, Praeserwatijff, Läkedomar och Befrijelser: Stält och Sammanfattadt*; Olof Olofsson Enæus: Strängnäs, Sweden, 1623.

44. Nyström, S. *I Apotekarens Trädgård: Medicinalväxtodling i Svenska Apotekarträdgårdar Med Fokus på 1700-Talet*; Institutionen för Växtförädling, Sveriges Lantbruksuniversitet, SLU: Alnarp, Sweden, 2021.

45. *Catalogus et Valor Medicamentorum in Officinis Pharmaceüticis Stockholmiensibus prostantium/Apothekare-TAXA, Uppå Alla de Medicamenter och Wahror, som på Apotheken i Stockholm Finnes Till Sahlu /Apotheker-Taxt, Aller Medicamenten undt Wahren, Welche in Denen Stockholmischen Apotheken zu Finden Seyn*; Stockholm, Sweden, 1698; Available online: https://libris.kb.se/bib/14227586 (accessed on 1 December 2022).

46. Malmsten, P.H.; Sandahl, O.T. *Pharmacopoea Svecica. Editio septima*; Norstedt: Stockholm, Sweden, 1869.

47. Retzius, A.J. *Försök till en Flora Oeconomica Sveciae Eller Swenska Wäxters Nytta och Skada i Hushållningen*; J.A. Lundblad: Lund, Sweden, 1806; Volume 1.

48. Pieroni, A.; Giusti, M.E. Alpine ethnobotany in Italy: Traditional knowledge of gastronomic and medicinal plants among the Occitans of the upper Varaita valley, Piedmont. *J. Ethnobiol. Ethnomed.* **2009**, *5*, 32. [CrossRef] [PubMed]

49. Petelka, J.; Plagg, B.; Säumel, I.; Zerba, S. Traditional medicinal plants in South Tyrol (northern Italy, south Alps). *J. Ethnobiol. Ethnomed.* **2020**, *16*, 74. [CrossRef]

50. Brioschi, C. *Botanical and Ethnobotanical Studies in Peucedanum ostruthium (Apiaceae) from the Upper Saastal: Variability of Morphology and Coumarin Components, and Use as a Medicinal Plant*; Institute of Systematic and Evolutionary Botany, University of Zürich: Zürich, Switzerland, 2020.

51. Svanberg, I. «The Lapps Chew This Root a Lot»: Milk Parsley (*Peucedanum palustre*) in Sami Plant Knowledge. In *Cultural Interaction between East and West. Archaeology, Artefacts and Human Contacts in Northern Europe*; Fransson, U., Ed.; Stockholm Studies in Archaeology; Department of Archaeology, University of Sweden: Stockholm, Sweden, 2007; pp. 328–330.

52. Färje., C.G. *C.G. Transtrands Flora. Med Artförteckning och Kommentar Samt de Dialektala Namnen*; Transtrands Hembygdsförening: Sälen, Sweden, 1961.

53. Eriksson, P.A. Växter. In *Lima och Transtrand. Ur Två Socknars Historia*; Björklund, S., Pettersson, T.J.-E., Eds.; Malungs Kommun: Malung, Sweden, 1987; Volume 2, pp. 69–108.

54. Enell, H.; Thedenius, H.; Luhr, R. *Pharmaca Composita*; C. & E Gernandts Förlag: Stockholm, Sweden, 1896.

55. Nygårds, L. *Rällsjö Brita är Mitt Namn: En Berättelse Från Självhushållets tid Färgad av Tre Sekel*; Ackedo: Stockholm, Sweden, 2017.

56. Behm, M. *Een ny och Ganska Nyttigh Läkiare-Book om Häste Läkedomar. Therutinnan Allehanda Wisse och Kostlege Recepter Äre Beskreffne och Experminenterade*; Henrich Keyer: Stockholm, Sweden, 1648.

57. Rålamb, Å.C. *Adelig Öfning, XIII: Hushåld och Åkerbruk eller Oeconomia; Läkebok för Hästar, Boskap och Andra Stora och små Creatur*; Niclas Wankijff: Stockholm, Sweden, 1690.

58. Samuelsson, S. *Den Förträffeliga och Tilförlåteliga Häste-Boken, af den Namnkunnoge Rusthållaren Sven Samuelsson i Hadelö Upgifwen; Innehållande Säkra Underrättelser om alla Häste-Kreaturen Tilstötande Sjukdomar Samt Ofelbara Præservativer och Botemedel Deremot, Samt Tecken til Rätta Sjukdomen. Jemte Några Recepter mot Smittosamma Boskaps Sjukdomar; af Honom Sjelf Med Egen Hand Anteknade under Desz Mångåriga Försök och Förträffeliga Curer. Samt nu på Många Respective Stånds-Personers och Fleres Åstundan til Allmänhetens Tjenst och Nytta til Trycket Befordrad*; Björckegren: Linköping, Sweden, 1775.

59. Heurgren, P.G. *Husdjuren i Nordisk Folktro*; Örebro Dagblad: Örebro, Sweden, 1925.

60. Anonymous. Om mästerroten m.m. *Skansvakten* **1949**, *34*, 22.

61. Steensland, L. *Älvdalska Växtnamn Förr och Nu*; Dialekt-Och Golkminnesarkivet: Uppsala, Sweden, 1994.

62. Pettersson, O. *Gamla Byar i Vilhelmina*; Generalstabens Litografiska Anstalt: Stockholm, Sweden, 1941; Volume 3.

63. Høeg, O.A. *Planter og Tradisjon: Floraen i Levende Tale og Tradisjon i Norge 1925–1973*; Universitetsforlaget: Oslo, Norway, 1974.

64. Marzell, H. Die Meisterwurz. Die Geschichte einer einst hochberühmten alpinen Heilpflanze. *Jahrb. Ver. Schutze Alp. Tiere* **1959**, *24*, 36–42.

65. Kamp, J. *Danske Folkeminder, Æventyr, Folkesagn, Gaader, Rim og Folketro*; Nielsen: Odense, Denmark, 1877.

66. Waller, P.J.; Bernes, G.; Thamsborg, S.M.; Sukura, A.; Richter, S.H.; Ingebrigtsen, K.; Höglund, J. Plants as de-worming agents of livestock in the Nordic countries: Historical perspective, popular beliefs and prospects for the future. *Acta Vet. Scand.* **2001**, *42*, 31–44. [CrossRef]

67. Sweden's Virtual Herbarium. Available online: http://herbarium.emg.umu.se/ (accessed on 22 October 2022).

68. Artportalen. Available online: www.artportalen.se (accessed on 22 October 2022).

69. Mattalia, G.; Belichenko, O.; Kalle, R.; Kolosova, V.; Kuznetsova, N.; Prakofjewa, J.; Stryamets, N.; Pieroni, A.; Volpato, G.; Sõukand, R. Multifarious trajectories in plant-based ethnoveterinary knowledge in northern and southern eastern Europe. *Front. Vet. Sci.* **2021**, *8*, 710019. [CrossRef]

70. Ingty, T. Pastoralism in the highest peaks: Role of the traditional grazing systems in maintaining biodiversity and ecosystem function in the alpine Himalaya. *PLoS ONE* **2021**, *16*, e0245221. [CrossRef] [PubMed]

71. FAO. *Global Plan of Action for the Conservation and Sustainable Utilization of Plant Genetic Resources for Food and Agriculture*; FAO: Rome, Italy, 1996.

Review

Iconic Arable Weeds: The Significance of Corn Poppy (*Papaver rhoeas*), Cornflower (*Centaurea cyanus*), and Field Larkspur (*Delphinium consolida*) in Hungarian Ethnobotanical and Cultural Heritage

Gyula Pinke [1,*], Viktória Kapcsándi [1] and Bálint Czúcz [2]

[1] Albert Kázmér Faculty of Mosomagyaróvár, Széchenyi István University, Vár 2., H-9200 Mosonmagyaróvár, Hungary

[2] European Commission, Joint Research Centre, Via Fermi 2749, 21027 Ispra, Italy

* Correspondence: pinke.gyula@sze.hu

Abstract: There are an increasing number of initiatives that recognize arable weed species as an important component of agricultural biodiversity. Such initiatives often focus on declining species that were once abundant and are still well known, but the ethnographic relevance of such species receives little recognition. We carried out an extensive literature review on the medicinal, ornamental, and cultural applications of three selected species, *Papaver rhoeas*, *Centaurea cyanus*, and *Delphinium consolida*, in the relevant Hungarian literature published between 1578 and 2021. We found a great diversity of medicinal usages. While *P. rhoeas* stands out with its sedative influence, *D. consolida* was mainly employed to stop bleeding, and *C. cyanus* was most frequently used to cure eye inflammation. The buds of *P. rhoeas* were sporadically eaten and its petals were used as a food dye. All species fulfilled ornamental purposes, either as garden plants or gathered in the wild for bouquets. They were essential elements of harvest festivals and religious festivities, particularly in Corpus Christi processions. *P. rhoeas* was also a part of several children's games. These wildflowers were regularly depicted in traditional Hungarian folk art. In poetry, *P. rhoeas* was used as a symbol of burning love or impermanence; *C. cyanus* was frequently associated with tenderness and faithfulness; while *D. consolida* regularly emerged as a nostalgic remembrance of the disappearing rural lifestyle. These plants were also used as patriotic symbols in illustrations for faithfulness, loyalty, or homesickness. Our results highlight the deep and prevalent embeddedness of the three iconic weed species studied in the folk culture of the Carpathian Basin. The ethnobotanical and cultural embeddedness of arable weed species should also be considered when efforts and instruments for the conservation of arable weed communities are designed.

Keywords: anthropology; arable weed conservation; charismatic species; cultural history; cultural symbols; ethnobotany; human–plants relations; medicinal plants; wild food plants

Citation: Pinke, G.; Kapcsándi, V.; Czúcz, B. Iconic Arable Weeds: The Significance of Corn Poppy (*Papaver rhoeas*), Cornflower (*Centaurea cyanus*), and Field Larkspur (*Delphinium consolida*) in Hungarian Ethnobotanical and Cultural Heritage. *Plants* **2022**, *12*, 84. https://doi.org/10.3390/plants12010084

Academic Editor: Renata Sõukand

Received: 21 November 2022
Revised: 18 December 2022
Accepted: 19 December 2022
Published: 23 December 2022

1. Introduction

Farmers and agronomists have been desperately engaged in reducing the adverse economic effects of arable weeds for a long time. Nevertheless, arable weeds may also exhibit beneficial properties [1,2] and they can also contribute to several important ecosystem services, for example, pest control and soil fertility improvement [3–6]. Moreover, weeds are often the basis of agricultural food webs providing food resources to many organisms, including numerous insect and bird species, so they are considered beneficial from a conservation or even from an agricultural point of view [7–9]. Weeds, which exhibit a low level of competition with crops and provide a considerable resource value for higher trophic groups, are sometimes distinguished as "good weeds" [10,11]. Several of these "good" arable weed species have become threatened by agricultural intensification in

Europe [12–14]. The decline of well-known, often colourful "emblematic", weed species has been recognized in several EU-level conservation initiatives, including arable plant sanctuaries [15,16]. Nevertheless, these iconic species are also deeply embedded in the local culture of the European rural regions and are accompanied by considerable traditional knowledge, and ethnobotanical and cultural heritage, which is also becoming endangered with their decline [17–19]. In turn, the regional cultural embeddedness of these species should also be considered as an important factor in the design of the conservation programmes that are aimed at protecting arable weed communities for the future generations.

In this paper, we aim to explore the cultural embeddedness of three emblematic arable weed species: corn poppy (*Papaver rhoeas* L., henceforward *poppy*), cornflower (*Centaurea cyanus* L., syn. *Cyanus segetum* Hill), and field larkspur (*Delphinium consolida* L., syn. *Consolida regalis* Gray, henceforward *larkspur*), in the culture of Hungarian-speaking communities in the Carpathian Basin in Eastern Europe from a diachronic perspective.

Although sacred plants [20], magical herbs [21,22], ritual [23] and long-lived [24] trees, antique fruits [25], peculiar food plants [26], and orchids [27,28] are frequently subjects of ethno-cultural botanical studies, very few studies focus on arable weed species with cultural significance [17–19]. The three weed species in the focus of this paper were introduced to Central Europe as archaeophytes [29] and their remains were found in archaeological sites, also in the Carpathian Basin, from the Copper Age until early modern times [30–32]. Former field observations [33], as well as contemporaneous findings, in adobe bricks [34] suggest that the three studied species were among the most abundant arable weeds by the end of the 19th century in the Carpathian Basin. Due to their brightly coloured flowers, all three species were popular wildflowers, and, despite their recent decline, they are still well known by the general public [35–38] (Figure 1). Because of their general recognition and charismatic nature, they could function as potential "flagship species" in conservation programmes [39] to combat the general decline of botanical interest and awareness (also known as "plant blindness" [40]), underlying many further recent global challenges [41].

Figure 1. The three studied weed species and their spectacular mass occurrences in arable fields: (**a**,**b**) *Papaver rhoeas* (Hegyeshalom, NW-Hungary, 2018); (**c**,**d**) *Centaurea cyanus* (Öskü, W-Hungary, 2011); (**e**,**f**) *Delphinium consolida* (Püski, NW-Hungary, 2020; all photographs by Gyula Pinke).

The poppy is still relatively frequent in Central Europe, probably due to its persistent seed bank [42], and can be a noxious weed in some crops, including the opium poppy (*Papaver somniferum* L.) [43] or in regions where herbicide-resistant poppy biotypes have recently been detected [44]. The cornflower was once common in many European countries, but it has largely declined and become threatened almost everywhere due to agricultural intensification [12]. Now, it is considered as an indicator species of low input cereal fields; thus, it is often addressed as a "flagship species" in conservation programmes [45,46]. The larkspur was previously widespread and is now regionally rare in some European countries [47]. It functions as an emblematic species for reintroduction projects of rare arable plants [48]. Currently, in the central part of the Carpathian Basin, the cornflower has only sporadic distribution and is still decreasing, while the larkspur is still frequent [49].

In order to gather and document information about the cultural embeddedness of the three studied species, we performed an extensive literature review, focusing on historical sources available in the Hungarian language, with which we aimed to create a comprehensive inventory of the occurrence of these species in traditional folk culture, including medicinal, nutritional, and ornamental applications, as well as their cultural roles in traditional festivals and children's games. Furthermore, we complement this overview with an outlook on the representation of these species in the visual arts and literature, providing further illustrations of the symbolic significance of these species. By doing so, we are opening up the "footprint" of these iconic weed species in Hungarian culture to a broad international audience, thus making this otherwise relatively inaccessible rich cultural heritage more accessible. We also hope that a better knowledge of this threatened cultural heritage can help improve and enrich the predominantly negative public discourse on arable weeds. Accordingly, to assign a weed as "beneficial" not only will its rarity status and importance in food chains be taken into account, but also its ethnobotanical and cultural relevance will be considered.

2. Methods

We performed a series of targeted literature searches in several Hungarian online databases, including Arcanum, Hungaricana Közgyűjteményi Portál, Matarka, Magyar Elektronikus Könyvtár, Elektronikus Periodika Adatbázis Archívum, and Erdélyi Magyar Elektonikus Könyvtár. These databases only contain works in Hungarian; accordingly, our study did not cover sources written in other languages of the Carpathian Basin. However, Hungarian articles that describe other ethnic groups living in the same area were included. As keywords, we used the names of the species in several forms: for scientific names we used the Plants of the World Online (POWO) database [50] as our primary reference (including also the main synonyms—see Introduction), whereas for Hungarian vernacular names (including regional and local folk names), we relied on the books of Wagner [33], Vörös [51], and Rácz [52] [i.e., *pipacs, pipats, pippancs, pipanc, papics, papantz, papcsik, pipók, vadmák, veres mák, lúdmák, czúczik, cucik, pitypalatyvirág* for *poppy*; *búzavirág, dődike, égi virág, kék virág, csüküllő, sukollat, vadpézsma, kékkonkoly, gabonavirág* for *cornflower*; and *szarkaláb, királyvirág, sarkvirág, sarkantyúfű, sarkasfű, dalisarkanytú, vitézi farkanytú* for *larkspur*]. We combined the species names with further search terms identifying possible cultural uses (e.g., the Hungarian terms for "ethnobotany", "medicinal", "remedy", "edible", "food", "fodder", "dye", "ornamental", "bouquet", "wreath", "garden", "toy", "game", "festival", "religion", "feast", "harvest", "folklore", "belief", "symbolism", "art", "motif", "handicraft", "painting", and "poetry"). For the literature databases which made this possible, we also extended the search to the whole text of the primary studies, and not just the title, abstract, and keyword fields. The studied historical sources are presented in Tables 1–3.

In a second step, our search expressions were translated into English, and we repeated the search in four selected major scientific literature databases (Web of Science, Scopus, Google Scholar, and ResearchGate). With this follow-up search, we aimed to place our results into a broader European context.

Most of the results are presented in a narrative format, but in the case of a few complex subtopics, that were particularly interesting and rich in details (medicinal uses, religious uses, and symbolic connotations related to human characters and feelings), we constructed tables to enumerate the results in a more structured form. For the interpretation of the archaic Hungarian names of diverse ailments, we used the book of Magyary-Kossa [53]. To present plant parts and modes of preparations, we followed the terminology of the American Botanical council [54].

In order to find relevant illustrations for our results, we made further ad-hoc searches on the websites of several Hungarian museums, other institutes, and online collections, applying the scientific and vernacular names of the target species as the main search terms. For some cultural uses without available original images, we created our own illustrations by reconstructing "animated scenes" depicting the activities (e.g., for children's games) or by using related contemporary items (e.g., for medicinal and food dying uses).

Table 1. Records on the medicinal and veterinary uses of *Papaver rhoeas*, *Centaurea cyanus*, and *Delphinium consolida* between 1578 and 2018 in the Carpathian Basin.

Year of Publication/Relevant Period	Source	Region (Current Country)	Species	Part Used	Mode of Preparation/Administration	Treated Disease(s)/Folk Medical Use(s)
1578	Melius Juhász Péter [55]	Hungary	*P. rhoeas*	Fructus ("Poppy heads")	Decoction (made with water or wine)/oral	Insomnia
				Not specified	Infusion/topical (mouthwash)	Mouth and gum diseases
					Infusion, poultice/topical (genitals)	Heavy menstruation bleeding
				Semen	Infused honey/oral	Intestinal pain
				Latex	Oral	Fever, throat, and tongue swelling
					Topical	"St Anthony's fire" (erysipelas)
					Poultice/topical	Nose- and liver-bleeding
1595	Beythe András [56]	Hungary	*P. rhoeas*			*Same as Melius Juhász (1578)*
1690	Pápai Páriz Ferencz [57]	Hungary	*P. rhoeas*	Latex	Oral	Insomnia (particularly after venesection), stomach pain, dysentery
				Flos, latex	Oral	Bleeding
Early 18th c.	Unknown physician [58]	Transylvania (Romania)	*D. consolida*	Herba	Infused vinegar/topical (nose)	Nose-bleeding
18–19th c.	Gulyás Éva [59]	East Hungary	*P. rhoeas*	Flos	Infusion/oral	Breast pain
18–19th c.	Novák László [60]	East Hungary	*C. cyanus*	Flos	Infused wine, poultice/topical	Eye inflammation
			P. rhoeas	Not specified	Oral	Stomach pain
1775	Csapó József [61]	Hungary	*P. rhoeas*	Flos	Infusion/oral	Catarrh, pleurisy
			C. cyanus	Flos	Poultice/topical	Eye inflammation
			D. consolida	Herba	Powder/oral	Heartburn
				Green herba	Pressed sap/topical (washing and bandage)	Fresh wounds
				Flos	Decoction (in rose-water), poultice/topical	Eye inflammation
1789	Zsoldos Xavér [62]	West Hungary	*P. rhoeas*	Flos (petals)	Infusion/oral	Panacea
1798	Veszelszki Antal [63]	Hungary	*D. consolida*	Semen	Infused wine/oral	Plague, intestinal pain, lithiasis
				Herba	Decocted wine/oral	Parasitic worms
			P. rhoeas			*same as Melius Juhász (1578)*
			C. cyanus	Flos	Smashed powder/oral	Jaundice
					Poultice/topical	Eye inflammation
					Pressed sap/topical (mouthwash)	Bad breath
1813	Diószegi Sámuel [64]	Hungary	*P. rhoeas*	Flos (petals)	Infusion and syrup/oral	Pain relief
			D. consolida	Flos	Not specified	Parasitic worms, epilepsy
1899	Temesváry Rezső [65]	Hungary	*P. rhoeas*	Latex	Added to milk/oral	Gynaecological bleeding
				Not specified	Infusion/oral	Pain relief during childbirth
			D. consolida	Flos	Topical (hot bath)	Gynaecological bleeding
					Decoction added to red wine/oral	Gynaecological bleeding
					Fumigation/topical (vulva)	Premature birth prevention
				Not specified	Breast plaster, fumigation, poultice/topical	Mastitis
				Not specified	Infusion/oral	Pain relief during childbirth
1902	Gönczi Ferenc [66]	West Hungary	*D. consolida*	Flos	Decocted beer/oral	Bleeding
1910	Gönczi Ferenc [67]	West Hungary	*C. cyanus*	Not specified	Dew collected from the plant/topical (face wash)	Freckles
1925	Darvas Ferenc [68]	Hungary	*D. consolida*	Flos	Not specified	Conjunctivitis, chronic constipation, menstruation disorders
				Semen	Not specified	Lice and other skin parasites
			C. cyanus	Flos	Infusion (blend component)/oral	Colour enhancer
					Fumigant blends/topical	Colour enhancer
			P. rhoeas	Flos	Infusion/oral	Pain and spasm relief
					Not specified	Syrups, cough drops, dyeing sugar solutions
1928	Relkovic Davorka [69]	West Hungary	*D. consolida*	Flos	Not specified	Bleeding

Table 1. *Cont.*

Year of Publication/ Relevant Period	Source	Region (Current Country)	Species	Part Used	Mode of Preparation/Administration	Treated Disease(s)/Folk Medical Use(s)
1932	Rapaics Raymund [70]	Hungary	*P. rhoeas*	Fructus ("Poppy heads")	Fresh poppy heads/oral (used as a "pacifier")	Babies crying too much (sedative)
1935	Luby Margit [71]	Northeast Hungary	*D. consolida*	Flos	Decoction/topical (footbath) Dried flowers/used as a shoe insert	Bleeding
1940	Réthelyi József [72]	Hungary	*P. rhoeas*	Flos (petals)	Infusion/oral	Panacea, exorcism (to expel bad illness)
1941	Vajkai Aurél [73]	West Hungary	*D. consolida*	Not specified	Infusion/oral	Bleeding
				Not specified	Decoction (with horse chestnut)/oral	Haematuria (veterinary: cattle)
1944	Greszné Czimmer Anna [74]	East Hungary	*P. rhoeas*	Not specified	Infusion/oral	Heavy menstruation, gynaecological bleeding
1945	Vargyas Lajos [75]	Central Hungary	*D. consolida*	Flos	Infusion/oral	Bleeding after childbirth
			C. cyanus	Not specified	Infusion/oral	Cough
			P. rhoeas	Not specified	Decocted red wine/oral	Bleeding
				Flos	Decocted wine/oral	Inducing abortion
				Not specified	Oral	Contraception
			D. consolida	Not specified	Not specified	Contraception
1968	Farkas József [76]	Northeast Hungary	*D. consolida*	Not specified	Decoction/topical (hot bath)	Bleeding
					Shoe insert/topical	
1969	Seregély György [77]	Hungary	*P. rhoeas*	Not specified	Oral	Inducing abortion
1976	Péntek János [78]	Transylvania (Romania)	*C. cyanus*	Flos	Infusion/poultice	Eye inflammation
			D. consolida	Flos	Infusion/oral	Diuretic therapy
					Infusion	Leucorrhoea
					Tincture (in brandy)/oral	Bleeding
1976	Szabóné Futó Rózsa [79]	North Hungary	*D. consolida*	Herba	Decoction, poultice/topical (bath)	Eczema
			P. rhoeas	Not specified	Infusion/oral	Pulmonary diseases, cough, bleeding
			C. cyanus	Not specified	Infusion/oral	Pulmonary diseases, bleeding
1979	Oláh Andor [80]	Southeast Hungary	*P. rhoeas*	Flos	Infusion/oral	Cough
1980	Ujváry Zoltán [81]	Hungary	*D. consolida*	Not specified	Topical (bath)	Rheumatism
			C. cyanus	Not specified	Decoction/topical (wash)	Ulcer
1983	Petercsák Tivadar [82]	North Hungary	*D. consolida*	Not specified	Infusion (mixed with milk)/oral	Against witchcraft (veterinary: cattle)
1984	Rácz Gábor [83]	Transylvania (Romania)	*C. cyanus*	Flos	Infusion (blend component)/oral	Diuretic therapy
			P. rhoeas	Flos (petals)	Infusion (blend component)/oral	Tea corrigent to improve colour
			D. consolida	Flos	Infusion (blend component)/oral	Not specified
1985	Péntek János [84]	Transylvania (Romania)	*D. consolida*	Herba	The plant was tied to the horn of the livestock on the opposite side of the sick eye	Cataract (veterinary: cattle)
1985	Kóczián Géza [85]	Transylvania (Romania) and Southwest Hungary	*P. rhoeas*	Flos (petals)	Infusion/oral	Sleep-inducing, pain relief for stomach pain
				Semen	Decoction/oral	Stomach pain, internal purifying therapy, smooth muscle spasm relief
			D. consolida	Herba	Infusion/oral	Cough, tuberculosis
					Tincture (in brandy)/oral	Heavy menstruation bleeding
					Fumigation/topical	Sick humans and livestock
1986	Tóth József [86]	West Hungary	*P. rhoeas*	Flos (petals)	Infusion/oral	Tranquilizer, throat rinse
			C. cyanus	Flos	Infusion/topical, oral	Eye inflammation, heart palpitations, high blood pressure
					Fumigation	Air disinfection
			D. consolida	Flos	Not specified	Cough sedative, vasodilator
1989	Tisovszki Zsuzsanna [87]	Central Hungary	*P. rhoeas*	Flos (petals)	Infusion/oral	Cough

Table 1. *Cont.*

Year of Publication/ Relevant Period	Source	Region (Current Country)	Species	Part Used	Mode of Preparation/Administration	Treated Disease(s)/Folk Medical Use(s)
1991	Gelencsér József [88]	Central Hungary	C. cyanus, D. consolida, P. rhoeas	Herba	Decoction/topical (bath)	Evil eyes (children)
1993	Lenkey István [89]	North Hungary	D. consolida	Herba	Infusion/oral	Cough, common cold, pneumonia, gastrospasm
2000	Gub Jenő [90]	Transylvania (Romania)	C. cyanus	Herba	Infusion/topical (wash)	Wound
2001	Bartha Júlia [91]	East Hungary	D. consolida	Not specified	Infused wine/oral	Heavy menstruation, vaginal discharge, venereal diseases, nervousness
2002	Ujváry Zoltán [92]	Upper Hungary (Slovakia)	D. consolida	Not specified	Not specified	Eye inflammation
2005	Szabó László Gy. [93]	Hungary	P. rhoeas	Not specified	Not specified	Bleeding (unspecified)
			C. cyanus	Flos	Not specified; Infusion (blend component)	Diuretic, throat rinse; Tea corrigent (to improve colour)
			D. consolida	Flos	Not specified; Infusion (blend component)	Laxative, vasodilator; Tea corrigent to improve colour
			P. rhoeas	Semen; Flos (petals)	Not specified; Not specified	Purgative, diuretic, vermifuge; Mild sedative, expectorant
2010	Horváth Katalin [94]	Transcarpathia (Ukraine)	P. rhoeas	Not specified	Infusion/oral	Bleeding (unspecified), cough
				Not specified	Infusion, poultice/topical	Eye inflammation
2011	Grynaeus Tamás [95]	Southeast Hungary	P. rhoeas	Flos (petals)	Not specified	Common cold, cough
2018	Papp Nóra [96]	Transylvania (Romania)	C. cyanus	Flos	Infusion/topical, oral	Eye inflammation, earache, hearing loss

Table 2. Records on the uses of arable wildflowers in harvest festivals from the 19th century in the Carpathian Basin.

Relevant Period	Source	Region (Current Country)	Date	Species	Name	Description
19th–20th c.	Kapronyi Teréz [97]	North and West Hungary	End of harvest (mid-July)	Arable wildflowers	Harvest feast	A harvest wreath was made of ears, arable wildflowers, colourful bandannas, and paper ribbons
1850s	Prónay Gábor [98]	Hungary	End of harvest (mid-July)	Arable wildflowers	Harvest feast	A wreath made of ears and arable wildflowers was taken to the landlord in a formal march
1850s	Bozena Nemcová [99]	North Hungary (Slovakia)	End of harvest (mid-July)	Arable wildflowers	Harvest feast	A harvest wreath made of ears and arable wildflowers was given to the landlord by the nicest couple among the harvesters
Mid 19th–late 20th c.	Kapronyi Teréz [97]	Hungary	Start of harvest (late June)	P. rhoeas, C. cyanus	Binding ceremony	A bunch of ears with a poppy and cornflower was tied to the hand of the landlord amidst good wishes
1870s	Ébner Sándor [100]	Transylvania (Romania)	Evenings during harvest (late June–mid-July)	Arable wildflowers	Harvest feast	Harvesters wore colourful wreaths made of arable wildflowers
1890s	Kovács Bálint [101]	Transylvania (Romania)	End of harvest (mid-July)	C. cyanus	Harvest feast	A wreath made of ears and the cornflower was placed on the head of the funniest harvester who carried it to the landlord
1900s	Platthy Adorján [102]	North Hungary	End of harvest (mid-July)	Arable wildflowers	Harvest feast	A harvest wreath made of ears and arable wildflowers was ceremonially given to the landlord

Table 2. *Cont.*

Relevant Period	Source	Region (Current Country)	Date	Species	Name	Description
1900s	Illés Péter [103]	West Hungary	End of harvest (mid-July)	Arable wildflowers	Harvest feast	A harvest chariot was decorated with ears and arable wildflowers
Early 20th c.	Manga János [99]	North Hungary	Start of harvest (late June)	Arable wildflowers	Binding ceremony	A bunch of ears with wildflowers was tied to the hand of the landlord who gave money in exchange for drinks
		North Hungary	End of harvest (mid-July)	Arable wildflowers	Harvest tradition	A bunch of ears with arable wildflowers was taken home and hung on a wooden beam and its seeds were used as sowing seeds in the following autumn
		West Hungary	End of harvest (mid-July)	Arable wildflowers	Harvest tradition	A wreath formed from ears and arable wildflowers was laid around the neck of one of the harvesters, then it was taken home and given to hens to increase egg laying
Early 20th c.	Gelencsér József [88]	Central Hungary	End of harvest (mid-July)	*P. rhoeas* and *C. cyanus*	Harvest feast	A wreath made of ears, the poppy, and cornflower was ceremonially given to the landlord
1930s	Faggyas István [104]	North Hungary	Start of harvest (late June)	*C. cyanus, D. consolida, Vicia* sp.	Binding ceremony	A bunch of ears with the cornflower, larkspur, and wild vetches was tied to the hand of the land steward by a young girl amidst good wishes
		North Hungary	End of harvest (mid-July)	Arable wildflowers	Harvest feast	A harvest wreath was made of ears, wildflowers, and colourful paper ribbons
1930s	Gyimesiné Gömöri Ilona [105]	North Hungary	End of harvest (mid-July)	*P. rhoeas, C. cyanus*	Harvest feast	A wreath made of ears, the poppy, and cornflower was ceremonially given to the landlord
1940s	Illés Péter [103]	West Hungary	End of harvest (mid-July)	*P. rhoeas* and *C. cyanus*	Harvest feast	A harvest wreath was created from ears and arable wildflowers, and the stage was also decorated with the remaining flowers.
Early 1950s	Illés Péter [103]	West Hungary	End of harvest (mid-July)	*P. rhoeas, C. cyanus, A. githago*	Harvest feast	Two wreaths from ears, the poppy, cornflower, and corncockle were made. One was taken to the grave of the previous landlord and the second was given to the new one

Table 3. Records on the uses of arable wildflowers in religious ceremonies from the mid-19th century in the Carpathian Basin.

Relevant Period	Source	Region (Current Country)	Date	Species	Name	Description
Mid 19th c.–today	Sz. Tóth Judit [106]	German (Swabian) communities near Budapest	21 May–25 June	*D. consolida* and other unspecified wild flowers and garden plants	Corpus Christi	Wreaths and bouquets from the plants are placed next to the altar or hung on the wall of the chapel to increase their remedial power
Mid 19th c.–today	Sz. Tóth Judit [107]	German (Swabian) communities near Budapest	21 May–25 June	*C. cyanus, Leucanthemum vulgare,* and other unspecified arable wild flowers, *Sambucus* sp., *Robinia* sp., *Sedum* sp., garden plants	Corpus Christi	Flowers are gathered and flower carpets are laid on the route of the procession. Flower wreaths are also made and hung on chapels and tents
Early 20th c.	Horváth Iván [108]	Croatian communities in West Hungary	21 May–25 June	*P. rhoeas* and other unspecified wildflowers	Corpus Christi	Petals of wildflowers, especially those of the poppy, were gathered by children and were thrown on the route of the procession

Table 3. *Cont.*

Relevant Period	Source	Region (Current Country)	Date	Species	Name	Description
Early 20th c.	Kalapis Zoltán [109]	Vojvodina (Serbia)	21 May–25 June	*P. rhoeas, D. consolida,* and unspecified grasses	Corpus Christi	Freshly mown poppy and larkspur together with grasses were transported from the fields and meadows with carts to decorate the streets
Early 20th c.	Demeter Zsófia [110]	Central Hungary	21 May–25 June	*P. rhoeas, C. cyanus, Leucanthemum vulgare*	Corpus Christi	Hay, poppy, cornflower, and marguerite were thrown on the route of the procession
Early 20th c.	Nagy Netta [111]	Southeast Hungary	21 May–25 June	*P. rhoeas, Rosa* sp.	Corpus Christi	Previously petals of the poppy, later those of the rose, were thrown by young girls during the procession
1930s	Császi Irén [112]	North Hungary	21 May–25 June	*A. githago, P. rhoeas, C. cyanus, L. tuberosus*	Corpus Christi	Petals of wildflowers, mainly those of the poppy, cornflower, corncockle, and tuberous pea were gathered and thrown by young girls during the procession
1930s	Illés Péter [103]	West Hungary	2 July	Arable wildflowers	Visitation of Our Lady	During the service of thanksgiving, the priest blessed the harvest tools that were decorated with ears and arable wildflowers
Early–late 20th c.	Bencsik János [113]	Romanian and Serbian communities in South Hungary	24 May–27 June	*P. rhoeas, C. cyanus,* green wheat	Wheat blessing during Orthodox Pentecost	Wreaths were prepared from the corn poppy, cornflower, and green wheat, after which they were placed into wells (to prevent water pollution) or fed to livestock (to prevent diseases)
		Romanian and Serbian communities in South Hungary	24 June	*G. verum, P. rhoeas, C. cyanus,* other wildflowers and garden plants	Ivana Kupala (John the Batist) day	After the religious ceremony, flowers were gathered by children to make wreaths with magical powers of fortune-telling

3. Results and Discussion

Altogether, we found 108 publications discussing the ethnobotanical uses of the three studied arable weed species in the relevant Hungarian literature published between 1578 and 2021. In terms of their medicinal usage, we found 100 records in 43 documents. Among these notes, 42 refer to *P. rhoeas*, 39 to *D. consolida*, and 19 to *C. cyanus*. There is a great variety, both in the therapeutic purposes and mode of application, of each species (Table 1). *P. rhoeas* stands out with its sedative influence, *D. consolida* was primarily applied to stop bleeding, while *C. cyanus* was most frequently used to cure eye inflammation. In relation to food items, *P. rhoeas* was mentioned in four publications, as a famine food, delicacy, or food dye; while *D. consolida* was mentioned in two papers, as a food colouring or a delicacy. Four papers asserted the melliferous potential of *C. cyanus*, and one paper suggested its potential for grazing livestock. Ten articles highlighted the importance of these species as ornamental plants. Twelve papers described the role of these wildflowers in rituals and traditions related to (cereal) harvest, while their role in religious festivities, especially in Corpus Christi, were reported in ten. The cultural significance of these species for children's games and toys, particularly that of *P. rhoeas*, was presented in 13 studies. We also found 20 studies that discussed the role of the studied species in folk art. These plants are also regularly depicted in the visual arts and literature—from which a few iconic ones will also be discussed to illustrate the symbolic meaning and metaphorical applications of these species and to highlight the deep cultural embeddedness of these charismatic arable weeds.

3.1. Medicinal Uses

The first written records between the 16th and 17th centuries concern only *P. rhoeas* and cover a very broad scale of remedial power from insomnia through gum diseases, intestinal pain, St Anthony's fire (erysipelas), and bleeding (Table 1) [55–57]. In the late 18th century, this plant was considered a panacea [62]. At the same time, physicians started to recommend it for respiratory disorders [61] and the plant later became a general cough reliever in Hungarian folk medicine [68,75,79,80,87,94,96]. *P. rhoeas* also used to be listed in many European dispensatories as 'syrupus rhoeados', a sweet infusion made of poppy petals that was most notably used as a red colouring in pharmaceutical mixtures [114]. Poppy heads were also widely used as a sedative-hypnotic tool to calm crying babies, who would suck on this "pacifier", producing sounds such as "peep-peep" while gradually sinking into a deep sleep [52]. According to Rácz [52], this practice was so widespread that these baby sounds can even be related to the etymologic origin of the Hungarian name of the poppy ("pipacs").

The earliest Hungarian hint of the medicinal application of *D. consolida* is in a set of handwritten margin notes from the very beginning of the 18th century: *"When the blood of your nose starts to run heavily, take the herb called larkspur growing among wheat, dry it, smash it, and blow it into the nose with vinegar."* [58]. Even though this plant was later suggested to treat many different ailments, its most important application was to halt diverse types of bleeding, including wounds and gynaecological complaints, as a topical treatment [65,66,69,71,73,76,84,85]. While most sources do not specify the type of bleeding, in many cases, the context suggests gynaecological problems. The plant was prepared and administered in diverse ways against bleeding. For example, its herb would have been boiled together with bathwater and served as hot as the patient could bear, or it could also be used "secretly", smuggled under shoe inserts to cure an unspecified bleeding [71,76]. In some cases, the larkspur was also administered internally when its decoction was mixed into red wine [65], brewed together with beer [66], or extracted in a brandy tincture [85]. In some Hungarian regions, fumigation with *D. consolida* was conducted to cure gynaecological bleeding or to prevent premature childbirth, and it was also applied as a component of breast plasters in the case of mastitis [65].

In Hungarian folk medicine, the flowers of *C. cyanus* were used as a poultice for eye inflammation since the 18th century [60,61,63], and this stayed in use as its most common application until the middle of the 20th century [78,86,92,94,96]. Table 1 shows that it was also utilized in diuretic therapies [78,79,83] and we also found a record demonstrating that it was used to mitigate heart palpitations [86].

Related to the collection of these herbs, there is an interesting observation from 1798 in a book by Veszelszki [63]: *"Hard-working fathers and mothers send their children, who are regarded too weak to bear heavy work, to pick poppy flowers in the fields in midsummer, which they sell, or dry in sites where there is no sunlight."* This suggests that this plant was gathered and processed mainly by children at the turn of the 18–19th century. Later, during the Great War, there was a special ministerial decree to oblige school-teachers to organise their pupils to harvest medicinal plants (particularly the poppy and cornflower) during the summer holiday, which was intended to mitigate the general scarcity of medicaments in the era [115]. According to the yearly Hungarian pharmaceutical bulletins of the era, the poppy, cornflower, and larkspur were among the most popular medicinal plants gathered from the wild, which could be sold at a relatively good price until the 1960s. While the direct pharmaceutical usage of these drugs was gradually declining, they were still important for the beauty industry, mainly as components for face creams [116]. Today, these plants are rarely used for medicinal purposes and the poppy is even regarded as obsolete in modern phytotherapy. However, all these plants are still used in herbal tea blends in small quantities as minor "corrigents" to improve the colour of the infusion (Figure 2) [117].

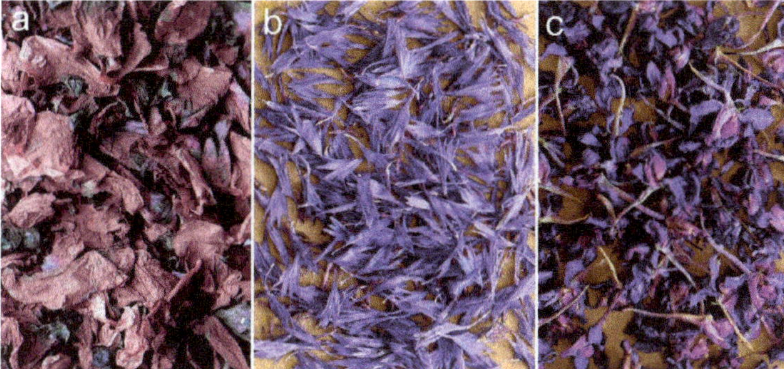

Figure 2. Traditional medicinal products from the three studied species, including (**a**) dried petals of corn poppy (*Rhoeados flos*); (**b**) ray florets of cornflower (*Cyani flos*); and (**c**) tepals of field larkspur (*Calcatrippae flos*) (Photographs by Gyula Pinke).

P. rhoeas was also extensively used as a soothing agent for various ailments in Poland [18], Serbia [118], Italy [119–127], Greece [128], Spain [129–132], and Turkey [133–135]. The poppy was also used outside Europe, e.g., in Morocco [136], Algeria [137], Tunisia [138], Iraq [139], and Iran [140]. In most cases, the flower, or less frequently its seeds, was prepared for traditional medicines. In Italy, a poppy was also administered to crying babies to induce sleeping but in the form of a decoction [120], while, in Kosovo, an infusion made of poppy seeds was used for a similar purpose [141]. The use of a larkspur as a remedy was also documented in Switzerland: if a woman suffered from discharge, she placed a larkspur in her shoes and kept it there for three days [142]. In Northwest Europe, chirurgeons also applied a larkspur on wounds and broken bones [142]. Wound treatments with *D. consolida* were also performed in Serbia [143] and Catalonia [144], and the plant was also used for diuretic therapies in Romania [145] and Italy [146]. The cornflower was also valued as a disinfectant for eyes and wounds in the Renaissance herbals of Western Europe [142], and it was used as a folk remedy for eye diseases in Belarus [147], the Polish–Lithuanian–Belarusian borderland [148], Kosovo [141], Bosnia and Herzegovina [149,150], Italy [122,151], Spain [144], and Armenia [152]. Furthermore, it was also used in diuretic therapies in Belarus [147], Ukraine [153], and Bulgaria [154], and to relieve heart palpitations in Italy [155]. The plant was also used for blood purification and cleansing the respiratory tract by traditional Polish herbalists [156], as well as in Lithuania [157] and Kosovo [141]. In Ukraine, cornflower tea was once considered as a panacea [158]. All these plants are still used in herbal tea blends in small quantities as minor "corrigents" in some European countries [142,159].

3.2. Food Uses

There are only a few records in Hungarian literature mentioning that the studied species were directly consumed as food items. The large buds of *P. rhoeas* were sporadically consumed as a famine food in the 19th century [160], but in some regions they may have also been consumed more regularly [161]. Two cases of accidental poppy poisoning of children were reported in the Austrian–Hungarian borderland in the early 20th century by Barsi [162], who suspected that more undocumented cases of fever and dazed sleepiness among children may have been caused by the consumption of these attractive, large, and apparently delicious buds. According to Barsi [162], poppy consumption may have spread to Western Hungary from neighbouring Styria and Carinthia, where the plant used to be consumed as a vegetable, even as pottage.

The petals of a poppy were gathered not only for folk remedies and pharmaceutical dye, but they were also used to colour cheese, cakes, and wine [163] (Figure 3). Similarly, the flowers of *D. consolida* were also used as a food dye; mixing the original green dye with alum could turn it blue, and both colours were utilized by confectioners [33]. This species was also picked by children, who sucked out the nectar from the long spurs of its flowers [164].

In many European countries, *P. rhoeas* used to have more significant culinary applications. However, in Italy, young poppy leaves are still eaten raw in mixed salads, or cooked in vegetable soups, omelettes, and pizzas [165–172]. Similar usages have also been reported from Croatia [173,174], Bulgaria [175], Greece [176], Spain [177,178], and Turkey [179,180]. From the latter country, there were even recent intoxication cases related to its consumption [181]. In Italy, poppy seeds were used to flavour bread and cookies [171,182]. The poppy was also used as a component of alcoholic drinks in Croatia and in Catalonia [132,174]. Additionally, poppy petals were used as a food colouring in Croatia [183] and as an ingredient for cosmetics, including lipsticks and cheek make-up in Italy [171,184,185]. The bright red colour of poppy petals was also utilized as a fabric dye in Italy [186]. Polish ethnobotanists reported that the very young shoots of *C. cyanus* were added to non-sour soups, while its flowers were combined with sugar to make wine and beer; moreover, they were also used to dye vinegar [147,173,187]. Cornflower infusion was also used for cosmetic purposes in Italy; it gave a special gloss and blue nuance to grey and white hair [185]. The green pigment from the flowers of *D. consolida* was also used in dying confectionery in many parts of Europe [142].

Figure 3. Food applications of the studied species, including poppy petals and cornflower flowers, used to dye and decorate homemade black elder (*Sambucus nigra* L.) syrup (**a–d**); handcrafted cheese (**e,f**); and cakes (**g,h**) coloured using poppy petals and petal extracts (Photographs by Gyula Pinke).

3.3. Fodder and Ethnoveterinary Uses

The arable fields stuffed with the poppy, cornflower, larkspur, and other weed species also provided important bee pastures [188], thus contributing indirectly to a further important food resource (honey). One of the earliest beekeeper books printed in Hungary [189] mentions that honeybees adored *C. cyanus* so much that they visited these flowers in mass quantities even in the period of linden (*Tilia* sp.) blossom. In the first part of the 19th century, this plant was regarded as one of the best melliferous plants, and bees could gather its nectar and pollen for up to six weeks [189], producing large amounts of greenish-yellow, delicious, monofloral honey [190]. Even in the middle of the 20th century, large amounts of *C. cyanus* pollen could still be detected in almost every type of summer floral honeys in Hungary [191]. Although *P. rhoeas* does not produce considerable amounts of nectar, it was also significant for beekeepers due to the large amount of pollen gathered by honeybees [190]. Among the three studied species, *C. cyanus* was also reported to have been used as an occasional livestock feed grazed on by sheep [192]. Occasionally, *P. rhoeas* was also foraged by animals, but it could be severely toxic, causing spasms and intestinal pain, sometimes with lethal consequences [193].

The larkspur was also applied for veterinary purposes. In West Hungary, its decoction with horse chestnut (*Aesculus hippocastanum* L.) was used to cure haematuria in cattle [73]. In North Hungary, infusion from a larkspur and other plants was mixed with foremilk and was given for newborn calves to secure fertility and milk-benefit, as well as to protect against witchcraft [82]. An interesting magical healing process was reported from Transylvania; in the case of a cataract, a larkspur was tied to the horn of the livestock on the opposite side of the sick eye [84].

The poppy was also used for animal feed in Turkey [179], particularly for rabbits and pigs. In Italy, poppy leaves were given to hens to increase egg laying [171], and, today, its seeds are still used as birdseed [120,186]. In Sicily, poppy flowers were once fed to farm animals in large quantities to stun them before slaughter [194].

3.4. Ornamental Uses

The first historical record of *C. cyanus* and *D. consolida* being cultivated in Hungary as ornamental garden plants comes from the 17th century [195], but these plants could have actually been grown for ornamental purposes since the 11–12th centuries [196]. *P. rhoeas* was also planted in gardens in the Renaissance era [70] and it was still cultivated sporadically in the 19th century [163]. In the late 20th century, when the arable flora suffered a significant decline [197], the ethnographer, Béla Gunda, observed in many villages that people were striving to "save" the disappearing cornflower: *"Village women gather the cornflower seeds which they sow in their small gardens in the autumn or spring, so that this flower, a folk favourite, could continue to thrive there"* [198]. Today, the seeds of all three species studied in this paper are commercially available in Hungary [199]. Nevertheless, they are relatively rare in gardens, where they have been replaced by more fashionable horticultural cultivars developed from their close relatives (Figure 4). They also used to be popular as cut flowers, providing a small temporary income source to impoverished women [73]. Monofloral and mixed bouquets of these wildflowers were regularly sold in flower markets as recently as the 1950s [200]. In Romania (Transylvania), they are still gathered for decoration in vases [96].

Figure 4. *Delphinium consolida* and *Delphinium ajacis* L. in a street garden (Markotabödöge, NW Hungary, 2020. Photograph by Gyula Pinke).

These colourful flowers have also been popular ornamentals in many other parts of Europe. Both *P. rhoeas* and *C. cyanus* have been cultivated since the 16th century throughout Central Europe as garden plants [201,202], which may partly be motivated by a high demand for wreaths [201]. The cornflower was a beloved species in the home gardens of Austria, where it seems to have maintained its high popularity until relatively recently [203]. The diverse horticultural varieties of *D. consolida* have also been common in many gardens of Central Europe [204]. *P. rhoeas* and *C. cyanus* have also been used to embellish bouquets in some regions of Spain [205], and bouquets and wreaths from *C. cyanus* are still sold in open-air markets in Poland [206].

3.5. Cutural Uses

3.5.1. Harvest Festivals

Gál [207] suggests in a short story that wildflowers were important emotional factors in setting the atmosphere of cereal harvests of the past, adding a little colour to the otherwise long and laborious days: *"The harvest goes on like a song, it has rhythm and melody. (…) The little flames of the poppies, the blue glitter of cornflowers and larkspurs. Every colour, every rhythm, every beauty all together"* (Figure 5a). These flowers were closely associated with harvest rituals and celebrations, which usually took place both at the beginning and end of the harvest period (Table 2). This period was typically opened with the ceremonial "binding" of the landlord (or steward) during their first visit to the harvesters when the landlord was symbolically tied up with a rope made of the first ears that were decorated with arable wildflowers [99,104] (Figure 5b). The first ears were often regarded to have magical powers, carrying God's blessings, which would ensure an abundant yield for the next year [97]. The first available report of wreaths made by the harvesters to celebrate the end of the harvest originates from 1806 [208]; however, this record still lacks any mention of wildflowers. Though, by the middle of the 19th century, these plants reportedly became key components of the wreaths [98,99].

Figure 5. The studied species in historical artworks: (**a**) cereal sheaves with corn poppy and corn-flower (postcard with good wishes for a newborn baby from May 1905, collection of Gyula Pinke); (**b**) ceremonial binding of the landlord (drawing by Mihály Szobonya, source: Vasárnapi Újság, 1888, 35 (28): 457; (**c**) harvesters with a harvest wreath (postcard, Ostoros, N Hungary, 1910–1920, © Zempléni Múzeum); (**d**) harvest festival (Kazár, N Hungary, 1940. Photograph by Géza Buzinka, courtesy of Fortepan).

According to a note from 1870, harvest workers spent their evenings together singing and wearing colourful wreaths made of various wildflowers during the harvest period [100]. This idyllic image of bucolic harvest celebrations became prominent in many subsequent documents in the late 19th and early 20th century. Nevertheless, it is good to keep in mind that, during this period, the Hungarian government started to recognize the untapped potential of rural traditions in building a national image, and harvest festivals became an important component in this new "country marketing". During this period, various forms of guidelines and recommendations were issued by various government agencies, some-times even prescribing the components of harvest wreaths. These recommendations were then mixed with pre-existing traditions at the local level, which makes it difficult to sepa-rate the genuine traditions from the new 'top-down' trends [209]. The recommendations were outlined in the following protocol: when the toughest part of the work, the reaping, was finished, a large wreath was made from the thickest ears, intertwined with poppy, cornflower, larkspur, corncockle (*Agrostemma githago* L.), and vetch (*Vicia* sp.) flowers, as well as fancy ribbons (Figure 5c). Then, the harvesters carried it through the farm in a solemn march, singing aloud (Figure 5d). The wreath was delivered and handed over to the landlord with a nice speech and, in exchange, he thanked the workers for all their hard effort. Subsequently, the harvest ball could be started with live music, traditional costumes, and a lot of dancing and shouting [88,97,99,101,103–105].

In some regions, a wreath or a bunch of ears with all these wildflowers was taken to the homes of the harvesters where it was hung on a wooden beam. These hanging decorations were then left in place for a long time, the maturing seeds of the drying wildflowers were saved and added to the sowing seeds to be used in the following autumn or given to hens to increase egg laying [99].

These flowers used to have similar roles in Czech, Romanian, and Russian harvest celebrations [99,210].

3.5.2. Religious Ceremonies and Rituals

The three studied wildflowers were also important props in several religious celebrations that took place during the summer months (Table 3). The most significant liturgical event related to arable wildflowers was the feast of Corpus Christi, which is celebrated two months after Easter in the Catholic calendar, typically in late May or early June. The event was celebrated with spectacular processions in several regions of Hungary. To prepare for this feast, children were sent out to the countryside to gather flower petals, especially those of the poppy [108]. Then, during the procession, the priest carrying the Holy Communion was followed by a group of young girls dressed in white holding the petals in small baskets and tossing them around in such an abundance that the ground was often fully covered in a floral carpet (Figure 6a,b). According to the reports of Horváth [108] and Demeter [110], the poppy was preferred because it produced the most spectacular floral carpet, but the flowers of the cornflower, corncockle, tuberous pea (*Lathyrus tuberosus* L.), and marguerite (*Leucanthemum vulgare* Lam.) were also used. Later, the petals of roses (*Rosa* sp.) and peonies (*Paeonia* sp.) became more frequent, and bouquets of garden flowers were also integrated into this celebration. The flowers from Corpus Christi, blessed with the Holy Communion, used to be one of the most respected paraliturgical items for Catholic Hungarians [112]. After the ceremony, these flowers were taken home by the churchgoers to save their houses from lightning strikes or to treat sick children and animals (e.g., via fumigation or infusion baths) [107–111].

Figure 6. Flower carpet for the procession of Corpus Christi in Budaörs, Hungary (**a**) 1940; by unknown photographer; (**b**) 1943; photograph by Carl Lutz; donated by Archiv für Zeitgeschichte ETH Zürich/Agnes Hirschi. Courtesy of Fortepan).

In communities that follow Eastern Christian traditions (Greek Catholic, Orthodox), the most important feast involving wildflowers was the birth of John the Baptist (Ivana Kupala). In the Greek-Catholic villages of Northern Hungary, this feast was celebrated with bunches of flowers from the fields, meadows, and gardens that typically included large amounts of the larkspur and cornflower. These flowers were then taken to the church and blessed by the priest at the end of the liturgy (Table 3), which endowed them with magical properties, and they were used for making decoctions, and vapour or fume treatments to heal children and livestock [211]. In the settlements of Southern Hungary, populated by ethnic Romanians and Serbians of the Orthodox faith, John the Baptist (Ivana Kupala) used to be celebrated on 24 June (Table 3). For this event, children and elderly women went to the fields and meadows to pick flowers, which were then bundled into large bunches and wreaths. The mainstay of these wreaths was yellow bedstraw (*Galium verum* L.) but the poppy and cornflower were also common components. These wreaths were flung up onto the thatched roofs and, if they fell off, it was considered to be a bad omen (e.g., prophesising death), while if they remained on the roof, it was believed to be a good sign (e.g., the young girl of the family would get married soon) [113].

The celebration of the Orthodox Pentecost, which usually followed its Catholic counterpart by several weeks, also involves some traditions related to arable wildflowers. In the Romanian and Serbian communities discussed above, Pentecost involved a traditional wheat blessing ceremony where wildflower wreaths were prepared and blessed. These wreaths were later placed into the wells (to prevent water pollution) or were fed to livestock (to prevent diseases) [113].

According to Luczaj [212], floral decorations and a petal toss were featured in Corpus Christi processions in many European countries until the 19th century. Poland seems to be the last refuge for the once widespread tradition of blessing floral wreaths for Corpus Christi in which the most important flowers used are roses; however, the cornflower and poppy can be found in them as well [212]. Cornflower bouquets were also used in Orthodox Pentecostal rituals in western Ukraine. In this region, the seedpods of *P. rhoeas* were used for decoration in other religious festivities (Easter, Feast of the Transfiguration) [213]. In certain regions of Spain, both *P. rhoeas* and *C. cyanus* were also the subject of magical and religious beliefs and practices [205].

3.5.3. Children's Culture

The relatively large and brightly coloured flowers of the poppy captivated the attention and imagination of children as well. Particularly in smallholder families, where the grandparents, parents, and larger children used to make various types of "poppy puppets" throughout the Carpathian Basin [164,214–219]. According to Ortutay [220], the starting point of making a poppy puppet was a poppy bud; first, the sepals were removed, then, the petals were folded down and tied with a blade of grass. This resulted in something that looked like a doll in a red robe (the petals), whose head was the ovary, and the stamens formed a collar. With a few finishing touches, these dolls could be turned into various figures, e.g., an elegant lady, a devil, or the baby of a larger rag doll, which could be used creatively while playing, e.g., as the participants of a wedding or a funeral. In a wedding game, a poppy with white petals was typically used to make the bride, while the red ones were used for the groom and the other guests. In a burial scene, a white poppy was selected for the decedent and the priest was red [164,214–219] (Figure 7a). Sometimes, these dolls were dressed up with further accessories, e.g., with a necklace woven of larkspur flowers [164,214].

Moreover, before unfolding a poppy bud, a colour-guessing game accompanied with the nursery rhyme of "*Is it beer, is it wine, is it brandy, or is it a pink ribbon?*" was often played by children in the Carpathian Basin. A poppy with rusty-reddish petals represented "beer", red was "wine", white was "brandy", and a pink poppy stood for the "ribbon". If the answer was right, the respondent received a treat (e.g., they could eat the petals or received a flower from the questioner), otherwise they had to pay (e.g., with another poppy flower) [161,217,221]. In other popular children's games, poppy petals were snapped with the lips [222] (Figure 7b) or poppy buds were hit on the back of the hand [70] (Figure 7c). Thus, similar to the German names (e.g., "Klatschmohn", "Klapperrose", and "Klatschrose"), the Hungarian name of this plant ("pipacs" and its archaic forms "pippancs" and "papics") could also be of onomatopoeic origin (closely related to the words "pattint" [snap] and "pacskol" [slap]) [52]. These words probably came into existence before the first half of the 16th century independently of other languages [223], which suggests a long-standing cultural relationship to this plant by the Hungarian ethnic populations living in the Carpathian Basin. The poppy was also frequently represented as a character in tales and poems for kids, often in a protagonist role as the "poppy king", "poppy lady", or a "little poppy". These fabulous heroes usually had a red face, wore red clothes, and made bright, flirtatious appearances [36].

Figure 7. Animated scenes with old-fashioned children's toys: (**a**) puppets made from poppy flower; (**b**) snapping poppy petals on the lips; (**c**) snapping a poppy bud on the back of the hand (Mosonmagyaróvár, NW Hungary, 2019); (**d**) a cornflower wreath and a bunch of larkspur (Halászi, NW Hungary, 2022) (Photographs by Gyula Pinke).

The cornflower was also used to make wreaths by young Hungarian peasant girls [63,84,215] and the decorations made of larkspur flowers were also highly appreciated in wedding games [224] (Figure 7d). On the day of Pentecost, the cornflower was used as a gift of love in some Hungarian regions where lads gave small bunches of the cornflower, sometimes accompanied with other wildflowers, to the girls they liked most [225].

Poppy dolls were also popular among children in Germany and Central Italy [171,204]. Italian children also used to play a colour-guessing game with the still closed flower buds, saying *"frate, monaca o cappuccino?"* (monk, nun, or capuchin?) [171]. In Italy, poppy ovaries and seed pods were also played with as "stamps" that left a nice mark on the skin [171,182]. One of its German folk names ("Tintenblume") indicates that a red ink could also be made from the petals by the children [204]. Folkloric records suggest a widespread use of the poppy in children's games within [226] and beyond Europe [227]. The cornflower was also listed in one of the earliest ethnobotanical inventories as a plant used in children's games in Germany and Upper Austria [228].

3.5.4. Visual Arts

Together with the rose, carnation (*Dianthus* sp.), and later tulip (*Tulipa* sp.), the cornflower has been one of the oldest and most archaic floral motifs in Hungarian folk art [229,230]. The cornflower used to be particularly popular as a Christian motif, appearing regularly in diverse religious contexts, e.g., on painted church ceilings [229] (dated 16th c.), [230] (dated 18th c.); embroidered church tablecloths (Figure 8), [231,232] (dated 17th c.), [233] (dated 1898); altar cloths [234] (dated 17th c.); and vestments [235] (dated 1792), [236] (dated 20th c.). Furthermore, from the 19th century, the cornflower has gradually infiltrated the decoration of household items, including sheets [231] (dated 19th c.); bonnets [237] (dated 19–20th c.); chests [238] (dated 1853); ceramic pots [239] (dated 1926–1929); and horn carvings [240] (dated 19–20th c.). From the 17th century, the motifs of the poppy also emerged, first on pewters [241] (dated 17th c.), but later it became a favourite element on diverse folk embroideries [242] ranging from liturgical tablecloths [243] (dated 1897), [244] (dated 1915) to evening dresses [245] (dated 20th c.). In some regions of Hungary, larkspur flowers also became popular floral motifs on embroideries as well as wall paintings [246].

While these wildflowers do not belong to the most common floral motifs of Hungarian folk art, they are present in several traditional ornamental styles. Cornflower, poppy, and larkspur motifs also appear among the famous patterns of Kalotaszeg (Țara Călatei) in West Transylvania, Romania, probably originating from the 18th century [247]. These wildflowers are also prominent motifs in the embroidery techniques from Torontálvásárhely (Debeljača) (Figure 9a) and Ada (Figure 9b), both developed during the 20th century in Vojvodina, Serbia [248–250]. Despite the fact that other, older folk-art styles used strongly stylised floral patterns and were largely detached from any concrete species [251], these new styles applied relatively easily recognizable naturalistic and recognizable figurative floral motifs. In the case of Ada, these motifs include the corncockle, marguerite, and buttercup (*Ranunculus* sp.), in addition to the three studied species [248]. Furthermore, some of these appear on traditional hand-embroidered slippers in nearby Szeged (Hungary, Figure 9c) [252,253]. Traces of these folk-art styles were adopted by more recent "souvenir folk art" designed and mass produced by business ventures trying to meet the demand of tourists [254].

Figure 8. Fragment of a panelled cover cloth (late 17th century, Calvinist church of Marosvécs [Brâncovenești], Transylvania, Romania). Cornflower is located in the centre of the motif from which carnations and tulips emerge. The central cornflower is a favourite stylistic element of Transylvanian embroideries (Museum of Applied Arts, Budapest. Photograph by Áment Gellért. © Iparművészeti Múzeum).

The cornflower is also one of the most popular motifs on a popular high-end product line of the Hungarian porcelain manufacture 'Zsolnay pottery' (Figure 10a) [255]. Furthermore, the three wildflowers examined are also often featured on other hand-painted ceramics (Figure 10b). The cornflower and poppy were popular elements of the Hungarian Art Nouveau at the turn of the 20th century (Figure 11). In Hungary, Pál Szinyei Merse is the most eminent painter in terms of depicting emblematic landscapes with vibrant poppies at the end of the 19th century [35]. He created those pictures on his provincial estate, which became masterpieces of Hungarian naturalism (Figure 12a). He was characterized by the art historian Antal Hekler [256] as, " … *a warm-hearted interpreter of the Hungarian reality aflame with poppies*". The studied three wildflowers were also frequently illustrated in genre paintings (Figure 12b) and still life pictures (Figure 12c).

Figure 9. Contemporary folk embroideries with wheat ears and arable wildflowers motifs: (**a**,**b**) textiles from Vojvodina (N Serbia) with motifs from Ada (**a**) and Torontálvásárhely [Debeljača] (**b**) (both manufactured by Veronika Serfőző, photographs by Ágnes Nagy Abonyi. Courtesy of Rozetta Kézműves Társaság, Zenta); (**c**) Slippers from Szeged (S Hungary, courtesy of Sallay Szegedi Papucs).

Figure 10. (**a**) Porcelain vase with cornflower motifs (Zsolnay factory); (**b**) Hand-painted wall plate ceramics (Photographs by Gyula Pinke).

Figure 11. Ceramics from the Hungarian Art Nouveau: (**a**) Vase with red poppies (ca. 1900, Zsolnay factory, Museum of Applied Arts, Budapest. Photograph by Jonatán Urbán and Dávid Kovács. © Iparművészeti Múzeum); (**b**) Cup with cornflowers (1896, Henrik Giergl company, Museum of Applied Arts, Budapest. Photograph by Ágnes Kolozs. © Iparművészeti Múzeum).

Figure 12. Hungarian paintings featuring arable wildflowers; (**a**) Meadow with poppies (Pál Szinyei Merse, 1896; Museum of Fine Arts, Budapest; © Szépművészeti Múzeum 2022); (**b**) Summer (The reproduction cards of the 'Publisher Könyves Kálmán'. Reproduced after: A hatted lady with a bouquet of arable wildflowers, Fülöp Szenes, 1913. Courtesy of Bedő Papírmúzeum); (**c**) Bouquet of arable wildflowers (Gizella Czeglédi, 1999. Courtesy of Gizella Czeglédi).

The three studied wildflowers have also been popular decorative elements in folk art in several other European countries [18,257,258]. According to Polish-Ukrainian beliefs, the cornflower was one of the favourite flowers of water-nymphs [259]. The cornflower was also often depicted in paintings during the Middle Ages and the Renaissance period, frequently seen in Christian frescos as a symbol of Mary or Christ [17,142]. In modern art, the French impressionist, Claude Monet, garnered worldwide fame with vibrant poppies in his paintings [260].

3.5.5. Literary Works

Sándor Petőfi, the most famous Hungarian patriotic poet of the 19th century, wrote: *"The fields are filled with flowers, you will find/Poppies grow in gay profusion/All genera, every kind"* (*"Szántóföld szépen virít,/Termi bőven a pipacsnak/Mindenféle nemeit"*). In this satirical poem, published in 1847, the thriving poppies refer to the negligence of lazy Hungarian landlords of the era. Another famous epic of Petőfi, titled *"John the Valiant"* (*"János vitéz"*, 1844), was transformed into an opera by the composer Pongrác Kacsóh in 1904. In this opera, the poppy emerges as a symbol of homesickness and patriotism. At the climax of this piece, the sound of a flute touches John's heart, who says: *"Back home, poppies and larkspurs have started to bloom, by the time I get home it will be time for harvest … "*.

In the first half of the 20th century, the poppy, sometimes also the cornflower and larkspur, appears as metaphorical illustrations of the Hungarian homeland and folk spirit [35]. *"Hungary, my beautiful mother country with poppy flowers and wheat ears"*—passionate exclamations, similar to this one written by Mici Gruber (1928), were quite common in public magazines of the era. In the early 20th century, this flaming red flower could also symbolize the increasingly popular revolutionary political movements. *"The rich fields are set ablaze in poppies/by the fiery wonder of the Hungarian summer"* (*"Pipacsot éget a kövér határra/A lángoló magyar nyár tűzvarázsa"*)—this is the iconic beginning of one of the most famous poems of Gyula Juhász in 1918, where the poppies in the glowing landscape became an impressionist personification of the growing societal tensions, possibly heading towards an imminent revolution [261].

One of the novels by Zsigmond Móricz, a famous writer of the early 20th century, titled *"Poppies on the sea"* (*"Pipacsok a tengeren"*) (1908), takes place on the Great Hungarian plain, which was flooded by the Tisza river, and the flowers were floating on the surface (Figure 13). In this story, a little boy, who is the writer himself, falls in love with a little peasant girl wearing a red skirt, and secretly calls her *"my little blood red poppy flower"* (*"Kis vérszín pipacsvirágom"*). This suggests that the plant can also be associated with emotional infatuation. Poems by Károly Szász (1930) make a similar association, connecting the poppy to burning love: *"Poppies were burning in the grass/blazing like your kisses/red like my blood/I stuck a poppy in your hair"* (*"Pipacsok égtek lángolva a fűben./Tüzesek, mint a csókod,/Pirosak, mint a vérem./Egy pipacsot én a hajadba tűztem"*). This passion is further intensified by Lajos Nechanszky (1932): *"Your lips, the trembling blood red poppies/(…) are whispering their glowing embers at me"* (*"Imbolygó, vérszínű pipacs a szád/(…) és rámsusogja forró parazsát."*).

Nevertheless, due to the short life and quick fall of its petals, this plant was also considered as a metaphor of impermanence, and the resulting sorrow and lovesickness. As Mihály Tompa illustrates (1853): *"Oh, its ornaments are so perishable/they blow in the morning and fall before the evening!"* (*"Ah, de dísze oly múlandó./Reggel nyílik, estig elhull!"*). In his poetry book titled *"When you were poppies"*, László Király (1982) makes a dramatic observation: *"Youth is gone in a flash of red poppies"* (*"Tovatűnt az ifjúság pipacsszínű lobbanása"*) [262].

Figure 13. Drawings by Rozi Békés for the illustration of Zsigmond Móricz's novel *"Poppies on the sea"* (1908), where a major flood on the Tisza river causes the red flowers to float (Courtesy of Rozi Békés).

The cornflower was often used with nostalgic intent, to invoke the intimate atmosphere of harvests from days gone by. For example, Ferenc Mátyás (1952) wrote: *"Cornflowers in the girls' hair/flames in their eyes and hearts/their songs tear sorrow apart/as they are binding wheat into sheaves."* (*"Búzavirág a lányok hajában,/szemükben is, szívükben is láng van./Száll a daluk a bút összetépve,/úgy kötözik a búzát kévébe."*). According to an old legend [37], possibly originating from Western Europe [142], God made this flower blue, so that bent-over peasants could still admire the colour of Heaven during their tough work in the fields. This legend can also be traced back to a metaphor by Ferenc Mátyás (1952): *"As if the sky broke into pieces/and it would shine down here"* (*"Mintha az ég darabokra törne,/s csupakéken itt lenn tündökölne"*). Sometimes, the flowers came with a stronger religious meaning, as illustrated by József Erdélyi (1935): *"Your blue colour, like the clean, almighty sky/is an ethereal virgin faraway!"* (*"Kék színed, mint a tiszta, magas ég,/a földöntúli, szűzi messzeség!"*). István Toronyi (1932) also refers to the divine origin of this plant: *"Holy water dropped to your nice blue clothes/when you became in holy baptism/cleansed from sin: the flower of God."* (*"Szenteltvíz hullott szép kék ruhádra/S akkor lettél te szent keresztségben/Bűntől tisztulva: Isten virága"*). Accordingly, the cornflower used to be a universal symbol of innocence, virginity, perseverance, and faithfulness [263–265]. At the same time, this plant was often used as a metaphor of gentle and tender love, as Jenő Dsida (1930) expresses: *"I silently sigh the blue love of cornflowers towards you"* (*"Csöndesen feléd sóhajtom a búzavirágok kék szerelmét"*).

The deep relationship between the larkspur and farmers is reflected in the works of several poets from a peasant descent [38]. As György Dénes (1961) illustrates: *"I am walking on peasant-fields again/ . . . larkspurs are gently guiding my path"* (*"Paraszt-mezőkön járok újra/(. . .) szarkaláb hajlik/szelíden útamra"*). This plant could also encapsulate nostalgic memories, suggesting that this flower could be an important element in making an imprint on youth, as it is in the case made by Imre Oravecz (1997): *"You also used to be a child/holding a quail chick in your hands/walking barefoot on soft grounds/picking larkspur at sheaves binding"* (*"Voltál gyermek,/tartottál kezedben fürjfiókát,/lépkedtél mezítláb a föld puha hátán, /szedtél szarkalábat marokveréskor)"*. Similarly, Dániel Hatvani (1965) also mentions this plant evoking an old romance: *"Only the stacked sheaves of wheat might/keep the larkspur-scented memories of first loves/ . . . threshing machines murmuring in the dust/glittering bodies of girls in the evening sun"* (*"Talán csak búzaasztagok őrzik/az első szerelmek szarkaláb-illatát/(. . .) mormoló cséplőgépek a por halmazában/izzadó leánytesteken csordult szét a nap"*).

3.5.6. Societal Symbols

In the previous sections, we have reviewed cultural and artistic applications of the three studied species among the Hungarian-speaking communities in the Carpathian

Basin. All these cultural applications, whether they are traditional ceremonies or artistic motifs, are based on the symbolic meanings that these species convey. In this section, we summarize these symbolic messages (Table 4), connecting them to further social movements and phenomena.

Table 4. Metaphorical connotations of the studied wildflowers symbolizing human characters and emotions (own synthesis based on the cultural uses presented in Section 3.5.1, Section 3.5.2, Section 3.5.3, Section 3.5.4, Section 3.5.5, Section 3.5.6 of this article).

Character, Emotion Being Symbolised	Species
Homesickness, nostalgia, bucolic reminiscence	*P. rhoeas, C. cyanus, D. consolida*
Patriotism	*P. rhoeas, C. cyanus, D. consolida*
Historical remembrance	*P. rhoeas, C. cyanus*
Passion, infatuation, burning love, lovesickness	*P. rhoeas*
Impermanence, transience, ephemerality, fragility	*P. rhoeas*
Purity, innocence, virginity	*C. cyanus*
Pertinence, faithfulness, loyalty	*C. cyanus*
Gentle and tender love	*C. cyanus*

As we discussed above, all three species have often symbolised a bucolic nostalgia and yearning toward a simple rustic life, the idyllic reminiscence of a lost homeland or youth. Nevertheless, in the early 20th century, the three studied arable weed species also became patriotic symbols of the consolidating Hungarian state and the Hungarians in it that were seeking their identity in the dualistic Austrian-Hungarian monarchy. At this time, the poppy and cornflower were seen as the most important components of wreaths and bouquets used in summer burial ceremonies to decorate a coffin, hearse, and even streets where the funeral procession of a prominent public figure passed by. These flowers symbolised the connection that tied the decedent to the Hungarians [35,37]. This period coincides with a Europe-wide renewal of symbolic systems, with a proliferation of new national symbols all over Europe [266]. Nevertheless, the poppy, cornflower, and larkspur remained hidden but popular national symbols in Hungary during the era of socialism (Figure 14b) and afterward [267].

After the First World War, the poppy became a particularly important symbol representing the blood shed by the soldiers, but also a hope for regeneration and renewal. This symbol probably has multiple roots: poppies were reportedly abundant in the disturbed landscapes of the battlefields [268,269]. Their colour allowed for an easy association with the blood shed and the short-lived flowers also provided a natural allegory of transience and fragility, which has also been documented in this study (Table 4). The poppy, as a metaphor, was also sporadically used in Hungarian war coverage from the Eastern front, as poppies growing on soldiers' graves resembled the blood drops of the fallen soldiers [35]. Moreover, the unusually high abundance of the poppy in Hungarian arable fields in 1916 was explained in a contemporary article as the *"blood of the Earth overflowing in sorrow (…), as its sons are falling in the Eastern and Italian fronts (…), far away from their motherland"* [35]. Accordingly, the poppy became an important symbol of the huge and heroic, possibly pointless losses, and this meaning is preserved in several national symbols of remembrance still actively used today, including the *"Flanders poppy"*, or the emblem of the Royal British Legion [268,270].

Figure 14. Stamps with cornflower motifs: (**a**) *"For the mothers with many children"*, a series of charity stamps issued by the *"cornflower-action"* movement (1929–1939) (Courtesy of Bedő Papírmúzeum); (**b**) a Hungarian postal stamp with arable wildflowers (1980) (Designer József Vertel, Courtesy of Magyar Posta).

The association of ephemerality with the poppy might go back to very ancient roots—as suggested by Beuchert [201], who identified the poppy wreaths found in the grave of a young Egyptian princess as an indication of "fragile and evanescent existence". In Ukrainian Carpathian folklore, this short-lived flower is also associated with transience, briefness of youth, and unfortunate love [271]. Moreover, the poppy was also seen as a symbol of pride (due to its impressive display) and of sleep (based on its popular medicinal use) [142,272].

As discussed in the previous sections, the symbolic meanings of the cornflower are much more connected to primary human and societal values including innocence, perseverance, faithfulness, and loyalty (Table 4). Accordingly, it is not surprising that the sky-coloured cornflower was so often used as a symbol in religious contexts. The colour of the cornflower is often explained to have a celestial origin: for example, according to a British legend, the sky sent bits of itself down to the fields, thus creating cornflowers [142]. The cornflower has also been used as a symbol of charitable movements, with a noble societal purpose. For example, the "cornflower-action" was a Hungarian charitable movement between 1929 and 1939, aimed at subsidizing Hungarian mothers with many children by selling paper cornflowers and stamps with a cornflower illustration (Figure 14a) [37,273,274]. Similarly, in Germany, where this plant has a remarkable cultural appreciation [275], artificial cornflowers were also prepared and sold to support old veterans of the Franco-Prussian War in 1870–1871 [37]. As the cornflower also thrived in the battlefield landscapes, this flower was also used as a symbol of remembrance [269]. After World War I, pin badges with a poppy (e.g., in Britain) and a cornflower (e.g., in France) were made (mainly) by disabled soldiers for the purpose of supporting war orphans and veterans [36,37,142]. These charismatic wildflowers can be spotted even in present-day national symbols: for example, the colours of the French tricolour flag are often linked to the poppy and cornflower [142], while the blue of the Estonian flag is also often linked to the cornflower, which is also one of the main national symbols of this relatively young state [276].

Similar to the poppy, the cornflower was also widely used as a symbol of love. Nevertheless, aligned with its general symbolic meanings (perseverance, faithfulness, purity) [201], the cornflower symbolized a slower, more permanent, reliable, and tender emotion. Not surprisingly, around the turn of the 20th century, *"Cornflower"* was a common code word in personal ads in Hungarian newspapers, and it was also one of the most frequently used nicknames in salutations and signatures of secret love messages (*"To my cornflower"*, *"Your cornflower"*, etc.) [37]. As Erdélyi [277] pointed out, the cornflower (as well as the flax—*Linum usitatissimum* L.; and the blackthorn—*Prunus spinosa* L.) appears conspicuously frequently in Hungarian folk poetry describing an ideal (or desired) eye colour. Independent of hair colour, blue eyes were always considered to be signs of tenderness, faithfulness, and serenity in girls. These subconscious idealistic images may explain why folk art so often depicts girls with blue eyes [277]. Cornflowers were also used in various traditions of foretelling love in Western Europe [142].

4. Conclusions

Our review explores the long-established and deep cultural embeddedness of the studied three iconic arable weed species, *P. rhoeas*, *C. cyanus*, and *D. consolida* in the Carpathian Basin. Ethnobotanical records and historical artefacts suggest that these species were used on a broad scale among the Hungarian populations of the Carpathian Basin from (at least) the 16–17th centuries until modern times as medicinal, food, ornamental, and cultural resources. These species were emotionally linked to the peasant lifestyle in many ways, providing inspiring symbols for Hungarian literature and visual arts, as well as broader society. Many aspects of these cultural connections peaked in the early 20th century. The drop in the number of records from the late 20th and the 21st century is probably attributable to a combination of the declining diversity of arable weeds and the disappearing interest and knowledge related to these plants. The long-term deep cultural embeddedness of the studied species could be capitalized on to obtain a stronger societal support for the idea of arable weed species conservation. More generally, ethnobotanical and cultural embeddedness should be considered more seriously when efforts and instruments for the conservation of arable weed communities are designed.

Author Contributions: G.P. conceived of the idea for the paper, made the literature reviews, and wrote the first draft of the manuscript. V.K. prepared cheese and cakes dyed with poppy and revised the article. B.C. revised, completed, and finalized the article. All authors have read and agreed to the published version of the manuscript.

Funding: This research received no external funding.

Institutional Review Board Statement: Not applicable.

Informed Consent Statement: Prior and written consent of images from museums and private collections have been obtained for publication.

Data Availability Statement: Not applicable.

Conflicts of Interest: The authors declare no conflict of interest.

References

1. Naylor, R.E.L.; Lutman, P.J. What is a weed? In *Weed Management Handbook*; Naylor, R.E.L., Ed.; Blackwell Science: Oxford, UK, 2002; pp. 1–15.
2. Fagúndez, J. The paradox of arable weeds: Diversity, conservation, and ecosystem services of the unwanted. In *Agroecology, Ecosystems, and Sustainability*; Benkeblia, N., Ed.; CRC Press: Boca Raton, FL, USA, 2014; pp. 139–149.
3. Blaix, C.; Moonen, A.C.; Dostatny, D.F.; Izquierdo, J.; Le Corff, J.; Morrison, J.; Von Redwitz, C.; Schumacher, M.; Westerman, P.R. Quantification of regulating ecosystem services provided by weeds in annual cropping systems using a systematic map approach. *Weed Res.* **2018**, *58*, 151–164. [CrossRef]
4. Smith, B.M.; Aebischer, N.J.; Ewald, J.; Moreby, S.; Potter, C.; Holland, J.M. The potential of arable weeds to reverse invertebrate declines and associated ecosystem services in cereal crops. *Front. Sustain. Food Syst.* **2020**, *3*, 118. [CrossRef]
5. Yvoz, S.; Cordeau, S.; Ploteau, A.; Petit, S. A framework to estimate the contribution of weeds to the delivery of ecosystem (dis)services in agricultural landscapes. *Ecol. Indic.* **2021**, *132*, 108321. [CrossRef]

6. Merfield, C.N. Redefining weeds for the post-herbicide era. *Weed Res.* **2022**, *62*, 263–267. [CrossRef]

7. Bretagnolle, V.; Gaba, S. Weeds for bees? A Review. *Agron. Sustain. Dev.* **2015**, *35*, 891–909. [CrossRef]

8. Rollin, O.; Benelli, G.; Benvenuti, S.; Decourtye, A.; Wratten, S.D.; Canale, A.; Desneux, N. Weed-insect pollinator networks as bio-indicators of ecological sustainability in agriculture. A review. *Agron. Sustain. Dev.* **2016**, *36*, 8. [CrossRef]

9. Morrison, J.; Izquierdo, J.; Hernandez Plaza, E.; Gonzalez-Andujar, J.L. The attractiveness of five common Mediterranean weeds to pollinators. *Agronomy* **2021**, *11*, 1314. [CrossRef]

10. Marshall, E.J.P.; Brown, V.K.; Boatman, N.D.; Lutman, P.J.W.; Squire, G.R.; Ward, L.K. The role of weeds in supporting biological diversity within crop fields. *Weed Res.* **2003**, *43*, 77–89. [CrossRef]

11. Storkey, J.; Westbury, D.B. Managing arable weeds for biodiversity. *Pest Manag. Sci.* **2007**, *63*, 517–523. [CrossRef]

12. Storkey, J.; Meyer, S.; Still, K.S.; Leuschner, C. The impact of agricultural intensification and land-use change on the European arable flora. *Proc. R. Soc. B* **2012**, *279*, 1421–1429. [CrossRef]

13. Meyer, S.; Wesche, K.; Krause, B.; Leuschner, C. Dramatic losses of specialist arable plants in central Germany since the 1950s/60s—A cross-regional analysis. *Divers. Distrib.* **2013**, *19*, 1175–1187. [CrossRef]

14. Fried, G.; Petit, S.; Dessaint, F.; Reboud, X. Arable weed decline in northern France: Crop edges as refugia for weed conservation? *Biol. Conserv.* **2009**, *142*, 238–243. [CrossRef]

15. Albrecht, H.; Cambecèdes, J.; Lang, M.; Wagner, M. Management options for the conservation of rare arable plants in Europe. *Bot. Lett.* **2016**, *163*, 389–415. [CrossRef]

16. Bergmeier, E.; Meyer, S.; Pape, F.; Dierschke, H.; Haerdtle, W.; Heinken, T.; Hoelzel, N.; Remy, D.; Schwabe, A.; Tischew, S.; et al. Arable vegetation of calcareous soils (*Caucalidion*). Plant community of the year 2022. *Tuexenia* **2021**, *41*, 299–350. [CrossRef]

17. Kandeler, R.; Ullrich, W.R. Symbolism of plants: Examples from European-Mediterranean culture presented with biology and history of art. *J. Exp. Bot.* **2009**, *60*, 3297–3299. [CrossRef]

18. Zemanek, A.; Zemanek, B.; Klepacki, P.; Madeja, J. The poppy (*Papaver*) in old Polish botanical literature and culture. In *Plants and Culture: Seeds of the Cultural Heritage of Europe*; Morel, P., Mercury, A., Eds.; Centro Europeo per I Beni Culturali Ravello, Edipuglia: Bari, Italia, 2009; pp. 217–226.

19. Pinke, G.; Dunai, É.; Czúcz, B. Rise and fall of *Stachys annua* (L.) L. in the Carpathian Basin: A historical review and prospects for its revival. *Genet. Resour. Crop Evol.* **2021**, *68*, 3039–3053. [CrossRef]

20. Prigioniero, A.; Scarano, P.; Ruggieri, V.; Marziano, M.; Tartaglia, M.; Sciarrillo, R.; Guarino, C. Plants named *"Lotus"* in antiquity: Historiography, biogeography, and ethnobotany. *Harv. Pap. Bot.* **2020**, *25*, 59–71. [CrossRef]

21. Fatur, K. "Hexing herbs" in ethnobotanical perspective: A historical review of the uses of anticholinergic *Solanaceae* plants in Europe. *Econ. Bot.* **2020**, *74*, 140–158. [CrossRef]

22. Kujawska, M.; Svanberg, I. From medicinal plant to noxious weed: *Bryonia alba* L. (*Cucurbitaceae*) in Northern and Eastern Europe. *J. Ethnobiol. Ethnomed.* **2019**, *15*, 22. [CrossRef]

23. Gruca, M.; van Andel, T.R.; Balslev, H. Ritual uses of palms in traditional medicine in Sub-Saharan Africa: A review. *J. Ethnobiol. Ethnomed.* **2014**, *10*, 60. [CrossRef]

24. Leroy, T.; Plomion, C.; Kremer, A. Oak symbolism in the light of genomics. *N. Phytol.* **2020**, *226*, 1012–1017. [CrossRef] [PubMed]

25. Savo, V.; Kumbaric, A.; Caneva, G. Grapevine (*Vitis vinifera* L.) symbolism in the ancient Euro-Mediterranean cultures. *Econ. Bot.* **2016**, *70*, 190–197. [CrossRef]

26. Teixidor-Toneu, I.; Kjesrud, K.; Kool, A. Sweetness beyond desserts: The cultural, symbolic, and botanical history of angelica (*Angelica archangelica*) in the Nordic Region. *J. Ethnobiol.* **2020**, *40*, 289–304. [CrossRef]

27. Kumbaric, A.; Savo, V.; Caneva, G. Orchids in the Roman culture and iconography: Evidence for the first representations in antiquity. *J. Cult. Herit.* **2013**, *14*, 311–316. [CrossRef]

28. Molnár, V.A.; Nagy, T.; Löki, V.; Süveges, K.; Takács, A.; Bódis, J.; Tökölyi, J. Turkish graveyards as refuges for orchids against tuber harvest. *Ecol. Evol.* **2017**, *7*, 11257–11264. [CrossRef] [PubMed]

29. Kästner, A.; Jäger, E.; Schubert, R. *Handbuch der Segetalpflanzen Mitteleuropas*; Springer Verlag: Wien, Austria, 2001.

30. Hartyányi, B.; Nováki, G.; Patay, Á. Növényi mag- és termésleletek magyarországon az újkőkortól a XVIII. századig. *Magy. Mg. Múz. Közlem.* **1967**, *4*, 5–84.

31. Pető, Á.; Kenéz, Á.; Gyulai, F.; Lisztes-Szabó, Z.; Molnár, M.; Saláta, D. *Régészeti Növénytan: Leletek, Módszerek és Értelmezés*; Archaeolingua Alapítvány: Budapest, Hungary, 2018.

32. Gyulai, F. *Archaeobotany in Hungary*; Archaeolingua: Budapest, Hungary, 2010; ISBN 978 963 8046 93 2.

33. Wagner, J. *Magyarország Gyomnövényei*; Pallas: Budapest, Hungary, 1908.

34. Henn, T.; Nagy, D.U.; Pál, R.W. Adobe bricks can help identify historic weed flora—A case study from South-Western Hungary. *Plant. Ecol. Divers.* **2016**, *9*, 113–125. [CrossRef]

35. Pinke, G. A pipacs szimbolikája. A lángoló magyar nyár tűzvarázsa. *Élet Tud.* **2018**, *73*, 742–744.

36. Pinke, G. A pipacs szimbolikája. Vénusz könnyeitől a flandriai csatamezőkig. *Élet Tud.* **2018**, *73*, 713–715.

37. Pinke, G. A kék búzavirág átlényegülése. Mintha az ég darabokra törne. *Élet Tud.* **2019**, *74*, 1040–1043.

38. Pinke, G. Egy mezei virág kultúrtörténeti hagyatéka. Szarkaláb hajlik szelíden utamra. *Élet Tud.* **2020**, *75*, 843–846.

39. Jepson, P.; Barua, M. A theory of flagship species action. *Conservat. Soc.* **2015**, *13*, 95–104. [CrossRef]

40. Pany, P.; Heidinger, C. Useful plants as potential flagship species to counteract plant blindness. In *Cognitive and Affective Aspects in Science Education Research: Selected Papers from the ESERA 2015 Conference*; Hahl, K., Juuti, K., Lampiselkä, J., Uitto, A., Lavonen, J., Eds.; Springer International Publishing: Cham, Switzerland, 2017; pp. 127–140. ISBN 978-3-319-58685-4.

41. Stroud, S.; Fennell, M.; Mitchley, J.; Lydon, S.; Peacock, J.; Bacon, K.L. The botanical education extinction and the fall of plant awareness. *Ecol. Evol.* **2022**, *12*, e9019. [CrossRef] [PubMed]

42. Krähmer, H.; Andreasen, C.; Economou-Antonaka, G.; Holec, J.; Kalivas, D.; Kolarova, M.; Novak, R.; Panozzo, S.; Pinke, G.; Salonen, J.; et al. Weed surveys and weed mapping in Europe: State of the art and future tasks. *Crop Prot.* **2020**, *129*, 105010. [CrossRef]

43. Pinke, G.; Pál, R.W.; Tóth, K.; Karácsony, P.; Czúcz, B.; Botta-Dukát, Z. Weed vegetation of poppy (*Papaver somniferum*) fields in Hungary: Effects of management and environmental factors on species composition. *Weed Res.* **2011**, *51*, 621–630. [CrossRef]

44. Stankiewicz-Kosyl, M.; Synowiec, A.; Haliniarz, M.; Wenda-Piesik, A.; Domaradzki, K.; Parylak, D.; Wrochna, M.; Pytlarz, E.; Gala-Czekaj, D.; Marczewska-Kolasa, K.; et al. Herbicide resistance and management options of *Papaver rhoeas* L. and *Centaurea cyanus* L. in Europe: A review. *Agronomy* **2020**, *10*, 874. [CrossRef]

45. Bellanger, S.; Guillemin, J.; Bretagnolle, V.; Darmency, H. *Centaurea cyanus* as a biological indicator of segetal species richness in arable fields. *Weed Res.* **2012**, *52*, 551–563. [CrossRef]

46. Guillemin, J.-P.; Bellanger, S.; Reibeil, C.; Darmency, H. Longevity, dormancy and germination of *Cyanus segetum*. *Weed Res.* **2017**, *57*, 361–371. [CrossRef]

47. Meyer, S.; Wesche, K.; Hans, J.; Leuschner, C.; Albach, D.C. Landscape complexity has limited effects on the genetic structure of two arable plant species, *Adonis aestivalis* and *Consolida regalis*. *Weed Res.* **2015**, *55*, 406–415. [CrossRef]

48. Himmler, D.; Lang, M.; Albrecht, H.; Kollmann, J. *Ackerwildkräuter für Bayerns Kulturlandschaft*; Bayerische KulturLandStiftung: München, Germany, 2016.

49. Király, G. (Ed.) *Új Magyar Füvészkönyv. Magyarország Hajtásos Növényei*; Aggteleki Nemzeti Park Igazgatóság: Jósvafő, Hungary, 2009; ISBN 978 963 87082 9 8.

50. Plants of the World Online. Facilitated by the Royal Botanic Gardens, Kew, UK. Available online: https://powo.science.kew.org/ (accessed on 16 December 2022).

51. Vörös, É. *A Magyar Gyógynövények Neveinek Történeti-Etimológiai Szótára*; A Debreceni Egyetem Magyar Nyelvtudományi Intézetének Kiadványai: Debrecen, Hungary, 2008.

52. Rácz, J. *Növénynevek Enciklopédiája*; Tinta Könyvkiadó: Budapest, Hungary, 2010.

53. Magyary-Kossa, G. *Magyar Orvosi Emlékek*; Magyar Orvosi Könyvkiadó Társulat: Budapest, Hungary, 1929.

54. American Botanical Council. Your source for reliable herbal medicine information. *Terminology*. Available online: https://www.herbalgram.org/resources/terminology-page/ (accessed on 30 August 2022).

55. Melius Juhász, P. *Herbarium az Faknac Fuveknec Nevekroel, Természetekroel, és Hasznairól / Magyar Nyelwre, és ez Rendre Hoszta az Doctoroc Koenyveiboel az Horhi Melius Peter*; Heltai Gaspárne Muehellyébe: Colosvar, (Hungary), Romania, 1578.

56. Beythe, A. *Fives Koenuev Fiseknek es Faknac Nevökröl, Termezetoekroel es Hasznokrul Irattatot, es Szoeroeztetoet Mag'ar Nyelvoen*; Manlius Ianos: Nymet-Vivaratt, (Hungary), Austria, 1595.

57. Pápai Páriz, F. *Pax Corporis, Az az: Az Emberi Testnek Belsö Nyavalyáinak Okairól, Fészkeiről, s' Azoknak Orvoslásának Módgyáról Való Tracta*; Tótfalusi Kis Miklós: Kolosváratt, (Hungary), Romania, 1695.

58. Vilmon, G. Régi magyar orvosi könyvtárak érdekességei: Egy 1695-ös Pax Corporis széljegyzetei. *Orv. Hetil.* **1991**, *132*, 85–86.

59. Gulyás, É. Adatok a Jászság és Nagykunság XVIII–XIX századi gyógyító gyakorlatához. In *Szolnok Megyei Múzeumi Évkönyv*; Tálas, L., Ed.; Damjanich János Múzeum: Szolnok, Hungary, 1990; Volume 7.

60. Novák, L. Egy XVIII század végi—XIX század eleji orvosló könyv Kecskemétről és korának gyógyító gyakorlata. In *Bács-Kiskun Megye Múltjából*; Iványosi-Szabó, T., Ed.; Bács-Kiskun Megyei Levéltár: Kecskemét, Hungary, 1982; Volume 4, pp. 469–509.

61. Csapó, J. *Új Füves és Virágos Magyar Kert, Mellyben Mindenik Fünek es Virágnak Neve, Neme, Ábrázatja, Természete és Ezekhez Képest Különbféle Hasznai Értelmessen Megjegyeztetnek*; Landerer: Posony, (Hungary), Slovakia, 1775.

62. Csepregi, K.; Papp, N. A Pannonhalmi Főapátság gyógyászati értékei: Gyógynövények Zsoldos Xavér herbáriumában (1788). *Bot. Közlem.* **2010**, *97*, 25–38.

63. Veszelszki, A. *A' Növevény-plánták' Országából Való Erdei, és Mezei Gyűjtemény, Vagy-is Fa- és Fűszeres Könyv, Mellyben Azoknak Deák, Magyar, Német, Frantz, Tseh, és Oláh Neveik, Külső, Belső, és Köz Hasznaikkal Egyetemben Máthiolusból "s más több Fa-, és Fűvész-Írókból a" Köz-rendű Hazafiak' Kedvekért Szálanként Egybe-Szedettek Veszelszki Antal Által*; Találtatik Trattner Mátyásnál, és Kiss István Nemzeti Könyv-Táros Bóltjábann: Pesthen, Hungary, 1798.

64. Diószegi, S. *Orvosi Fűvészkönyv, Mint a' Magyar Füvész Könyv Praktika Része*; Csáthy Gy: Debrecen, Hungary, 1813.

65. Temesváry, R. Előítéletek, népszokások és babonás eljárások a szülészet körében Magyarországon. *Orv. Hetil.* **1899**, *10*, 1–36.

66. Gönczi, F. Az emberi betegségek gyógyítása a göcseji népnél. *Ethnografia* **1902**, *13*, 214–224.

67. Gönczi, F. A természeti elemek kultuszának maradványai a göcseji s hetési népnél. *Ethnografia* **1910**, *21*, 284–291.

68. Darvas, F.; Magyary-Kossa, G. *Hazai Gyógynövények*; Athenaeum: Budapest, Hungary, 1925.

69. Relkovic, D. Adalékok a Somló-vidék folklorejához. *Ethnographia* **1928**, *39*, 94–105.

70. Rapaics, R. *A Magyarság Virágai*; Királyi Magyar Természettudományi Társulat: Budapest, Hungary, 1932.

71. Luby, M. Gyógynövények és gyógyító eljárások a szatmár megyei parasztság körében. *Debr. Szle.* **1935**, *9*, 360–364.

72. Réthelyi, J. Orvosi babonák, varázslatok és kuruzslások. *Gyógyszerészi Hetil.* **1940**, *79*, 309–311.

73. Vajkai, A. A gyűjtögető gazdálkodás Cserszegtomájon. *Néprajzi Értesítő* **1941**, *33*, 231–258.
74. Greszné, A. Adatok a Tiszántúl népies orvoslásához. *Debr. Szle.* **1943**, *17*, 8–19.
75. Vargyas, L. Adatok a makádi néphithez II. *Ethnographia* **1945**, *56*, 60–65.
76. Farkas, J. Bakos Ferenc a mátészalkai parasztorvos. In *A Nyíregyházi Jósa András Múzeum Évkönyve*; Csallány, D., Ed.; Múzeumi Ismeretterjesztő Központ: Budapest, Hungary, 1968; Volume 10, pp. 145–180.
77. Seregély, G. Népszokások, babonák és tévhitek az abortívumok és fogamzásgátlók köréből Magyarországon. *Egészség* **1969**, *63*, 147–150.
78. Péntek, J.; Szabó, A. Egy háromszéki falu népi növényismerete. *Ethnographia* **1976**, *87*, 203–225.
79. Szabóné, R. Növények szerepe a népi gyógyászatban Taktaszadán. In *A Herman Ottó Múzeum Évkönyve*; Szabadfalvi, J., Ed.; Herman Ottó Múzeum: Miskolc, Hungary, 1976; Volume 15, pp. 249–261.
80. Oláh, A. „Eddig dolgom után, ezután hitem után élek"—A magyar népi gerontológia tanulságai. *Orv. Közlemények* **1979**, *12*, 179–206.
81. Ujváry, Z. Virág a Magyar Népi Kultúrában. *A Hajdú-Bihar Megyei Múzeumok Közleményei* **1980**, *35*, 383–424.
82. Petercsák, T. *Népi Szarvasmarhatartás a Zempléni Hegyközben*; Borsodi Kismonográfiák; Herman Ottó Múzeum: Miskolc, Hungary, 1983; Volume 17.
83. Rácz, G.; Rácz-Kotilla, E.; Laza, A. *Gyógynövényismeret*; Ceres: Bukarest, Romania, 1984.
84. Péntek, J.; Szabó, A. *Ember és Növényvilág. Kalotaszeg Növényzete és Népi Növényismerete*; Kriterion: Bukarest, Romania, 1985.
85. Kóczián, G. *A Hagyományos Parasztgazdálkodás Termesztett, a Gyűjtögető Gazdálkodás vad Növényfajainak Etnobotanikai Értékelése*; Nagyatádi Kulturális és Sport Központ: Nagyatád, Hungary, 2014.
86. Tóth, J. Gyógynövények és hasznosításuk Óladon. *Vasi Honism. Helytörténeti Közlemények* **1986**, *1–2*, 82–90.
87. Tisovszki, Z. Népi orvoslás, növényismeret Esztergom-Szentgyörgymezőn. In *Komárom-Esztergom Megyei Múzeumok Közleményei*; Fűrészné, A., Ed.; Komárom Megyei Múzeumok Igazgatósága: Tata, Hungary, 1989; Volume 3, pp. 41–52.
88. Gelencsér, J.; Lukács, L. *Szép Napunk Támadt. Népszokások Fejér Megyében*; István Király Múzeum: Székesfehérvár, Hungary, 1991.
89. Lenkey, I. Népi gyógyítás Alsószuhán. In *A Miskolci Herman Ottó Múzeum Közleményei*; Viga, G., Ed.; Herman Ottó Múzeum: Miskolc, Hungary, 1993; Volume 28, pp. 195–210.
90. Gub, J. Borogatók, kenőcsök, sebtapaszok a Sóvidéken. *Hazanéző* **2000**, *11*, 27–29.
91. Bartha, J. Népi orvoslás a Nagykunságon. *Tisicum* **2001**, *12*, 283–292.
92. Ujváry, Z. *Gömöri Magyar Néphagyományok*; Herman Ottó Múzeum: Miskolc, Hungary, 2002.
93. Szabó, L. *Gyógynövény-Ismereti Tájékoztató*; Melius Alapítvány: Pécs, Hungary, 2005.
94. Horváth, K. Gyógynövények párhuzamos megnevezései a kárpátaljai magyar nyelvjárásokban. *Acta Hung.* **2010**, *20*, 177–186.
95. Grynaeus, T. Makó és környéke hagyományos orvoslása III. *Stud. Ethnogr.* **2011**, *7*, 197–240.
96. Papp, N. „*A virágok mindegyik orvosság*"—*Hagyományok és népi orvoslás Lövétén*; Lövéte Közbirtokossága: Lövéte, Romania, 2018.
97. Kapronyi, T. Arató-szokások a summáséletben. *Zenetudományi Dolg.* **1980**, *1*, 37–57.
98. Prónay, G. *Vázlatok Magyarhon Népéletéből*; Geibel Armin: Pest, Hungary, 1855.
99. Manga, J. Aratószokások, aratóénekek. In *Népi Kultúra—Népi Társadalom*; Ortutay, G., Ed.; Akadémiai Kiadó: Budapest, Hungary, 1977; pp. 241–276.
100. Ébner, S. Gabonáinkban előforduló gaznemek. *Erdélyi Gazda* **1870**, *2*, 13–14.
101. Kovács, M. A zágoni nép szokásai és babonái. *Ethnographia* **1895**, *6*, 396–398.
102. Platthy, A. Hevesmegye mátrai része. In *Az Osztrák–Magyar Monarchia Írásban és Képben*; Magyar Királyi Államnyomda: Budapest, Hungary, 1900; Volume 6.
103. Illés, P. Vasi aratóünnepek és aratókoszorúk. *Vasi Szle.* **2008**, *62*, 954–970.
104. Faggyas, I. Aratás a keleméri földesúri birtokon. In *Néprajzi Dolgozatok Borsod-Abaúj-Zemplén Megyéből: Válogatás az Önkéntes Néprajzi Gyűjtők Pályamunkáiból*; Viga, G., Ed.; Miskolci Herman Ottó Múzeum: Miskolc, Hungary, 1965; pp. 79–97.
105. Gyimesiné, I. Koszorúba fontuk, amit isten adott—Aratási hálaünnepek Hevesen. In *Annales Musei Agriensis*; Veres, G., Ed.; Dobó István Vármúzeum: Eger, Hungary, 2007; Volume 43, pp. 327–338.
106. Sz. Tóth, J. Virágkoszorúk a Buda környéki nemzetiségek hagyományaiban. In *Népi Vallásosság a Kárpát-Medencében*; Pilipkó, E., Ed.; Laczkó Dezső Múzeum: Veszprém, Hungary, 2013; Volume 8, pp. 148–168.
107. Sz. Tóth, J. Virágszőnyeg és virágkoszorú. In *"Leborulva Áldlak"—Az Oltáriszentség és az Úrvacsora a Magyarországi Vallási Kultúrában*; Barna, G., Ed.; Szent István Társulat & Szent István Tudományos Akadémia: Budapest, Hungary, 2020; pp. 295–309.
108. Horváth, I. *A Naptári Ünnepekhez Fűződő Szokások és Hiedelmek a Nyugat-Magyarországi Horvátoknál*; Folklór Archívum; Magyar Tudományos Akadémia Néprajzi Kutatócsoportja: Budapest, Hungary, 1978; Volume 10.
109. Kalapis, Z. A változó életmóddal kivesző szokásokról és hiedelmekről Tóbán. *Létünk* **1989**, *19*, 639–683.
110. Demeter, Z.; Gelencsér, J.; Lukács, L. *Palotavárosi Emlékek*; Az István Király Múzeum Közleményei: Székesfehérvár, Hungary, 1990.
111. Nagy, N. Virág- és kertkultúra három Szeged környéki településen. *Stud. Ethnogr.* **2003**, *4*, 133–156.
112. Császi, I. A vallásos naptári és egyházi ünnepi szokások szerepe a gyermeknevelésben. In *Annales Musei Agriensis*; Petercsák, T., Veres, G., Eds.; Dobó István Vármúzeum: Eger, Hungary, 2000; Volume 36, pp. 398–406.
113. Bencsik, J. A tejoltó galaj a román és szerb Iván napi szokásokban. In *Paraszti és Mezővárosi Kultúra a XVIII-XX Században*; Viga, G., Ed.; Miskolci Herman Ottó Múzeum: Miskolc, Hungary, 1993; pp. 112–118.
114. Balogh, K. *A Magyar Gyógyszerkönyv Kommentárja*; M. Orv. Könyvkiadó Társ.: Budapest, Hungary, 1879.
115. Vajticzky, E. A gyógyító növények és a háború. *Néptanítók Lapja* **1916**, *31*, 5–7.
116. Benkő, J. Gyógynövények. *Magyarország* **1969**, *6*, 41.

117. Rácz, G.; Rácz-Kotilla, E.; Szabó, L. *Gyógynövények Ismerete*; Galenus Kiadó: Budapest, Hungary, 2012.

118. Matejic, J.S.; Stefanovic, N.; Ivkovic, M.; Zivanovic, N.; Marin, P.D.; Dzamic, A.M. Traditional uses of autochthonous medicinal and ritual plants and other remedies for health in Eastern and South-Eastern Serbia. *J. Ethnopharmacol.* **2020**, *261*, 113186. [CrossRef]

119. Signorini, M.A.; Piredda, M.; Bruschi, P. Plants and traditional knowledge: An ethnobotanical investigation on Monte Ortobene (Nuoro, Sardinia). *J. Ethnobiol. Ethnomed.* **2009**, *5*, 6. [CrossRef]

120. Idolo, M.; Motti, R.; Mazzoleni, S. Ethnobotanical and phytomedicinal knowledge in a long-history protected area, the Abruzzo, Lazio and Molise National Park (Italian Apennines). *J. Ethnopharmacol.* **2010**, *127*, 379–395. [CrossRef]

121. Tuttolomondo, T.; Licata, M.; Leto, C.; Savo, V.; Bonsangue, G.; Gargano, M.L.; Venturella, G.; La Bella, S. Ethnobotanical investigation on wild medicinal plants in the Monti Sicani Regional Park (Sicily, Italy). *J. Ethnopharmacol.* **2014**, *153*, 568–586. [CrossRef] [PubMed]

122. Menale, B.; De Castro, O.; Cascone, C.; Muoio, R. Ethnobotanical investigation on medicinal plants in the Vesuvio National Park (Campania, Southern Italy). *J. Ethnopharmacol.* **2016**, *192*, 320–349. [CrossRef]

123. Fortini, P.; Di Marzio, P.; Guarrera, P.M.; Iorizzi, M. Ethnobotanical study on the medicinal plants in the Mainarde Mountains (Central-Southern Apennine, Italy). *J. Ethnopharmacol.* **2016**, *184*, 208–218. [CrossRef]

124. Mautone, M.; De Martino, L.; De Feo, V. Ethnobotanical research in Cava de' Tirreni Area, Southern Italy. *J. Ethnobiol. Ethnomed.* **2019**, *15*, 50. [CrossRef] [PubMed]

125. Fontefrancesco, M.F.; Pieroni, A. Renegotiating situativity: Transformations of local herbal knowledge in a Western Alpine valley during the past 40 years. *J. Ethnobiol. Ethnomed.* **2020**, *16*, 58. [CrossRef] [PubMed]

126. Motti, R.; de Falco, B. Traditional herbal remedies used for managing anxiety and insomnia in Italy: An ethnopharmacological overview. *Horticulturae* **2021**, *7*, 523. [CrossRef]

127. Grauso, L.; de Falco, B.; Motti, R.; Lanzotti, V. Corn poppy, *Papaver rhoeas* L.: A critical review of its botany, phytochemistry and pharmacology. *Phytochem. Rev.* **2021**, *20*, 227–248. [CrossRef]

128. Mathianaki, K.; Tzatzarakis, M.; Karamanou, M. Poppies as a sleep aid for infants: The *"Hypnos"* remedy of Cretan folk medicine. *Toxicol. Rep.* **2021**, *8*, 1729–1733. [CrossRef]

129. Benitez, G.; Gonzalez-Tejero, M.R.; Molero-Mesa, J. Pharmaceutical ethnobotany in the western part of Granada Province (Southern Spain): Ethnopharmacological synthesis. *J. Ethnopharmacol.* **2010**, *129*, 87–105. [CrossRef]

130. Gonzalez, J.A.; Garcia-Barriuso, M.; Amich, F. Ethnobotanical study of medicinal plants traditionally used in the Arribes Del Duero, Western Spain. *J. Ethnopharmacol.* **2010**, *131*, 343–355. [CrossRef]

131. Carrio, E.; Valles, J. Ethnobotany of medicinal plants used in Eastern Mallorca (Balearic Islands, Mediterranean Sea). *J. Ethnopharmacol.* **2012**, *141*, 1021–1040. [CrossRef] [PubMed]

132. Gras, A.; Serrasolses, G.; Valles, J.; Garnatje, T. Traditional knowledge in semi-rural close to industrial areas: Ethnobotanical studies in Western Girones (Catalonia, Iberian Peninsula). *J. Ethnobiol. Ethnomed.* **2019**, *15*, 19. [CrossRef] [PubMed]

133. Polat, R.; Cakilcioglu, U.; Kaltalioglu, K.; Ulusan, M.D.; Turkmen, Z. An ethnobotanical study on medicinal plants in Espiye and its surrounding (Giresun-Turkey). *J. Ethnopharmacol.* **2015**, *163*, 1–11. [CrossRef] [PubMed]

134. Polat, R. Ethnobotanical study on medicinal plants in Bingol (City Center) (Turkey). *J. Herb. Med.* **2019**, *16*, 100211. [CrossRef]

135. Emre, G.; Dogan, A.; Haznedaroglu, M.Z.; Senkardes, I.; Ulger, M.; Satiroglu, A.; Emmez, B.C.; Tugay, O. An ethnobotanical study of medicinal plants in Mersin (Turkey). *Front. Pharmacol.* **2021**, *12*, 664500. [CrossRef]

136. Merzouki, A.; Ed-Derfoufi, F.; Mesa, J.M. Contribution to the knowledge of Rifian traditional medicine. II: Folk medicine in Ksar Lakbir District (NW Morocco). *Fitoterapia* **2000**, *71*, 278–307. [CrossRef]

137. Ouelbani, R.; Bensari, S.; Mouas, T.N.; Khelifi, D. Ethnobotanical investigations on plants used in folk medicine in the regions of Constantine and Mila (North-East of Algeria). *J. Ethnopharmacol.* **2016**, *194*, 196–218. [CrossRef]

138. Salah, M.; Barhoumi, T.; Abderrabba, M. Ethnobotanical study of medicinal plant in Djerba Island, Tunisia. *Arab J. Med. Arom. Plants* **2009**, *5*, 67–97.

139. Mohammed, F.; Akgül, H. Ethnobotanical analysis of cultivated and indigenous plants in Duhok Province in Iraq. *Turk. J. Agric. Food Sci. Technol.* **2018**, *6*, 1191. [CrossRef]

140. Kheirollahi, A.R.; Mahmoodnia, L.; Khodadustan, E.; Kazemeini, H.; Hasanvand, A.; Hatamikia, M. Medicinal plants for kidney pain: An ethnobotanical study on Shahrekord city, West of Iran. *Plant Sci. Today* **2019**, *6*, 328–332. [CrossRef]

141. Mustafa, B.; Hajdari, A.; Pulaj, B.; Quave, C.L.; Pieroni, A. Medical and food ethnobotany among Albanians and Serbs living in the Shterpce/Strpce area, South Kosovo. *J. Herb. Med.* **2020**, *22*, 100344. [CrossRef]

142. De Cleene, M.; Lejeune, M. *Compendium of Symbolic and Ritual Plants in Europe*; Man & Culture Publishers: Ghent, Belgium, 2003; Volume 2, Herbs.

143. Popović, Z.; Smiljanic, M.; Kostić, M.; Nikic, P.; Jankovic, S. Wild flora and its usage in traditional phytotherapy (Deliblato Sands, Serbia, South East Europe). *Ind. J. Trad. Know.* **2014**, *13*, 9–35.

144. Gras, A.; Garnatje, T.; Ibanez, N.; Lopez-Pujol, J.; Nualart, N.; Valles, J. Medicinal plant uses and names from the herbarium of Francesc Bolos (1773-1844). *J. Ethnopharmacol.* **2017**, *204*, 142–168. [CrossRef] [PubMed]

145. Tita, I.; Mogosanu, G.D.; Tita, M.G. Ethnobotanical inventory of medicinal plants from the South-West of Romania. *Farmacia* **2009**, *57*, 141–156.

146. Leporatti, M.L.; Ivancheva, S. Preliminary Comparative Analysis of medicinal plants used in the traditional medicine of Bulgaria and Italy. *J. Ethnopharmacol.* **2003**, *87*, 123–142. [CrossRef] [PubMed]

147. Soukand, R.; Hrynevich, Y.; Vasilyeva, I.; Prakofjewa, J.; Vnukovich, Y.; Paciupa, J.; Hlushko, A.; Knureva, Y.; Litvinava, Y.; Vyskvarka, S.; et al. Multi-functionality of the few: Current and past uses of wild plants for food and healing in Liuban Region, Belarus. *J. Ethnobiol. Ethnomed.* **2017**, *13*, 10. [CrossRef] [PubMed]

148. Kujawska, M.; Klepacki, P.; Luczaj, L. Fischer's plants in folk beliefs and customs: A previously unknown contribution to the ethnobotany of the Polish-Lithuanian-Belarusian borderland. *J. Ethnobiol. Ethnomed.* **2017**, *13*, 20. [CrossRef]

149. Redzic, S.S. The ecological aspect of ethnobotany and ethnopharmacology of population in Bosnia and Herzegovina. *Coll. Anthropol.* **2007**, *31*, 869–890.

150. Saric-Kundalic, B.; Dobes, C.; Klatte-Asselmeyer, V.; Saukel, J. Ethnobotanical survey of traditionally used plants in human therapy of East, North and North-East Bosnia and Herzegovina. *J. Ethnopharmacol.* **2011**, *133*, 1051–1076. [CrossRef]

151. Bellia, G.; Pieroni, A. Isolated, but transnational: The glocal nature of Waldensian ethnobotany, Western Alps, NW Italy. *J. Ethnobiol. Ethnomed.* **2015**, *11*, 37. [CrossRef]

152. Nanagulyan, S.; Zakaryan, N.; Kartashyan, N.; Piwowarczyk, R.; Luczaj, L. Wild plants and fungi sold in the markets of Yerevan (Armenia). *J. Ethnobiol. Ethnomed.* **2020**, *16*, 26. [CrossRef] [PubMed]

153. Stryamets, N.; Elbakidze, M.; Ceuterick, M.; Angelstam, P.; Axelsson, R. From economic survival to recreation: Contemporary uses of wild food and medicine in rural Sweden, Ukraine and NW Russia. *J. Ethnobiol. Ethnomed.* **2015**, *11*, 53. [CrossRef] [PubMed]

154. Ivancheva, S.; Stantcheva, B. Ethnobotanical inventory of medicinal plants in Bulgaria. *J. Ethnopharmacol.* **2000**, *69*, 165–172. [CrossRef] [PubMed]

155. Danna, C.; Poggio, L.; Smeriglio, A.; Mariotti, M.; Cornara, L. Ethnomedicinal and ethnobotanical survey in the Aosta valley side of the Gran Paradiso National Park (Western Alps, Italy). *Plants* **2022**, *11*, 170. [CrossRef] [PubMed]

156. Spalek, K.; Spielvogel, I.; Prockow, M.; Prockow, J. Historical ethnopharmacology of the herbalists from Krummhubel in the Sudety Mountains (seventeenth to nineteenth century), Silesia. *J. Ethnobiol. Ethnomed.* **2019**, *15*, 24. [CrossRef] [PubMed]

157. Pranskuniene, Z.; Dauliute, R.; Pranskunas, A.; Bernatoniene, J. Ethnopharmaceutical knowledge in Samogitia Region of Lithuania: Where old traditions overlap with modern medicine. *J. Ethnobiol. Ethnomed.* **2018**, *14*, 70. [CrossRef]

158. Mattalia, G.; Stryamets, N.; Pieroni, A.; Soukand, R. Knowledge transmission patterns at the border: Ethnobotany of Hutsuls living in the Carpathian Mountains of Bukovina (SW Ukraine and NE Romania). *J. Ethnobiol. Ethnomed.* **2020**, *16*, 41. [CrossRef]

159. Und, I.; Schönfelder, P. *Das neue Handbuch der Heilpflanzen*; Kosmos-Verlag: Stuttgart, Germany, 2004.

160. Juhász, A. A falu társadalma. In *Néprajzi Tanulmányok Apátfalváról*; Bárkányi, I., Ed.; Móra Ferenc Múzeum: Szeged, Hungary, 2015; pp. 89–138.

161. Szigeti, G. Az apátfalvi nép táplálkozása. In *A Móra Ferenc Múzeum Évkönyve*; Trogmayer, O., Ed.; Móra Ferenc Múzeum: Szeged, Hungary, 1971; Volume 1, pp. 211–229.

162. Barsi, D. Pipacsmérgezés. *Bud. Orv. Új. Tud. Köz.* **1903**, *1*, 1–6.

163. Bokor, J.; Gerő, L. *A Pallas Nagy Lexikona*; Pallas Irodalmi és Nyomdai Rt: Budapest, Hungary, 1893.

164. Bartha, K. Játék. In *A Magyarság Néprajza*; Királyi Magyar Egyetemi Nyomda: Budapest, Hungary, 1937; pp. 453–498.

165. Guarrera, P.M. Food medicine and minor nourishment in the folk traditions of Central Italy (Marche, Abruzzo and Latium). *Fitoterapia* **2003**, *74*, 515–544. [CrossRef]

166. Di Tizio, A.; Luczaj, L.J.; Quave, C.L.; Redzic, S.; Pieroni, A. Traditional food and herbal uses of wild plants in the ancient South-Slavic diaspora of Mundimitar/Montemitro (Southern Italy). *J. Ethnobiol. Ethnomed.* **2012**, *8*, 21. [CrossRef]

167. Sansanelli, S.; Tassoni, A. Wild food plants traditionally consumed in the area of Bologna (Emilia Romagna Region, Italy). *J. Ethnobiol. Ethnomed.* **2014**, *10*, 69. [CrossRef] [PubMed]

168. Sansanelli, S.; Ferri, M.; Salinitro, M.; Tassoni, A. Ethnobotanical survey of wild food plants traditionally collected and consumed in the middle Agri Valley (Basilicata Region, Southern Italy). *J. Ethnobiol. Ethnomed.* **2017**, *13*, 50. [CrossRef]

169. Biscotti, N.; Bonsanto, D.; Del Viscio, G. The traditional food use of wild vegetables in Apulia (Italy) in the light of Italian ethnobotanical literature. *IB* **2018**, *5*, 1–24. [CrossRef]

170. Geraci, A.; Amato, F.; Di Noto, G.; Bazan, G.; Schicchi, R. The wild taxa utilized as vegetables in Sicily (Italy): A traditional component of the Mediterranean diet. *J. Ethnobiol. Ethnomed.* **2018**, *14*, 14. [CrossRef] [PubMed]

171. Lucchetti, L.; Zitti, S.; Taffetani, F. Ethnobotanical uses in the Ancona district (Marche Region, Central Italy). *J. Ethnobiol. Ethnomed.* **2019**, *15*, 9. [CrossRef] [PubMed]

172. Motti, R.; Bonanomi, G.; Lanzotti, V.; Sacchi, R. The contribution of wild edible plants to the Mediterranean diet: An ethnobotanical case study along the coast of Campania (Southern Italy). *Econ. Bot.* **2020**, *74*, 249–272. [CrossRef]

173. Luczaj, L.; Koncic, M.Z.; Milicevic, T.; Dolina, K.; Pandza, M. Wild vegetable mixes sold in the markets of Dalmatia (Southern Croatia). *J. Ethnobiol. Ethnomed.* **2013**, *9*, 2. [CrossRef]

174. Luczaj, L.; Jug-Dujakovic, M.; Dolina, K.; Jericevic, M.; Vitasovic-Kosic, I. The ethnobotany and biogeography of wild vegetables in the Adriatic Islands. *J. Ethnobiol. Ethnomed.* **2019**, *15*, 18. [CrossRef]

175. Ivanova, T.; Bosseva, Y.; Chervenkov, M.; Dimitrova, D. Enough to feed ourselves!—Food plants in Bulgarian rural home gardens. *Plants* **2021**, *10*, 2520. [CrossRef]

176. Vasilopoulou, E.; Georga, K.; Grilli, E.; Linardou, A.; Vithoulka, M.; Trichopoulou, A. Compatibility of computed and chemically determined macronutrients and energy content of traditional Greek recipes. *J. Food Compos. Anal.* **2003**, *16*, 707–719. [CrossRef]

177. Tardio, J.; Pardo-De-Santayana, M.; Morales, R. Ethnobotanical review of wild edible plants in Spain. *Bot. J. Linnean Soc.* **2006**, *152*, 27–71. [CrossRef]

178. Gras, A.; Valles, J.; Garnatje, T. Filling the Gaps: Ethnobotanical study of the Garrigues district, an arid zone in Catalonia (NE Iberian Peninsula). *J. Ethnobiol. Ethnomed.* **2020**, *16*, 34. [CrossRef]

179. Akgul, A.; Akgul, A.; Senol, S.G.; Yildirim, H.; Secmen, O.; Dogan, Y. An ethnobotanical study in Midyat (Turkey), a city on the silk road where cultures meet. *J. Ethnobiol. Ethnomed.* **2018**, *14*, 12. [CrossRef] [PubMed]

180. Sargin, S.A. Plants used against obesity in Turkish folk medicine: A Review. *J. Ethnopharmacol.* **2021**, *270*, 113841. [CrossRef] [PubMed]

181. Günaydın, Y.K.; Dündar, Z.D.; Çekmen, B.; Akıllı, N.B.; Köylü, R.; Cander, B. Intoxication due to *Papaver rhoeas* (corn poppy): Five case reports. *Case. Rep. Med.* **2015**, *2015*, 321360. [CrossRef] [PubMed]

182. Ranfa, A.; Bodesmo, M. An ethnobotanical investigation of traditional knowledge and uses of edible wild plants in the Umbria region, Central Italy. *J. Appl. Bot. Food Qual.* **2017**, *90*, 246–258. [CrossRef]

183. Kosic, I.V.; Juracak, J.; Luczaj, L. Using Ellenberg-Pignatti values to estimate habitat preferences of wild food and medicinal plants: An example from Northeastern Istria (Croatia). *J. Ethnobiol. Ethnomed.* **2017**, *13*, 31. [CrossRef]

184. Di Sanzo, P.; De Martino, L.; Mancini, E.; De Feo, V. Medicinal and useful plants in the tradition of Rotonda, Pollino National Park, Southern Italy. *J. Ethnobiol. Ethnomed.* **2013**, *9*, 19. [CrossRef]

185. Motti, R.; Bonanomi, G.; Emrick, S.; Lanzotti, V. Traditional herbal remedies used in women's health care in Italy: A Review. *Hum. Ecol.* **2019**, *47*, 941–972. [CrossRef]

186. Guarrera, P.M. Household dyeing plants and traditional uses in some areas of Italy. *J. Ethnobiol. Ethnomed.* **2006**, *2*, 9. [CrossRef]

187. Luczaj, L.; Szymanski, W.M. Wild vascular plants gathered for consumption in the Polish countryside: A review. *J. Ethnobiol. Ethnomed.* **2007**, *3*, 17. [CrossRef]

188. Bathó, E. *A Méhészet Szerepe a Jászság Paraszti Gazdálkodásában*; Studia Folkloristica et Ethnographica; Kossuth Lajos Tudományegyetem: Debrecen, Hungary, 1995; Volume 36.

189. Kalo, P. *A Méhtartásnak Külömbféle Tartományokra, Környékekre, és Esztendőkre Alkalmaztatott Igen Könnyű, Hasznos, és Gyönyörűséges Modgya*; Nyomt Érseki Fő Oskolák: Eger, Hungary, 1816.

190. Nyárády, A. *A Méhlegelő és Növényei*; Földművelésügyi és Erdészeti Minisztérium: Bukarest, Romania, 1958.

191. Hazslinszky, B. Adatok a méz pollenanalítikai vizsgálatához. *Mg. Kut.* **1938**, *11*, 144–159.

192. Molnár, Z.; Hoffmann, K. A hortobágyi pásztorok növény- és növényzetismerete II. A telkes helyek, útmezsgyék, csatornapartok és szántóföldek növényei, valamint a nem őshonos fásszárúak. *Debr. Déri Múzeum Évkönyve* **2011**, *82*, 29–43.

193. Kóssa, G. Házi állatainkon gyakrabban előforduló mérgezések. *Állategészség* **1898**, *2*, 2–12.

194. Leto, C.; Tuttolomondo, T.; La Bella, S.; Licata, M. Ethnobotanical study in the Madonie regional park (Central Sicily, Italy)—Medicinal use of wild shrub and herbaceous plant species. *J. Ethnopharmacol.* **2013**, *146*, 90–112. [CrossRef] [PubMed]

195. Lippay, J. *Posoni Kert*; Nagyszombat-Bécs, Slovakia, 1664.

196. Vaszaray, M. Nagyanyáink kertjei. *A Kert* **1913**, *22*, 690–693.

197. Pinke, G. The status of arable plant habitats in Eastern Europe. In *The Changing Status of Arable Habitats in Europe*; Hurford, C., Wilson, P., Storkey, J., Eds.; Springer: Cham, Switzerland, 2020; pp. 75–87.

198. Gunda, B. Kulturális ökológiai megfigyelések a növénytermesztés kezdeteiről. In *Herman Ottó Múzeum Évkönyve*; Dobrossy, I., Viga, G., Eds.; Herman Ottó Múzeum: Miskolc, Hungary, 1988; Volume 26, pp. 467–481.

199. Ecserei, K.; Mosonyi, I.D.; Mándy, A.T.; Honfi, P. Decorational value and herbicide sensibility of some ephemeral annual ornamental plants. *Sci Pap. Ser. B Hortic.* **2015**, *59*, 341–346.

200. Timár, L. Vadvirágok Szeged piacain. *Agrártudományi Egy. Kert Szőlőgazdaságtudományi Kar Évkönyve* **1952**, *15*, 109–116.

201. Beuchert, M. *Symbolik der Pflanzen*; Insel Verlag: Frankfurt, Germany, 2004.

202. Jagel, A. *Papaver rhoeas*—Klatsch-mohn (*Papaveraceae*), Blume des Jahres 2017. *Jahrb. Bochumer Bot. Ver.* **2018**, *9*, 269–276.

203. Vogl-Lukasser, B.; Vogl, C.R. The changing face of farmers' home gardens: A diachronic analysis from Sillian (Eastern Tyrol, Austria). *J. Ethnobiol. Ethnomed.* **2018**, *14*, 63. [CrossRef]

204. Hegi, G. *Illustrierte Flora von Mittel-Europa*; Verlag Paul Parey: Berlin/Hamburg, Germany, 1975.

205. Gras, A.; Garnatje, T.; Angels Bonet, M.; Carrio, E.; Mayans, M.; Parada, M.; Rigat, M.; Valles, J. Beyond food and medicine, but necessary for life, too: Other folk plant uses in several territories of Catalonia and the Balearic Islands. *J. Ethnobiol. Ethnomed.* **2016**, *12*, 23. [CrossRef] [PubMed]

206. Kasper-Pakosz, R.; Pietras, M.; Luczaj, L. Wild and native plants and mushrooms sold in the open-air markets of Southeastern Poland. *J. Ethnobiol. Ethnomed.* **2016**, *12*, 45. [CrossRef] [PubMed]

207. Gál, S. Az egy és az egész. *Irodalmi Szemle* **2006**, *49*, 16–35.

208. Kultsár, I. Kassáról 6. August. *Hazai Tudósítások* **1806**, *15*, 123.

209. Balassa, I. *Az Aratómunkások Magyarországon 1848-1944*; Akadémiai Kiadó: Budapest, Hungary, 1985.

210. Murgoci, A. The "Cununa": A Transilvanian harvest festival. *Folklore* **1929**, *40*, 245–261. [CrossRef]

211. Bartha, E. Virágszentelés Észak-Magyarországon. *Miskolci Herman Ottó Múzeum Közleményei* **1982**, *20*, 109–115.

212. Luczaj, L.J. A relic of medieval folklore: Corpus Christi Octave herbal wreaths in Poland and their relationship with the local pharmacopoeia. *J. Ethnopharmacol.* **2012**, *142*, 228–240. [CrossRef]
213. Stryamets, N.; Fontefrancesco, M.F.; Mattalia, G.; Prakofjewa, J.; Pieroni, A.; Kalle, R.; Stryamets, G.; Soukand, R. Just beautiful green herbs: Use of plants in cultural practices in Bukovina and Roztochya, Western Ukraine. *J. Ethnobiol. Ethnomed.* **2021**, *17*, 12. [CrossRef] [PubMed]
214. Hamvai, V. Játék és pedagógia. *Pedagógiai Szemle* **1968**, *18*, 441–450.
215. Zomborka, M. Gyermekjátékok. In *Szolnok Megye Népművészete*; Bellon, T., Szabó, L., Eds.; Európa Könyvkiadó: Budapest, Hungary, 1987; pp. 201–218.
216. Bárkányi, I. Gyermekjátékok Csongrád tanyavilágában az 1910 -1930-as Években. In *Móra Ferenc Múzeum Évkönyve*; Juhász, A., Ed.; Móra Ferenc Múzeum: Szeged, Hungary, 1992; pp. 239–262.
217. Nagy Abonyi, Á. A népi gyermekjátékszerek Zentán és környékén. *Létünk* **1996**, *26*, 102–121.
218. Császi, I. Bábuk és bubák. Babakészítés Észak-Magyarországon. In *Annales Musei Agriensis*; Petercsák, T., Veres, G., Eds.; Dobó István Vármúzeum: Eger, Hungary, 2001; Volume 37, pp. 403–419.
219. Schőn, M. *Hajósi Sváb Népi Elbeszélések*; Cumania Alapítvány: Kecskemét, Hungary, 2005.
220. Ortutay, G. *Magyar Néprajzi Lexikon*; Akadémiai Kiadó: Budapest, Hungary, 1977; Volume I.
221. Kerényi, G. *Gyermekjátékok*; A magyar népzene tára; Magyar Tudományos Akadémia: Budapest, Hungary, 1951; Volume 1.
222. Czuczor, G.; Fogarasi, J. *A Magyar Nyelv Szótára*; Athenaeum: Budapest, Hungary, 1870.
223. Benkő, L. Pipacs. Egy hangutánzó szó etimológiai problémái. *Magy. Nyelv* **1961**, *57*, 162–169.
224. Rind, M. Topolyai gyermekjátékok Zöldy Pál gyűjteményében. In *Szavak Szivárványa*; Bárth, J., Ed.; Bács-Kiskun Megyei Önkormányzat Múzeumi Szervezete: Kecskemét, Hungary, 2006; pp. 163–173.
225. Benkő, É. Az ajándékozás szokásai a Szuha-völgyében. In *A Hajdú-Bihar Megyei Múzeumok Közleményei*; Balassa, I., Ujváry, Z., Eds.; Debrecen, Hungary, 1982; Volume 39, pp. 631–640.
226. Tomasini, S.; Theilade, I. Local knowledge of past and present uses of medicinal plants in Prespa National Park, Albania. *Econ. Bot.* **2019**, *73*, 217–232. [CrossRef]
227. Iqbal, I.; Hamayun, M. Studies on the traditional uses of plants of Malam Jabba Valley, District Swat, Pakistan. *Ethnob. Leafl.* **2004**, *2004*, 15.
228. Blümml, A.; Rott, E. Die Verwendung der Pflanzen durch die Kinder in Deutschböhmen und Niederösterreich. *Z. Des Ver. Volkskd.* **1901**, *11*, 49–94.
229. Huszka, J. *Magyar Diszitő Styl*; Deutsch M.-féle Művészeti Intézet: Budapest, Hungary, 1885.
230. Malonyai, D. *A Magyar Nép Művészete*; Franklin-Társulat: Budapest, Hungary, 1909; Volume 2.
231. László, E. *Magyar Reneszánsz és Barokk Hímzések*; Iparművészeti Múzeum: Budapest, Hungary, 2001.
232. Mikó, Á.; Verő, M. *Mátyás Király Öröksége. Késő Reneszánsz Művészet Magyarországon*; Magyar Nemzeti Galéria: Budapest, Hungary, 2008.
233. Felhősné, S. Úrasztali Terítők a gömöri egyházmegye keleti részén. In *Népi Vallásosság a Kárpát-Medencében*; Pilipkó, E., Ed.; Laczkó Dezső Múzeum: Veszprém, Hungary, 2013; Volume 8, pp. 617–640.
234. Szendrei, J. Magyar diszítmények. *Művészi Ip.* **1891**, *6*, 73–96.
235. Genthon, I. *Esztergom Műemlékei*; Műemlékek Országos Bizottsága: Budapest, Hungary, 1948.
236. Lanszki-Széles, G. Egyházi öltözékek, miseruhák Gölle és Kisgyalán községekben a 18-21. században. *Kaposvári Rippl Rónai Múzeum Közleményei* **2020**, *7*, 307–320. [CrossRef]
237. Zentai, J. Baranya magyar főkötői. *Janus Pannon. Múzeum Évkönyve* **1974**, *14–15*, 233–257.
238. Szendrődiné, Á. Egy 19. századi debreceni típusú láda restaurálása. In *A Szabadtéri Néprajzi Múzeum Évkönyve*; Cseri, M., Füzes, E., Eds.; Szabadtéri Néprajzi Múzeum: Szentendre, Hungary, 2003; Volume 16, pp. 193–210.
239. Takács, B. A sárospataki kőedénygyártás. In *A Herman Ottó Múzeum Évkönyve*; Komáromy, J., Ed.; Herman Ottó Múzeum: Miskolc, Hungary, 1965; Volume 5, pp. 333–346.
240. Varga, L. Ung-vidéki szarufaragványok. *Új Mindenes Gyűjtemény* **1984**, *3*, 75–83.
241. Barabás, H. Az egykori Kézdiszék református egyházainak feliratos és díszített ónedényei. In *Népi Vallásosság a Kárpát-Medencében*; Lackovits, E., Gazda, E., Eds.; Székely Nemzeti Múzeum: Sepsiszentgyörgy, Romania; Veszprém Megyei Múzeumi Igazgatóság: Veszprém, Hungary, 2007; Volume 7, pp. 441–452.
242. Leitner, M. A magyar néphímzések ismertetése. In *A Magyar Királyi Állami Nőipariskola Ékönyve*; Resch, E., Ed.; Újvidék, Serbia, 1941; Volume 1.
243. Felhősné, S.; Küllős, I.; Molnár, A.; Szalay, E. *Magyar Református Egyházak Javainak Tára. Kárpátaljai Református Egyház I-IV*; Országos Református Gyűjteményi Tanács: Budapest, Hungary, 2001.
244. Agárdi, L. Agárdi, L. A Szoboszlói református egyház klenódiumai. In *A Bocskai István Múzeum Évkönyve*; Bihari-Horváth, L., Ed.; Bocskai István Múzeum: Hajdúszoboszló, Hungary, 2016; Volume 3, pp. 87–128.
245. Farnadi, I. Magyar estély. *Muskátli* **1935**, *4*, 76.
246. Szamosi, J. Pestvármegye népművészete. In *Pest-Pilis-Solt-Kiskun Vármegye és Kecskemét th. Jogu Város Adattára*; Csatár, I., Ed.; Pécs, Hungary, 1939; pp. 154–166.
247. Péntek, J. *A Kalotaszegi Népi Hímzés és Szókincse*; Kriterion: Bukarest, Romania, 1979.
248. Burány, B. *Dél-Tisza Vidéki Vadvirágok. Vajdasági Hímzésminták*; Rubicon: Temerin, Serbia, 1991.
249. Nagy, I. *Debellácsi Hímzések*; Forum Könyvkiadó: Újvidék, Serbia, 1996.

250. Nagy, I. *Mai Debellácsi Hímzések*; Szerző Kiadása: Debelyacsa, Serbia, 2008.
251. Kresz, M. *Virág és Népművészet*; Népművelési Propaganda Iroda: Budapest, Hungary, 1979.
252. Szögi, C.; Tomán, Z. *A Szegedi Papucs Készítésének és Viselésének élő Hagyománya*; Szellemi Kulturális Örökség Igazgatóság: Szentendre, Hungary, 2018.
253. Cseh, F. Lokális tárgyalkotás mint nemzeti örökség. Az örökségképzés stratégiái a höveji varrott csipke és a szegedi papucs példáján. In *Ethno-Lore*; Balogh, B., Ed.; Néprajztudományi Intézet: Budapest, Hungary, 2019; Volume 36, pp. 27–49.
254. Tarján, G. *Folklór, Népművészet, Népies Művészet: Kalauz a Magyar Népművészethez*; Nemzeti Tankönyvkiadó: Budapest, Hungary, 2005.
255. Fucskár, Á.; Fucskár, J. *The Book of Hungarikums: Hungary's Treasures*; Alexandra Kiadó: Pécs, Hungary, 2020.
256. Hekler, A. A 19. század magyar művészetének vázlata. *A Kisfaludy-Társaság Évlapjai* **1936**, *59*, 251–262.
257. Peters, D. The decoration and decorators of late 18th century sevres porcelain in the Bowes-Museum. *Burlingt. Mag.* **1991**, *133*, 306–311.
258. Kumpikaite, E.; Kot, L.; Tautkute-stankuviene, I. Ornamentation of Lithuanian ethnographic home textiles. *Fibres Text. East. Eur.* **2016**, *24*, 100–109. [CrossRef]
259. Kujawska, M.; Luczaj, L.; Typek, J. Fischer's Lexicon of Slavic beliefs and customs: A previously unknown contribution to the ethnobotany of Ukraine and Poland. *J. Ethnobiol. Ethnomed.* **2015**, *11*, 85. [CrossRef] [PubMed]
260. Zentai, É. *Monet*; Corvina Kiadó: Budapest, 1985.
261. Baróti, D. Pipacspirossal zendüljön a Világ! *Tiszatáj* **1976**, *30*, 53–59.
262. Bertha, Z. Király László: Amikor pipacsok voltatok. *Alföld* **1983**, *34*, 70–72.
263. Hoppál, M.; Jankovics, M.; Nagy, A.; Szemadám, G. *Jelképtár*; Helikon: Budapest, Hungary, 2000.
264. Surányi, D. Virágok, virágénekek és a virágnyelv. *Valóság* **2019**, *62*, 40–50.
265. Kriachko, I. Ukrainian clothing in history, social life and art. In *Perspectives of Comparative World Literature and Cultural Studies*; Korolenko National Pedagogical University: Poltava, Ukraine, 2022; Volume 2, pp. 80–82.
266. Selmeczi-Kovács, A. *Nemzeti Jelképek a Magyar Népművészetben*; Cser Kiadó, Budapest, Hungary; Néprajzi Múzeum: Budapest, Hungary, 2014.
267. Kapitány, Á.; Kapitány, G. *Magyarságszimbólumok*; Európai Folklór Intézet: Budapest, Hungary, 2002.
268. Lewis-Stempel, J. *Where poppies blow: The British soldier, nature, the Great War*; Weidenfeld & Nicolson: London, UK, 2017.
269. Wearn, J.A.; Budden, A.P.; Veniard, S.C.; Richardson, D. The flora of the Somme battlefield: A botanical perspective on a post-conflict landscape. *First World War Stud.* **2017**, *8*, 63–77. [CrossRef]
270. Iles, J. In remembrance: The Flanders Poppy. *Mortality* **2008**, *13*, 201–221. [CrossRef]
271. Cherniavska, A. Floronym mak / poppy in the National World conception of Ukrainian and English speakers. *SJPU* **2021**, *42*, 8–18. [CrossRef]
272. Farcas, C.; Cristea, V.M.; Farcas, S.; Ursu, T.M.; Roman, A. The symbolism of garden and orchard plants and their representation in paintings (I). *Contr. Bot.* **2015**, *50*, 189–200.
273. Németh, I. Ülést tartott a magyar anyák Nemzetvédő Bizottsága. *Magy. Országos Tudósító* **1936**, *18*, 84.
274. Debreczenyi, J. Sokgyermekes veszprémi családok anno. *Veszprémi Szle.* **2021**, *23*, 79–90.
275. Schubert, P.; Steinecke, H. Die Kornblume—Lieblingsblume der Preußen. *Palmengarten* **2013**, *77*, 29–33. [CrossRef]
276. Berg, T. Geography and vexillology: Landscapes in flags. In Proceedings of the 25th International Congress of Vexillology, Rotterdam, The Netherlands, 5–9 August 2013; Volume 8, pp. 1–10.
277. Erdélyi, Z. Adatok a magyar népköltészet színszimbolikájához (III.). *Ethnographia* **1961**, *72*, 583–596.

Article

Botanical Analysis of the Baroque Art on the Eastern Adriatic Coast, South Croatia

Nenad Jasprica [1], Vinicije B. Lupis [2] and Katija Dolina [1,*]

[1] Institute for Marine and Coastal Research, University of Dubrovnik, Kneza Damjana Jude 12, HR 20000 Dubrovnik, Croatia; nenad.jasprica@unidu.hr

[2] Institute of Social Sciences Ivo Pilar, Regional Center in Dubrovnik, HR 20000 Dubrovnik, Croatia; vinicije.lupis@pilar.hr

* Correspondence: katija.dolina@unidu.hr

Abstract: The analysis of plants featured in Baroque artworks on the eastern Adriatic coast has not previously been the subject of an in-depth study. The study of plant iconography in Baroque sacred artworks, which are mostly paintings, was carried out in eight churches and monasteries on the Pelješac peninsula in southern Croatia. Taxonomic interpretation of the painted flora on 15 artworks led to the identification of 23 different plant taxa (species or genera) belonging to 17 families. One additional plant was identified only by family taxonomic rank. The number of plants was relatively high, and most species were considered non-native (71%, "exotic" flora) phanerophytes. In terms of geographic origin, the Palaearctic region (Eurasia) and the American continent were identified as the main areas of plant origin. *Lilium candidum*, *Acanthus mollis*, and *Chrysanthemum* cf. *morifolium*, were the most common species. We think that the plants were selected for decorative and aesthetic reasons, as well as for their symbolic significance.

Keywords: art; Baroque; floral elements; NE Mediterranean; Pelješac peninsula; sacral heritage

1. Introduction

Plant science research is accelerating at a rapid pace in Croatia [1]. New technologies and expanding infrastructure development opened the door to cutting-edge research on a large scale. Despite this noteworthy growth, access to old artworks is not evenly distributed across the country. In fact, we identified a research gap in the plant study of artworks in the eastern Adriatic region. Actually, there is no place or artwork that was previously studied scientifically in such detail. This study may be classified as an inter-disciplinary approach designed to enable the interpretation of botanical species and facilitate a better understanding of the context of Baroque art in the area of interest. Besides the scientific problem of accurately defining the typology of the plants represented, an attempt should be made toward decoding the message underlying the decoration. In general, botanical analysis of artworks considers their physical–natural, historical, and ideological aspects as they change throughout history. This perspective can contribute to a more objective enhancement of this complex cultural heritage in which nature and culture are intertwined.

The importance of the study of plant iconography and, in general, the floristic richness of the artworks in the Mediterranean was highlighted in the last decade. The floristic richness in Roman iconography and the plants carved in the fountain (Rome, mid-17th century) were analyzed on the basis of iconographic and historical documents [2,3]. Hosseini and Caneva [4] emphasized the lack of a general methodological approach for the simultaneous evaluation of historical, structural (i.e., composition), and botanical features, as well as for revalorization the natural components of the lost gardens from antiquity. Images of date palms (*Phoenix dactylifera* L.) on coins were analyzed from agricultural, botanical, and geographic perspectives, particularly with respect to their relationship with the climatic conditions that were favorable for their cultivation [5].

Citation: Pinke, G.; Kapcsándi, V.; Czúcz, B. Botanical Analysis of the Baroque Art on the Eastern Adriatic Coast, South Croatia. *Plants* **2023**, *12*, 2080. https://doi.org/10.3390/plants12112080

Academic Editors: Renata Söukand and Raivo Kalle

Received: 17 April 2023
Revised: 12 May 2023
Accepted: 22 May 2023
Published: 23 May 2023

Baroque is a style of architecture, painting, sculpture, and other arts that followed Renaissance and Mannerist art and preceded Rococo (often referred to as "late Baroque") and Neoclassicism [6]. The Baroque era began in Rome, Italy, in the early 17th century and then spread rapidly to France, Northern Italy, Western Europe (Spain, Portugal), and other countries. In general, the Baroque style developed in different regions at different times, but was established throughout Europe by 1620 [7]. In Europe, the Baroque style influenced all aspects of the visual and performing arts in the 17th and 18th centuries [8]. In some areas (e.g., the Iberian Peninsula), it continued along with new styles in the first decade of the 19th century.

The Baroque style introduced a variety of thematic innovations to the visual arts, such as vedute, still life, battle scenes, magnificent landscapes, plants, fruits, etc. [9]. Many painters used a very realistic style. There was much attention to detail, albeit not with the scientific precision and detail of Renaissance art (1400–1540). In contrast to the Renaissance, the Baroque landscape did not focus on human figures, while nature was given prominence in the composition. However, representations of plants and, more generally, of natural elements were not merely decorative or chosen for aesthetic reasons, but often pursued a specific symbolic purpose [10]. We agree with Caneva [11] and Caneva et al. [12], who argue that people in the past were able to understand these symbolic meanings thanks to their deep connection with and understanding of their environment.

At that time, the entire region along the Dalmatian coast (the eastern Adriatic), with the exception of the city-state Republic of Dubrovnik (or Republic of Ragusa), belonged to the Republic of Venice. The entire area, from Istria in the north to the Bay of Kotor in the south, was predominantly under the influence of the Venetian school of art, while artists from central and southern Italy, especially Naples, left visible traces in Dubrovnik and its surroundings [13,14].

The Baroque period on the Pelješac peninsula was an extremely intense period of building and art acquisition, because it was the time when the rich class of sailors and shipowners, especially those located in the western part of the peninsula, started building more sacral monuments [15]. Among the sacred works of art, altarpieces with very popular motifs dominated in this period. These motifs are mostly associated with the cult and veneration of the Mother of God and various saints, from martyrs from early Christianity to 'new' local patron saints of Croatian coastal communities and dioceses (St. Anthony of Padua, St. Blasius, St. Anne, etc.). In addition, floral patterns are found on these paintings and architectural elements in Baroque churches and monasteries in the region [16,17].

The objectives of our study were as follows: (i) to analyse the presence of floral elements in Baroque sacred art on the Pelješac peninsula; (ii) to determine the relationship between the local flora and the flora recognisable on the artworks; and (iii) to contribute to a better understanding of the relationship between man and the environment in this area.

2. Materials and Methods

2.1. Study Area

The Pelješac peninsula (area 355 km^2, max. altitude 961 m a.s.l.) is located on the eastern Adriatic coast in southern Croatia (Figure 1). Archaeological findings from the western part of the peninsula indicate continuous human settlement for several millennia [18]. However, the earliest known historical records of Pelješac date back to ancient Greece. After the Illyrian Wars (220 to 219 BC), the area became part of the Roman province of Dalmatia. Human activities have affected the environment for thousands of years [19,20]. In the mid-17th century, Pelješac was located quite far from the nearest major urban center (Dubrovnik) and had about 8000 inhabitants [21]. The population fluctuated from the 15th to the 17th century due to the immigration of Christian refugees from Bosnia and Herzegovina, epidemics, the Cretan War (1645–1669), the earthquake of 1667, and emigration. In general, the majority of the population was poor and engaged in fishing and agriculture, while shipping and international maritime trade increased in the 16th and 17th centuries.

Phytogeographically, the peninsula belongs to the Mediterranean Region, the Eastern Mediterranean Subregion, Adriatic Province, and the Epiro-Dalmatian Sector (*sensu* [22]). It is predominantly composed of carbonate rocks. The climate in this area is Mediterranean with mild, humid, and rainy winters and dry and hot summers (*Csa* subtype of Mediterranean climate, *sensu*) [23,24]. This climate enables the development of eu-Mediterranean vegetation dominated by evergreen shrubs and sclerophyllous trees (maquis), with the most important tree species being the holm oak (*Quercus ilex* L.). Today, the Pelješac peninsula is one of the Important Plant Areas (IPAs) in Croatia and has a high structural diversity of vegetation [25,26]). On the peninsula, there are sites rich in endemic flora [27], while the larger part of the peninsula is covered by the NATURA 2000 network of protected areas in Croatia [28,29].

Figure 1. Map of Pelješac peninsula and its location on southeastern Adriatic coast (SE Europe). Numbers indicate location of churches and monasteries where artworks were examined: 1—St. Anne Chapel, Žukovac, near village of Kućište; 2—Franciscan Monastery and Church of the Great Lady, Podgorje; 3—Church of Our Lady of Carmel, village of Carmel, Podgorje; 4—Church of Christian's helpers, Orebić; 5—St. Anthony of Padua Church, Trpanj; 6—St. Blasius Church, Janjina; 7—St. Martin Church, Žuljana; 8—Church of Our Lady of the Rosary, Tomislavovac, near village of Putniković (for a detailed description see Section 2.2). Abbreviations: IT—Italia, SL—Slovenia, HR—Croatia, BiH—Bosnia and Herzegovina, MN—Montenegro, RS—Serbia, RKS—Kosovo, AL—Albania, NMK—North Macedonia. The circle on the map in the lower left corner indicates the research area in the SE European context.

2.2. Methods

The study of the plant iconography of the artworks was carried out in eight churches and monasteries on the Pelješac peninsula (Figure 1). A total of 15 artworks were analysed.

The criterion for the selection of the artworks was based on territorial and representative principles. For this purpose, we first studied the documents of the Museum Documentation Center (MDC), i.e., the Register of Museums, Galleries, and Collections in the Republic of Croatia, which contains relevant information about the collections and their artworks. However, a significant part of the cultural heritage is owned by religious communities, which keep the artworks in their collections and treasuries. Although not all collections owned by religious communities were included in the Register, it is clear that there are many more collections than are officially recorded [16,17]. In addition, we had only partial insight into the list of artworks in Catholic parishes on the peninsula, as in many cases such lists are missing. Therefore, we visited all available churches that we knew contained artwork and attempted to cover the entire peninsula area. Access to some artworks in churches and monasteries was not possible for various reasons (restoration, loan to other parties, etc.). Our main focus was on artworks with floral motifs located on the Pelješac peninsula that date from the Baroque period, whose creators were local artists (e.g., Filippo Naldi) who skillfully contributed to the development of new motifs in artistic expression and decorated Baroque interiors with plant motifs. The artworks considered in this study are an exclusive example of folk Baroque on the Pelješac peninsula, which emerged during the period of the Dubrovnik Republic. All the artworks considered come from places located on the peninsula, most of which correspond to settlements where a newly enriched maritime folk class with its own cultural needs began to develop.

The artworks are listed in Table 1 (see also Table 2).

Table 1. List of studied artworks with codes, name, and location of church/monastery, and year (period) of its construction.

Code	Artwork	Shown on Figures
1.	St. Anne and Our Lady with Jesus Christ and Our Father [unknown Baroque painter St. Anne Chapel, Žukovac near the village of Kućište, 1625].	Figure 2A–C.
2.	The altar of St. Francis of Assisi [The Franciscan Monastery and Church of the Great Lady, Podgorje, late 15th century].	Figure 3A–C.
3.	The altar St. Anthony of Padua [ibid].	Figure 3D,E.
4.	The altar of St. Anthony of Padua [The church of Our Lady of Carmel, the village of Carmel, Podgorje, near Orebić, 1470].	Figure 4A.
5.	The altar of St. Anthony of Padua [ibid].	Figure 4C,D.
6.	The altar of St. Anthony of Padua [ibid].	Figure 4F.
7.	The altar of St Joseph [ibid].	Figure 4E.
8.	The Archangel Gabriel [ibid].	Figure 4B.
9.	The Escape to Egypt [unknown Baroque painter Church of Christian's helpers, Orebić, 1853–1886].	Figure 2D.
10.	Health-related Votive Tablet [ibid].	Figure 2E.
11.	The antependium [painted by Filippo Naldi, mid-18th century, St. Anthony of Padua Church, Trpanj, 1695].	Figure 5A.
12.	The Virgin with Child, St. Blasius, and St. Nicholas [painted by Filippo Naldi, oil on canvas, 198 × 100 cm, mid-18th century, the parish church of St Blasius, Janjina, after 1774].	Figure 2F.
13.	The antependium from altar of Our Lady of Mercy [painted by Filippo Naldi and St. John the Baptist, mid-18th century, the parish church of St. Martin, Žuljana, 1556].	Figure 5B.
14.	The antependium from altar of St. John the Baptist [ibid].	Figure 5C.
15.	Birth and Death with the Seven Holy Sacraments [painted by Filippo Naldi, oil on canvas, 700 × 100 cm, wooden fence at the choir, mid-18th century, the parish church of Our Lady of the Rosary, Tomislavovac near the village of Putniković, 1569].	Figure 5D–F.

The species presented were identified on the basis of the most diagnostic morphological aspects, such as the general form of the plant (habit), typology, shape, size, and color of the flowers and fruits, if present, and the morphology and arrangement of the leaves. The correct number of single parts was more or less easy to identify in real individuals (specimens), though this was not always possible in painted or engraved elements. This identification became even more difficult when time-related damage was added to the sometimes-poor accuracy of the painter in depicting plants. Therefore, in the absence of precise diagnostic elements, an assignment based on considerations related to habitat and probable abundance in adjacent natural contexts was proposed (see Caneva and

Bohuny [30], Caneva et al. [3]). When interpretation was too doubtful or ambiguous, identification was restricted to a general (higher taxonomic) level.

Various floras were used to determine the plants according to their diagnostic elements and their ecological and biogeographical aspects (for details, see Jasprica and Milović [31], Milović et al. [32], and Jasprica et al. [33], as well as references therein). Matthioli [34,35] was also consulted for iconographic analysis. The nomenclature of plant taxa follows the Plants of the World Online database [36]. The floristic list below (Table 2) includes the following aspects: the updated scientific name, the common name in English and Croatian (in parentheses), the structure expressed by the biological forms, and the chorology (geographical origin). Plants in the floristic list are given in alphabetical order. The frequency of their occurrence in the paintings and the elements that led to their identification are also indicated. Croatian common names, mainly used on the Pelješac peninsula, were identified using the Nomenclator botanicus Croaticus [37] and the Flora Croatica Database [1].

3. Results

The taxonomic interpretation of the painted flora led to the identification of 23 different plant taxa (species or genera) belonging to 17 families. One additional plant was identified only via family taxonomic rank (Table 2, Figures 2–5).

The most common plants (occurring in at least four artworks) were *Lilium candidum*, *Acanthus mollis*, and *Chrysanthemum* cf. *morifolium*.

Prunus and *Rosaceae* were the most represented genera and families, respectively (Table 2). The analysis of plant life forms showed that the artworks were dominated by phanerophytes (54%), followed by therophytes (21%).

Most species were non-native (71%, also referred to as "exotic" flora) and originated mainly from Asia. They were largely cultivated for their nutritional value (fruits, vegetables) and as ornamentals. Among the native taxa, Mediterranean floral element (i.e., *Centaurea cyanus*, taxa from the Orchidaceae family), which were mostly circum-Mediterranean plants, predominate. Although some plants (*Acanthus mollis*, *Vitis vinifera*) were signed as originating from outside the Mediterranean region, they have been cultivated since ancient times and, thus, have become an integral part of the local wild flora in the Mediterranean region.

Table 2. Identification and distribution of floristic elements painted on artworks of Pelješac peninsula. Corresponding family and Croatian name of genera or species are given in parentheses. Abbreviations, life-form: G—geophytes, H—hemicryptophytes, P—phanerophytes, T—therophytes; chorotype (biogeographic element): EA—Eurasian, AS—Asian, AF—African, Medit.—Mediterranean, AM—Americas, EU—European, Cos—Cosmopolitan. It is also indicated whether plant grows naturally (native) in Croatia or is non-native. Codes of artworks in which plant appears are given in Table 1 in Section 2.2.

Proposed Identification		Painted Part of the Plant	Life-Form	Chorotype, Native or Non-Native	Frequency of Occurrence (Code of Artworks in Which the Plant Appears)
Scientific Name	**Common Name**				
Acanthus mollis L. (Acanthaceae)	Common bear's breech (meki primog)	Leaves	H	AF, Medit., non-native	4 (1, 7, 11, 12)
Campsis radicans (L.) Bureau (Bignoniaceae)	trumpet vine (tekoma)	Flowers, inflorescence	P, liana	AM, non-native	1 (15)
Centaurea cyanus L. (Asteraceae)	Cornflower, bachelor's button (različak, zečina)	Flowers, inflorescence	T	Medit., native	1 (11)
Chrysanthemum cf. *morifolium* (Ramat.) Hemsl. [incl. *C. indicum*] (Asteraceae)	Florist's daisy, garden mum (krizantema)	Flowers, inflorescence	T	AS, non-native	4 (9, 13–15)
Citrullus lanatus (Thunb.) Matsum. and Nakai (Cucurbitaceae)	Watermelon (sađena lubenica)	Fruit	T	AF, non-native	1 (1)
Dianthus cf. *caryophyllus* L. (Caryophyllaceae)	Carnation (pitomi klinčić)	Flowers	H	EU, native	1 (11)
Hedera helix L. (Araliaceae)	Common ivy (obični bršljan)	Leaves	P, liana	EU, native	1 (9)
Hydrangea macrophylla (Thunb.) Ser. (Hydrangeaceae)	bigleaf hydrangea (velikolistna hortenzija)	Flowers, inflorescence	P	AS, non-native	1 (15)
Justicia carnea Lindl. (Acanthaceae)	Brazilian plume flower, jacobinia (dubrovački gospar)	Flowers, inflorescence	P	AM, non-native	1 (11)
Knautia/*Dipsacus* (Caprifoliaceae)	Widow flower/teasel (prženica/češljugovina)	Flowers, inflorescence	H	EA/EA, AF, native	1 (13)
Leonotis leonurus (L.) R.Br. (Lamiaceae)	Lion's tail, wild dagga (lavlji rep)	Flowers, inflorescence	P	AF, non-native	1 (11)
Lilium candidum L. (Liliaceae)	White lily, Madonna lily (bijeli ljiljan)	Flower, inflorescence	G	EA, native	7 (3, 4, 6–10)
Malus domestica (Suckow) Borkh. (Rosaceae)	Apple (obična jabuka)	Fruit	P	AS, non-native	1 (1)
Orchidaceae	Orchid (orhideje, kačunovice)	Flower shape and structure	G	Cos, native	2 (2, 3)
Paeonia sp. (Paeoniaceae)	Peony (božur)	Flowers	G	EA, AF, AM, non-native	1 (13)
Papaver rhoeas L. (Papaveraceae)	Common poppy (divlji mak)	Flowers	T	EA, native	2 (11, 12)
Passiflora caerulea L. (Passifloraceae)	Blue passionflower (krunica gospodinova)	Flowers, leaves	P, liana	AM, non-native	1 (2)
Phoenix dactylifera L. (Arecaceae)	Date palm (obični datuljevac)	Entire plant	P	AS, non-native	1 (9)
Prunus domestica L. (Rosaceae)	European plum (obična šljiva)	Fruit	P	EA (Türkiye), non-native	1 (1)
Prunus persica (L.) Batsch (Rosaceae)	Peach (breskva)	Fruit	P	AS, non-native	1 (1)
Pyrus communis L. (Rosaceae)	Common pear (obična kruška)	Fruit	P	EA, non-native	1 (1)
Rosa sp. (Rosaceae)	Rose (ruža)	Flowers	P	EA, AM, AF, non-native	2 (12, 13)
Solanum melongena L. (Solanaceae)	Eggplant (balančana)	Fruit	T	AS, non-native	1 (1)
Vitis vinifera L. (Vitaceae)	Grapevine (vinova loza)	Leaves, fruit	P, liana	EA, non-native	2 (1, 5)

Figure 2. (**A**)—St. Anne and Our Lady with Jesus Christ and Our Father in Chapel of St. Anne in Kućište; (**B**,**C**)—details from lower left and right parts of paintings; (**D**)—The Escape to Egypt, and (**E**)—Health-related Votive Tablet in Church of Christian's helpers in Orebić; (**F**)—The Virgin with Child, St. Blasius, and St. Nicholas in parish church of St. Blasius in Janjina.

Figure 3. The altars of St. Francis of Assisi (**A**) with details on *Passiflora caerulea* L. (**B**,**C**); St. Anthony of Padua (**D**) and details on *Lilium candidum* L. (**E**) in Franciscan Monastery and Church of the Great Lady in Podgorje.

Figure 4. (**A,C,D,F**)—The altars of St. Anthony of Padua; (**B**)—The Archangel Gabriel, (**E**)—St. Joseph in Podgorje.

Figure 5. (**A**)—Antependium in Church of St. Anthony of Padua in Trpanj; (**B,C**)—Antependia in Church of St. Martin in Žuljana; (**D**)—Polyptych Birth and Death with the Seven Holy Sacraments on wooden fence by choir in Church of Our Lady of the Rosary in Putniković; (**E,F**)—Details of first and final parts of polyptych.

4. Discussion

In this study, the number of plants is relatively high, and most species were considered non-native ("exotic" flora) phanerophytes, most of which are cultivated for various purposes. The main areas of origin of the plants were identified as the Palaearctic and the Americas [38]. Despite the great diversity of plants, a repetitive trend in the occurrence of species can be seen. *Lilium candidum*, *Acanthus mollis* and *Chrysanthemum* cf. *morifolium*, followed by *Vitis vinifera*, *Papaver rhoeas*, orchids and various *Rosa* spp., were the most common species.

Both *A. mollis* and *V. vinifera* (grapevine) are widely used in paintings, mosaics, and classical sculptures [2,39]. In general, these species, including some others found in this study (e.g., *Phoenix dactylifera*), are very common in sacred artworks due to their strong association with mythological and religious symbolic meanings. The frequent occurrence of *A. mollis*, which is a wild species in northwest Africa and the Mediterranean region, is related to the idea of "rebirth" and its symbolism [40,41]. The morphological characteristics of the grapevine (i.e., the single fruit is round, fleshy, and bears more than one seed) evoke specific symbolic meanings associated with ideas of wealth, fertility, and prosperity. However, the primary symbolic meaning of the grapevine is actually associated with notions of life and vitality [10].

The historical data prove the centuries-old tradition of cultivation of different varieties of *V. vinifera* on the Pelješac peninsula, which has been persevered until today [42]. In the case of *A. mollis*, the interpretation of its presence in this area must include an analysis of the activities of the Republic of Dubrovnik when it was an important thalassocracy. In that period, especially in the late Baroque, the ships and emissaries of the Republic often returned to port with exotic plants that they had collected along the Mediterranean coast and beyond (for a review, see Đurasović [43].

In this study, *Lilium candidum* was the "attribute" of St. Anthony of Padua, as well as of St. Joseph and the Archangel Gabriel. The white lilies (also called "Madonna lilies") bloomed at Easter, sprang from the staff of St. Joseph, and were carried by the Angel of the Annunciation [40,44]. To this day, St. Anthony is one of the most venerated and popular saints of the Catholic Church in Croatia, because his life was a constant struggle to face the ups and downs of life [45]. However, in church symbolism, the "lily of purity" is particularly suitable to represent the Virgin and adorn her altars [46], which was not the case in this study.

Although *Chrysanthemum* cf. *morifolium*, which was commercially introduced to Europe from China in the late 18th century, has been used as a medicinal, food, and ornamental plant for at least 2200 years [47], in our case, it has more cultural significance. In Croatian tradition, chrysanthemum is strongly associated with death [48]. Chrysanthemums can send a message of remembrance to the deceased, but they also convey to the living the family's commitment to the memory of a deceased loved one; this cultural belief is also reported in other European countries [49]. However, Moore [50] noted that there are several inconsistencies in defining floral meanings across generations, i.e., from funerals, mourning, and condolences to homecomings and celebrations. Morphologically, *Chrysantemum* cf. *morifolium* has some similarities with *C. indicum*. Therefore, due to the very complex taxonomic implications, we included *C. indicum* in addition to *Chrysanthemum* cf. *morifolium* in the floristic list (Table 2).

Chrysanthemums and various *Rosa* spp. are common plant species on the wooden antependia and fence at the choir (see Figure 5), which was painted by Filippo Naldi (?–1783), who was from Florence, Italy, and belonged to the group of artists who influenced 18th century Dalmatian art [17,51]. His depictions of sacred artwork on the Dalmatian coast were rich in floral patterns, while the figures are enriched with representations of floral ornaments covering the clothing [14]. However, echoes of Venetian painting of the Seicento and Settecento can also be seen in his works [14,51,52].

Meagher [53] emphasises that the presence of fruits (e.g., *Solanum melongena*, *Citrullus lanatus*, etc.) reflects the influence of the current stream of profane still life painting in

Southern Europe. In the present study, some plants (*Papaver rhoeas*, *Centaurea cyanus*, *Passiflora caerulea*, etc.) were found to be used in traditional or official medical practice in the eastern Adriatic islands [54]. Of all the islands, the longest list of medicinal plants used was recorded on the neighbouring island of Korčula, and most of these plants already appear in ancient and mediaeval herbal books [55]. For example, *P. rhoeas* is characterised by its sedative effect, and *C. cyanus* was mainly used to cure eye inflammations [56]. On the other hand, the presence of these two species could indicate more intensive agriculture in that period on the Pelješac peninsula. Nowadays, *C. cyanus*, which is a weed of cereal fields and olive groves, does not occur in this area, probably due to environmental changes, i.e., habitat loss or possible impact of management practices in olive groves (for a literature review, see *Flora Croatica Database* [1]). Finally, Pinke et al. [56] pointed out the deep cultural embeddedness of these charismatic arable weeds and their symbolic connotations related to human characters and feelings (patriotism, historical remembrance, virginity, loyalty, etc.).

The methodological limitations of the study must be emphasised. The highest number of plants was determined based on their flowers and fruits. The habit and the typical morphology of the leaves were used as good diagnostic elements for the herbaceous species. However, naturalistic descriptions of species are not rigorous, and their identification is often based on a few diagnostic elements, e.g., trees and shrubs are often identifiable based on their fruits. In the case of *Rosa* spp. and several other plants, the lack of details in habit or organs (e.g., flowers with petals, basal leaf rosette, etc.) made it impossible to identify them to species level. This problem was also emphasised by other studies, e.g., Caneva and Bohuny [30] and Caneva et al. [3]. In addition, the species represented a lack of seasonal consistency, in that some are species depicted as growing in springtime while some others are growing in autumn. For example, *A. mollis* has a specific phenology: it appears dead in summer but regrows after the first fall. *Vitis vinifera* is without leaves in winter and appears to be dead, but comes back to life in the growing season. The latter phenomenon is connected with the ideas of life and death and rebirth and regeneration.

Although our floristic list (Table 2) includes plant species mentioned in the Bible [57] and identified in very old artworks [2,30,58], surprisingly, some common Mediterranean plants, such as *Ficus carica* L., *Laurus nobilis* L., *Paliurus spina-christi* Mill., and *Tulipa* spp., were not found in the present study. In addition, only plant species from the Orchidaceae family can be considered native to Mediterranean small tree vegetation (maquis) or dry grassland habitats.

It is not known how much influence the regiment had on the appearance or content of the artwork. The Pelješac peninsula was a very rural area, where the inhabitants lived in poverty, and, in general, there was no need to show native plants, which people knew well anyway. However, it is important to emphasise that, for the first time, the inhabitants had the opportunity to see the appearance of the figures of Christ, the Mother of God, and the saints, which until then they had only heard about in church services [59].

Regardless of the poverty of the local population, all the artworks studied were donated by the faithful for sacred monuments and were in situ at the time of acquisition. The regiment freely commissioned the artworks for the churches, which the Church accepted. The authors of some artworks are not known, although the literature emphasises that they were under the influence of central and southern Italian artists [16]. Filippo Naldi painted almost as a rule for the poor who lived after the withdrawal of the Ottoman Turks in a wide area of what is now southern Croatia (Dalmatia), including the Pelješac peninsula [59]. His training in painting is unclear: he served in the Venetian army and was a port administrator in a small town not far from the northern coast of the Pelješac peninsula.

The floristic list offered here for the first time for the Croatian coastal region cannot be completed without an adequate base of artworks from other parts of the eastern and western Adriatic coast. However, a comparison and analysis could be made, at least in part, with the flora listed in a botanical database of ancient Roman paintings and sculptures, which includes a dataset of about 420 artworks [2]. The floristic study of Roman iconography included a large number of botanical elements (168 species, 78 families, and 159 genera) and shows some similarities to the most common plants (*Acanthus mollis*, *Vitis vinifera*) found in our study. However, a high proportion of phanerophytes and geophytes, as well as the presence of some taxonomic (e.g., pteridophytes) or functional (e.g., macrophytes) groups, were not found in our case.

Some similarities in the plant record are found in Islamic art. Although Islamic art is not art of a specific religion, time, place, or of a single medium (it spans ca. 1400 years, covers many lands and populations, and includes a range of artistic fields), the Ottomans (1299–1923) not only brought a new level of naturalism and detail to the design of flowers in Islamic artwork, especially in ornaments, but also introduced tulip and hyacinth to already developed floral motifs, such as lotus, lily, peony, chrysanthemum, and carnation [60]. In general, floral motifs in Islamic art avoid a focus on concepts of realism, such as growth or life [61]. Certain types of flowers or plants can have theological meanings; for example, the cypress often represents humility before God. From the late 16th to the mid-18th century, classical Ottoman artwork, especially in architecture, gradually lost ground to emerging western Baroque influences, and Baroque ornamentation became dominant even in famous religious buildings (e.g., Laleli Madrasa, Istanbul). In the religions of Buddhism and Hinduism, the lotus (genus *Nelumbo*) is the most commonly depicted plant and is associated with purity and beauty. Even a variety of colors are associated with different aspects of Buddhism; for example, the blue lotus flower is associated with the victory of the spirit over that of wisdom, intelligence, and knowledge [62]. In our study, the lotus flower (*Nelumbo nucifera* Gaertn.) was not found, though it was presented in Italian artworks [2].

5. Conclusions

A high proportion of non-native plants, mostly small trees or shrubs, from the Palaearctic and Americas was noted. The most common plant species recorded in the study are found not only in the artworks of the Baroque period, but also in artworks from various historical periods and, sometimes, from other religions. We assumed that the plants were selected for decorative and aesthetic reasons, as well as for their symbolic significance.

We believe that the most important result of this work lies in the information discovered about the botanical biodiversity of Baroque iconography in Croatia. However, considering the relatively small area and the artworks studied, the results should be read and analysed in the context of a better understanding of the cultural heritage, natural history, and knowledge of people from the Baroque period and possible higher agricultural land use in this part of the Mediterranean in the last few centuries.

Author Contributions: N.J. and V.B.L. conceived the study; N.J. wrote the manuscript; N.J. and K.D. analysed and interpreted data for the work. All authors critically revised the manuscript. All authors have read and agreed to the published version of the manuscript.

Funding: This research received no external funding.

Data Availability Statement: Data are contained within the article.

Conflicts of Interest: The authors declare no conflict of interest.

References

1. Flora Croatica Database. Available online: http://hirc.botanic.hr/fcd (accessed on 24 November 2022).
2. Kumbaric, A.; Caneva, G. Updated outline of floristic richness in Roman iconography. *Rend. Lincei Sci. Fis. Nat.* **2014**, *25*, 181–193. [CrossRef]
3. Caneva, G.; Altieri, A.; Kumbaric, A.; Bartoli, F. Plant iconography and its message: Realism and symbolic message in the Bernini fountain of the four rivers in Rome. *Rend. Lincei Sci. Fis. Nat.* **2020**, *31*, 1011–1026. [CrossRef]
4. Hosseini, Z.; Caneva, G. Lost gardens: From knowledge to revitalization and cultural valorization of natural elements. *Sustainability* **2022**, *14*, 2956. [CrossRef]
5. Rivera, D.; Obón, C.; Alcaraz, F.; Laguna, E.; Johnson, D. Date-palm (*Phoenix*, Arecaceae) iconography in coins from the Mediterranean and West Asia (485 BC–1189 AD). *J. Cult. Herit.* **2019**, *37*, 199–214. [CrossRef]
6. Martin, J.R. *Baroque*; Harper & Row: New York, NY, USA, 1977.
7. Snodin, M.; Nigel, L. (Eds.) *Baroque: Style in the Age of Magnificence 1620–1800*; V. & A. Publishing: London, UK, 2009.

8.   Hills, H. Introduction: Rethinking the Baroque. In *Rethinking the Baroque*; Hills, H., Ed.; Ashgate Publishing Ltd.: Aldershot, UK, 2011; pp. 1–10.

9.   Dahiya, B. Baroque art history. *J. Res. Humanit. Soc. Sci.* **2022**, *10*, 55–63.

10. Savo, V.; Kumbaric, A.; Caneva, G. Grapevine (*Vitis vinifera* L.) symbolism in the ancient Euro-Mediterranean cultures. *Econ. Bot.* **2016**, *70*, 190–197. [CrossRef]

11. Caneva, G. *The Augustus Botanical Code—Ara Pacis: Speaking to the People through the Images of Nature*; Gangemi: Rome, Italy, 2010.

12. Caneva, G.; Savo, V.; Kumbaric, A. Big messages in small details: Nature in Roman archeology. *Econ. Bot.* **2014**, *68*, 109–115. [CrossRef]

13. Prijatelj, K. Barok u Dalmaciji. In *Barok u Hrvatskoj*; Goldstein, S., Mirić, M., Čičin-Šain, V., Ivančić, Ž., Eds.; Sveučilišna naklada Liber, Odjel za povijest umjetnosti Centra za povijesne znanosti, Društvo povjesničara umjetnosti Hrvatske, Grafički zavod Hrvatske, Kršćanska sadašnjost: Zagreb, Croatia, 1982; pp. 651–869.

14. Tomić, R.; Marković, D. (Eds.) *Sveto i Profano, Slikarstvo Talijanskog Baroka u Hrvatskoj*; Galerija Klovićevi dvori: Zagreb, Croatia, 2015.

15. Vekarić, S.; Vekarić, N. Tri stoljeća pelješkog brodarstva (1600–1900). *Pelješki Zbornik* **1987**, *4*, 1–385.

16. Lupis, V. *Sakralna Baština Stona i Okolice*; Matica hrvatske: Ston, Croatia, 2000.

17. Lupis, V. Filippo Naldi—Dopune likovnoj baštini zapadnog Pelješca. In *Fra Jeronim Šetka Hrvatski Franjevac, Svećenik, Profesor, Jezikoslovac i Književnik*; Šešelj, S., Ed.; Neretvanska riznica umjetnina i inih vrijednosti; HKZ, Hrvatsko slovo d.o.o.: Zagreb, Croatia, 2016; pp. 269–279.

18. Forenbaher, S. Nakovana Cave Sanctuary. In *Archaeology and Speleology: From the Darkness of the Underground to the Light of Knowledge*; Janković, I., Drnić, I., Paar, D., Eds.; Arheološki Muzej: Zagreb, Croatia, 2021; pp. 87–98.

19. Trinajstić, I. Kronološka klasifikacija antropohora s osvrtom na helenopaleofite jadranskog primorja Jugoslavije. *Biosistematika* **1975**, *1*, 79–85.

20. Maslek, J. *Zemlja i Ljudi: Vinogradarstvo na Poluotoku Pelješcu u 19. i 20. Stoljeću*; Hrvatska Akademija Znanosti i Umjetnosti, Zavod za Povijesne Znanosti u Dubrovniku: Zagreb-Dubrovnik, Croatia, 2016.

21. Vekarić, N. The population of the Dubrovnik Republic in the fifteenth, sixteenth, and seventeenth centuries. *Dubrov. Ann.* **1998**, *2*, 7–28.

22. Rivas-Martínez, S.; Penas, A.; Diaz, T.E. Mapa Biogeografico de Europa. Available online: http://webs.ucm.es/info/cif/form/maps.htm (accessed on 20 November 2022).

23. Köppen, W.; Geiger, R. *Klima der Erde*; Justus Perthes: Darmstadt, Germany, 1954.

24. Sträßer, M. *Klimadiagramme zur Köppenschen Klimaklassifikation*; Klett-Perthes: Gotha-Stuttgart, Germany, 1998.

25. Jasprica, N.; Kovačić, S. Pelješac. In *Botanički Važna Područja Hrvatske*; Nikolić, T., Topić, J., Vuković, N., Eds.; Prirodoslovno-matematički fakultet Sveučilišta u Zagrebu, Školska knjiga d.o.o.: Zagreb, Croatia, 2010; pp. 335–340.

26. Jasprica, N.; Kovačić, S. The diversity of vegetation on the Pelješac peninsula. In *Zbornik Radova u Čast Ivice Žile*; Lupis, B.V., Ed.; Matica hrvatska—Ogranak Dubrovnik: Dubrovnik, Croatia, 2011; pp. 263–282.

27. Nikolić, T.; Milović, M.; Bogdanović, S.; Jasprica, N. (Eds.) *Endemi U Hrvatskoj Flori*; Alfa d.d.: Zagreb, Croatia, 2015.

28. Jasprica, N. Vegetation outline and NATURA 2000 habitats of the Pelješac peninsula, South Croatia. In Proceedings of the "Nature Protection in XXI Century", Žabaljak, Montenegro, 20–23 September 2011.

29. Anonymous. Uredba o ekološkoj mreži i nadležnostima javnih ustanova za upravljanje područjima ekološke mreže. *Off. Gaz. (OG)* **2019**, *80*. Available online: https://narodne-novine.nn.hr/clanci/sluzbeni/2019_08_80_1669.html (accessed on 20 November 2022).

30. Caneva, G.; Bohuny, L. Botanic analysis of Livia's villa painted flora (Prima Porta, Roma). *J. Cult. Herit.* **2003**, *4*, 149–155. [CrossRef]

31. Jasprica, N.; Milović, M. The vegetation of the islet of Badija (south Croatia), with some notes on its flora. *Nat. Croat.* **2016**, *25*, 1–24. [CrossRef]

32. Milović, M.; Kovačić, S.; Jasprica, N.; Stamenković, V. Contribution to the study of Adriatic island flora: Vascular plant species diversity in the Croatian Island of Olib. *Nat. Croat.* **2016**, *25*, 25–54. [CrossRef]

33. Jasprica, N.; Milović, M.; Dolina, K.; Lasić, A. Analyses of flora of railway stations in the Mediterranean and sub-Mediterranean areas of Croatia and Bosnia and Herzegovina. *Nat. Croat.* **2017**, *26*, 271–303. [CrossRef]

34. Matthioli, P.A. *I Discorsi di Pietro Andrea Matthioli su de Materia Medica di Dioscoride*; Presso Marco Ginammi: Venetia, Italy, 1568.

35. Matthioli, P.A. *De Plantis Epitome Utilissima Petri Andrea Matthioli*; J. Feyrabend: Francoforte, Germany, 1586.

36. Plants of the World Online. Available online: http://powo.science.kew.org (accessed on 28 April 2023).

37. Šugar, I. *Nomenclator Botanicus Croaticus*; Matica hrvatska: Zagreb, Croatia, 2008.

38. Vavilov, N.I. *Origin and Geography of Cultivated Plants*; Cambridge University Press: Cambridge, UK, 2009.

39. Gago, P.; Santiago, J.L.; Boso, S.; Alonso-Villaverde, V.; Martinez, M.C. Grapevine (*Vitis vinifera* L.): Old varieties are reflected in works of art. *Econ. Bot.* **2009**, *63*, 67–77. [CrossRef]

40. Zohary, M. *Plants of the Bible*; Cambridge University Press: Cambridge, UK, 1982.

41. Vandi, L. *La Trasformazione del Motivo Dell'acanto Dall'antichità al XV Secolo-Ricerche di Teoria e Storia Dell'ornamento*; Europäische Hochschulschriften, Kunstgeschichte 28; Peter Lang: Bern, Switzerland, 2002; Volume 386.

42. Andabaka, Ž.; Stupić, D.; Karoglan, M.; Marković, Z.; Preiner, D.; Maletić, E.; Karoglan Kontić, J. Historic development of the most important autochthonous dalmatian grapevine cultivars (*Vitis vinifera* L.). *Glas. Zašt. Bilja* **2016**, *39*, 14–20.

43. Đurasović, P. Unošenje egzotičnog drveća i grmlja na dubrovačko područje. *Šumar. List* **1997**, *121*, 277–289.

44. Skinner, C.M. *Myths and Legends of Flowers, Trees, Fruits and Plants in All Ages and in All Climes*, 3rd ed.; J.B. Lippincott Company: Phliladelphia, PA, USA, 1911.

45. Zlodi, Z. *Sveti Antun Padovanski, Životopis*; Verbum: Zagreb, Croatia, 2015.

46. Kandeler, R.; Ullrich, W.R. Symbolism of plants: Examples from European-Mediterranean culture presented with biology and history of art. *J. Exp. Bot.* **2009**, *60*, 1893–1895. [CrossRef]

47. Hao, D.-C.; Song, Y.; Xiao, P.; Zhong, Y.; Wu, P.; Xu, L. The genus Chrysanthemum: Phylogeny, biodiversity, phytometabolites, and chemodiversity. *Front. Plant Sci.* **2022**, *13*, 973197. [CrossRef]

48. Šilić, Č.; Mrdović, A. *Atlas Ukrasnih Vrtnih Biljaka*; Fram Ziral d.o.o: Mostar, Bosnia and Herzegovina, 2013.

49. Mathis, R. The evolution of *Chrysanthemum* in Europe. *Acta Hortic.* **1982**, *125*, 13–20. [CrossRef]

50. Moore, D.R. Flower Meanings: Are They Relevant Today? Master's Thesis, Murray State University, Murray, KY, USA, 2018. Available online: https://digitalcommons.murraystate.edu/etd/125 (accessed on 13 March 2023).

51. Tomić, R. Slikar Filippo Naldi (II). *Ars Adriat.* **2012**, *2*, 217–224. [CrossRef]

52. Fiocco, G. *Venetian Painting of the Seicento and the Settecento*; Pantheon: Florence, Italy, 1929.

53. Meagher, J. Still-Life Painting in Southern Europe, 1600–1800. In *Heilbrunn Timeline of Art History*; The Metropolitan Museum of Art: New York, NY, USA, 2008. Available online: http://www.metmuseum.org/toah/hd/sstl/hd_sstl.htm (accessed on 13 March 2023).

54. Łuczaj, Ł.; Jug-Dujaković, M.; Dolina, K.; Jeričević, M.; Vitasović-Kosić, I. Insular pharmacopoeias: Ethnobotanical characteristics of medicinal plants used on the Adriatic islands. *Front. Pharmacol.* **2021**, *12*, 623070. [CrossRef]

55. Tsioutsiou, E.E.; Giordani, P.; Hanlidou, E.; Biagi, M.; De Feo, V.; Cornara, L. Ethnobotanical study of medicinal plants used in Central Macedonia, Greece. *Evid. Based Complement. Alternat. Med.* **2019**, *1*, 4513792. [CrossRef]

56. Pinke, G.; Kapcsándi, V.; Czúcz, B. Iconic arable weeds: The significance of corn poppy (*Papaver rhoeas*), cornflower (*Centaurea cyanus*), and field larkspur (*Delphinium consolida*) in Hungarian ethnobotanical and cultural heritage. *Plants* **2023**, *12*, 84. [CrossRef]

57. Dafni, A.; Böck, B. Medicinal plants of the Bible—Revisited. *J. Ethnobiol. Ethnomed.* **2019**, *15*, 57. [CrossRef]

58. Dash, M. *Tulipomania—The Story of the World's Most Coveted Flower and the Extraordinary Passions It Aroused*; Gollancz: London, UK, 1999.

59. Tomić, R. Slikar Filippo Naldi. *Rad. Inst. za Povij. Umjet.* **2006**, *30*, 163–183.

60. Abdullahi, Y.; Embi, M.R. Evolution of abstract vegetal ornament sin islamic architecture. *Archnet-IJAR* **2015**, *9*, 31–49. [CrossRef]

61. Ghasemzadeh, B.; Fathebaghali, A.; Tarvirdinassab, A. Symbols and signs in Islamic architecture. *Eur. Rev. Artist. Stud.* **2013**, *4*, 62–78. [CrossRef]

62. Siddiqui Shahid, K. Significance of Lotus depiction in Gandhara Art. *J. Pak. Hist. Soc.* **2012**. Available online: https://www.academia.edu/31703261/ (accessed on 27 April 2023).

Article

Remnants from the Past: From an 18th Century Manuscript to 21st Century Ethnobotany in Valle Imagna (Bergamo, Italy)

Fabrizia Milani [1,2], Martina Bottoni [1,2], Laura Bardelli [1,2], Lorenzo Colombo [1,2], Paola Sira Colombo [1,2], Piero Bruschi [3,*], Claudia Giuliani [1,2] and Gelsomina Fico [1,2]

1   Department of Pharmaceutical Sciences, University of Milan, Via Mangiagalli 25, 20133 Milan, Italy; fabrizia.milani@unimi.it (F.M.); martina.bottoni@unimi.it (M.B.); laura.bardelli@studenti.unimi.it (L.B.); lorecolo.93@gmail.com (L.C.); pasico19@virgilio.it (P.S.C.); claudia.giuliani@unimi.it (C.G.); gelsomina.fico@unimi.it (G.F.)
2   Ghirardi Botanic Garden, Department of Pharmaceutical Sciences, University of Milan, Via Religione 25, 25088 Toscolano Maderno, Italy
3   Department of Agricultural, Environmental, Food and Forestry Science and Technology, University of Florence, 50144 Florence, Italy
*   Correspondence: piero.bruschi@unifi.it

Abstract: Background: This project originated from the study of an 18th century manuscript found in Valle Imagna (Bergamo, Italy) which contains 200 plant-based medicinal remedies. A first comparison with published books concerning 20th century folk medicine in the Valley led to the designing of an ethnobotanical investigation, aimed at making a thorough comparison between past and current phytotherapy knowledge in this territory. Methods: The field investigation was conducted through semi-structured interviews. All data collected was entered in a database and subsequently processed. A diachronic comparison between the field results, the manuscript, and a 20th century book was then performed. Results: A total of 109 interviews were conducted and the use of 103 medicinal plants, belonging to 46 families, was noted. A decrease in number of plant taxa and uses was observed over time, with only 42 taxa and 34 uses reported in the manuscript being currently known by the people of the valley. A thorough comparison with the remedies in the manuscript highlighted similar recipes for 12 species. Specifically, the use of agrimony in Valle Imagna for the treatment of deep wounds calls back to an ancient remedy against leg ulcers based on this species. Conclusions: The preliminary results of this study allow us to outline the partial passage through time fragments of ancient plant-based remedies once used in the investigated area.

Keywords: ethnobotanical investigation; diachronic comparison; historical sources; medicinal plants

Citation: Milani, F.; Bottoni, M.; Bardelli, L.; Colombo, L.; Colombo, P.S.; Bruschi, P.; Giuliani, C.; Fico, G. Remnants from the Past: From an 18th Century Manuscript to 21st Century Ethnobotany in Valle Imagna (Bergamo, Italy). *Plants* **2023**, *12*, 2748. https://doi.org/10.3390/plants12142748

Academic Editors: Renata Sõukand and Raivo Kalle

Received: 30 June 2023
Revised: 20 July 2023
Accepted: 20 July 2023
Published: 24 July 2023

1. Introduction

Diachronic approach in ethnobotanical studies has recently gained the attention of several authors. The use of historical sources allows researchers to collect valuable information about the mechanisms of transmission of Local Ecological Knowledge (LEK) and the nature of its transformation over time [1–3]. However, the interpretation of the results from diachronic studies is complicated by the intrinsic nature of LEK consisting of a complex multifaceted corpus of different practices, beliefs, and traditions having different origin and developmental patterns, molded by different socio-ecological factors. Some studies have shown that historical written records and iconography played a key role in shaping the modern Western medicinal knowledge as of the earliest manuscript copies of Dioscorides' De Materia Medica. According to Leonti et al. (2009; 2010) [4,5], the long-lasting influence of Pietro Andrea Mattioli (1501–1577) on the medicinal plant uses in Southern Italy clearly transpires through the review of contemporary ethnobotanical literature. On the one side, the continuous use of some medicinal taxa through the millennia can be explained by their therapeutic efficacy supporting the concept of social validation [6]. However, different explanations, beyond the efficacy of the remedy, have been invoked to account for the maintenance, erosion, or reshaping of ethnobotanical knowledge in the modern cultural contexts. Among the factors identified: the abundance and the accessibility of plants [7], the influence of the educational systems [8,9], the role of urbanization processes [10], the impact of climate change on the distribution of medicinal species [11,12], the effects of economic and government policies on the flow of information [13,14], the role of media in popularization of plant uses [15,16], and the exchange between different cultures [6].

In 2021, Milani and Fico published a book titled '*Raccolta di diversi rimedj a varj mali*' (Collection of various remedies to various pathologies), an in-depth ethnobotanical study of a pocket-size 18th century manuscript found in Valle Imagna (Bergamo, Lombardy, Italy) which contains approximately 200 plant-based complex recipes with therapeutic purposes collected and written down by an anonymous author [17] (Figure 1a,b). Some of these medicinal remedies had been retrieved from almanacs and medical textbooks and supplemented with personal notes from the author themselves. The manuscript was found in the private library of a 17th century house in Corna Imagna, one of the towns of the Valle Imagna (Bergamo, Italy), and was written in vulgar Italian with some references to Latin and to the vernacular tongue of Lombardy and the territory of Bergamo.

(**a**)

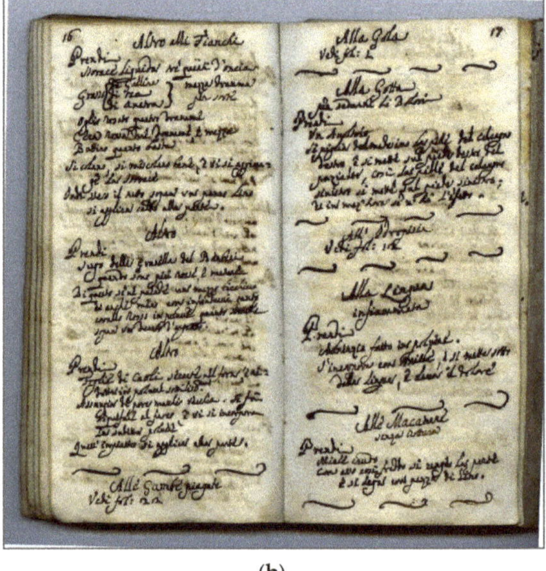

(**b**)

Figure 1. Pages taken from the 18th century manuscript: (**a**) remedies against oropharyngeal problems; (**b**) remedies against musculoskeletal and oropharyngeal disorders.

The remedies were considered useful to treat almost 80 different types of pathologies that were recognized by the official Western Medicine of that time (with the most treated pathologies being hemorrhoids, renal disorders, and ophthalmic problems) and were made of plant, mineral, animal, and human ingredients. Specifically, 205 plant species (21 of which are still unidentified) belonging to 70 botanical families were mentioned in the book, either used as they were or as derivatives. Today, most of these preparations can only be considered picturesque at best, because of the medical theories that the remedies were based on, the mix of ingredients used, the steps described for the preparations (which were often bound to religious rituals and superstitious beliefs), and the way of administering them to the patients. For example, in order to '*clear the eyes from the humors that cloud them*', the author suggests to pierce the earlobes and put a piece of root of *Helleborus niger* L.

in the holes as earrings; some of the preparations need to cook for '*as long as it is needed to run a Miserere*'; a preparation against kidney stones requires the collection of snails on a full moon night as the main ingredient [17]. However, bibliographic research in modern scientific literature showed that the use of some of the medicinal species involved and their active compounds could be potentially justified, to some extent, as treatment against some of the pathologies mentioned [17]. Admittedly, by the second half of the 19th century, improvements made in scientific research methodologies brought about the exclusion of ancient remedies, such as the ones described in the manuscript, from official Western Medicine standards. New pharmaceutical drugs (i.e., the first antibiotics, vaccines, semi-synthetic drugs, and others) contributed in no small part to this transformation [18]. However, in smaller and more isolated villages, such as the ones in Valle Imagna, this process was much slower and, even decades later, people frowned upon modern medicine, clinging to the 'old ways' instead, as reported by Dr. Giovanni Maconi MD in 2006 in his book '*La medicina popolare in Valle Imagna*' (Folk medicine in Valle Imagna) [19]. As a family doctor, born at the beginning of the 20th century and raised in Valle Imagna, Maconi collected from family and personal experience, as well as from professional encounters and from some of the elderly of the valley, information regarding magical, ritualistic, and empirical components of folk medicine practices since early 20th century Valle Imagna, thus depicting a picture of the traditional knowledge of the territory at that time.

The results obtained from the analyses of the manuscript prompted us to deepen our investigation on traditional uses of spontaneous medicinal species still surviving nowadays in Valle Imagna, and on whether part of this knowledge could be potentially traced back to centuries past. Thus, we conducted an ethnobotanical survey to investigate current knowledge of medicinal plants in the upper Valle Imagna. Then, we performed a thorough comparison between the plants and their traditional uses currently known and possibly practiced by the people of the valley and the ancient remedies included in the 18th century manuscript described in Milani and Fico (2021). The book by Maconi (2006) was considered too, in the attempt to sketch a potential *fil rouge* across the centuries, from the 18th century medicine to the 21st century folk medicine of Valle Imagna.

2. Results and Discussion

2.1. Current Ethnobotanical Survey

Between November 2021 and November 2022, a total of 109 informants were interviewed in the field, 93 of whom provided information concerning the medicinal use of spontaneous and cultivated plants of Valle Imagna. Out of the total 3849 citations, 669 referred to the Medicinal sector and involved 103 plant species, belonging to 46 botanical families. Four of these species were recalled only by their vernacular name and the descriptions provided by the informants during the interviews were not sufficient to allow for prompt identification. For this purpose, further investigation, both on the field and bibliographic, is currently taking place.

Seventy-eight percent of the citations involved species collected in the wild, 17% cultivated species, and 1% species that were collected both spontaneous and cultivated (i.e., *Malva sylvestris* L., which is commonly found spontaneous in Valle Imagna, but was also sometimes grown by the informants in their own gardens). Finally, 5% of the citations referred to species that could not be found throughout the territory, but that were usually kept in the house and became common ingredients for herbal remedies (i.e., *Syzygium aromaticum* (L.) Merr. & L.M.Perry dried flower buds and *Zingiber officinale* Roscoe rhizomes).

The most recurrent botanical families were Malvaceae (n. citations = 188; n. species = 3), Asteraceae (100; 13), Lamiaceae (78; 7), and Rosaceae (68; 12), while the most cited species were *Malva sylvestris* (n. informants = 67; number of citations = 135), *Tilia cordata* Mill. (32; 52), *Taraxacum* sect. *Taraxacum* (21; 26), *Hypericum perforatum* L. (16; 27), *Sambucus nigra* L. (12; 20), *Laurus nobilis* L. (12; 19), *Rosa canina* L. (11; 15), *Agrimonia eupatoria* L. (10; 27), *Rosmarinus officinalis* L. (9; 14), and *Citrus x limon* (L.) Osbeck. fil. (9; 12). It can be observed that Malvaceae was the most mentioned family only due to the high number of citations concerning *M. sylvestris* and, to a lesser extent, *T. cordata*, while the citations for the others were distributed among a higher number of different plant species. It is interesting to highlight that during the preliminary bibliographic research conducted on the territory and the local flora [20–23], Asteraceae, Rosaceae, and Lamiaceae resulted also among the families with the higher number of spontaneous species distributed in the area with 27, 19, and 13 species, respectively. Ranunculaceae, with 19 species catalogued, was also the second most widespread family in Valle Imagna, although its species were rarely mentioned by the informants (only 5 out of the total citations), supposedly because of their common toxicity.

Concerning the 103 plant species mentioned by the informants, they were prepared mainly as infusions (n. citations = 340), used raw (i.e., externally applied as they were; n = 144), macerated in oil (n = 62), or boiled in decoctions (n = 48). Leaves (n. citations = 291; n. of species = 46) and flowers/inflorescences/flowered aerial parts (185; 26) were the most recurrent plant parts. Mostly, the reported uses were personally experienced (84.0% of citations) and still currently of concern (94.0%).

The species were distributed among different categories of pathologies, according to their use (see Figure 2).

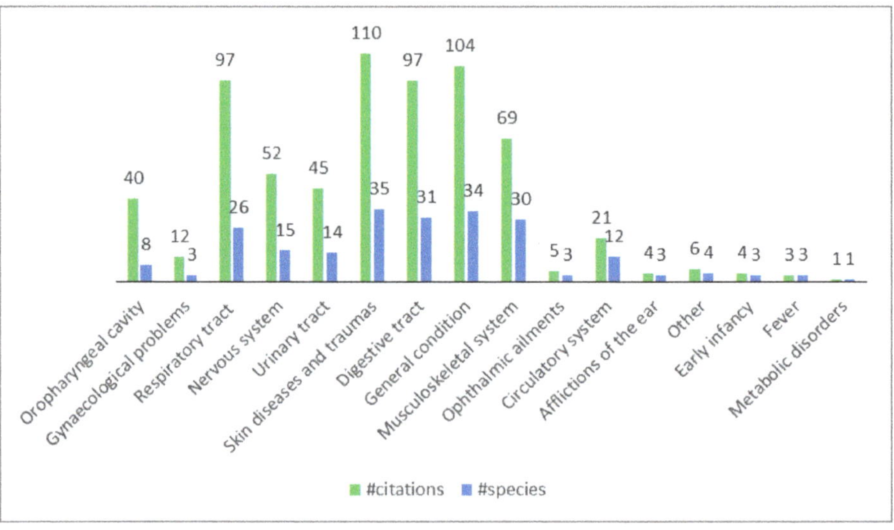

Figure 2. Categories of use (pathologies treated) in Valle Imagna. According to the number of citations (#citations) and number of species (#species).

Among the 35 species used for the treatment of skin problems, flowers and aerial parts of *Hypericum perforatum* (n. citations = 21) were macerated for at least 3 weeks in vegetable oil. This oleolite was then applied on skin as an anti-inflammatory, wound healing, and soothing agent for wounds, sunburn, and burns. Compresses prepared with the leaves of *Malva sylvestris* (n = 11) were used mainly as anti-inflammatory and soothing for skin irritations. Leaves of *Agrimonia eupatoria* (n = 10) were applied fresh externally or as compresses of the infusion, mainly as a wound healing agent for wounds that were, as some of the informants reported, 'hard to treat otherwise'.

In the category General condition, the infusion of leaves of *Malva sylvestris* (n = 38) was drunk for its general anti-inflammatory properties on the whole body, while the infusion of rosehips (*Rosa canina*, n = 8) was considered rich in vitamin C and useful as tonic and for the prevention of colds. The infusion of leaves and flowers of *Salvia officinalis* L. (n = 8) was drunk or added to the bath water as a tonic agent.

For the treatment of digestive problems, the infusion of leaves or flowers, or sometimes the whole above ground parts of *M. sylvestris* (n = 20) was drunk as laxative, or to ease abdominal pain. The leaves, rarely the flowerheads or the underground parts, of *Taraxacum* spp. (n = 16) were drunk in infusion or eaten mainly as liver depurative. The juice of *Citrus x limon* fruits (n = 7) was added to an infusion of *Thymus* spp., *Lavandula angustifolia* Mill., and *Zingiber officinale* to improve its digestive and antispastic properties. The juice was once added to *Ricinus communis* L. seeds oil and sugar as a laxative preparation. Finally, the infusion of leaves of *Artemisia absinthium* L. (n = 5) was taken for its digestive and carminative activities.

As for the category of respiratory tract infections, the most cited species was *Tilia cordata* (n = 30): its flowers were commonly collected and mainly used in infusions once dried for the treatment of cough, colds, and sore throat. Syrups of flowers (sometimes fruits) or infusions of flowers of *Sambucus nigra* (n = 13) were administered against cough for their expectorant properties, while a syrup of pinecones or fresh tops of *Pinus mugo* Turra (n = 8) was prepared for cough and sore throat. It is important to note that this last species can rarely be found throughout the territory of Valle Imagna and grows mainly at the higher altitudes of Mount Resegone (the highest mountain of the valley). For this reason, the informants who mentioned this use often collected the herbal ingredients for this preparation in the neighboring valleys, such as Val Taleggio and Val Brembana.

Among the 30 species used for musculoskeletal pains and inflammations, the most cited one was *Arnica montana* L. (n = 9). Its flowerheads were macerated in oil or in alcohol and the preparation used as it was or mixed with fat (i.e., bee wax) and rubbed externally on contusions or aching joints or muscles. For the same purpose, fresh leaves of *Brassica oleracea* L. (n = 7) were applied on the affected area. Moreover, the macerated oil of flowered aerial parts of *Achillea millefolium* L. (n = 5) was massaged (sometimes after mixing it with bee wax to obtain an ointment) on joints or on contusions and bruises.

The infusions of flowers of *T. cordata* (n = 11), flowered aerial parts of *L. angustifolia* (n = 8), and leaves of *Melissa officinalis* L. (n = 7) were the most mentioned preparations for the treatment of nervous system disorders, specifically as sedative remedies to facilitate and improve sleep. Additionally, the oleolite of lavender was massaged on the temples for the same purpose.

The most used species for the urinary tract was without doubt *M. sylvestris* (n = 21). The infusion of its leaves was drunk or used as vaginal douche for diuretic or anti-inflammatory purposes, specifically for cystitis and irritations. Another remedy against cystitis was the decoction of the whole above ground parts of *A. eupatoria* (n = 5), also drunk (in this case with the addition of cherry seeds) or used externally. The informants who reported this specific use of agrimony emphasized the fact that this remedy was once recommended by the midwives of the valley.

For the treatment of gingivitis and other inflammations of the oropharyngeal tract, the most used remedy was prepared once again with leaves (sometimes also flowers or underground parts) of *M. sylvestris* (n = 30). Its infusion was commonly used as anti-inflammatory mouthwashes. Less common was the similar use of the whole above ground parts of *A. eupatoria* (n = 2; decoction), the leaves of *Blitum bonus-henricus* (L.) Rchb. (n = 2; infusion), and the leaves of *Salvia officinalis* (n = 2; infusion or leaf applied on gums and teeth).

Concerning the circulatory system, leaves infusions of *Laurus nobilis* (n = 5) were considered useful mainly as a hypotensive, while the infusion of *Crataegus monogyna* Jacq. (n = 3) was drunk as a hypotensive as well, but also as an antiarrhythmic. Finally, decoctions of *Taraxacum* (n = 3) leaves or sometimes the leaves and underground parts eaten raw were considered a powerful blood depurative.

Figure 3a–d shows some of the preparations observed and collected during the fieldwork. For further information concerning the uses, please see Table S1 of the Supplementary Materials.

Figure 3. Some herbal preparations observed during the field work: (**a**) dried herbal teas of some of the most cited species, such as *M. sylvestris* and *T. cordata*; (**b**) decoction of aerial parts of *A. eupatoria*; (**c**) ointment obtained mixing extra virgin olive oil and bee wax; (**d**) decoction of rosehips (*R. canina*).

Finally, extensive bibliographic research in scientific literature was conducted on the 99 identified medicinal species mentioned by the informants in order to validate or refute their traditional uses. This bibliographic research highlighted that scientific evidence could be found for at least one of the uses concerning 62 taxa and detected during the field work, though it is important to note that the part of the plant and preparations analysed in literature rarely matched the traditional ones. Moreover, in vitro and in vivo studies could be more easily found, while clinical trials performed on human subjects were definitely lacking. Phytochemical investigations related to the potential biological activities were found only for 42 out of the 62 aforementioned taxa. For some of these, for example *Achillea millefolium*, *Calendula officinalis* L., *Malva sylvestris*, and *Sambucus nigra*, the literature concerning the active compounds involved in their activity and their potential mechanisms of action was relatively extensive, while in most cases the information was limited. The complete results of this research can be found in Table S1 of the Supplementary Materials.

Additionally, the analysis of knowledge distribution according to gender found that female and male informants presented different knowledge concerning medicinal plants, with females reporting a significant higher number of species (Mann–Whitney Test: $U = 649$; $Z = -3.07$; $p < 0.01$) and uses (Mann–Whitney Test: $U = 703$; $Z = -2.64$; $p < 0.01$) (Table 1). This finding was expected since women in Alpine areas represented the primary healthcare providers for the family until the recent past [8].

Table 1. Mean number (±sd) of species and uses according to the gender and age classes of informants.

Gender	#Informants	Species		Citations	
		Mean	±(sd)	Mean	±(sd)
Male	37	3.57	4.65	5.51	8.21
Female	56	5.25	4.67	8.44	9.29
Age					
20–29	5	4.30	7.30	9.37	5.90
30–39	6	4.66	5.31	6.16	7.65
40–49	10	5.30	3.79	8.10	7.40
50–59	16	5.69	7.31	9.37	10.36
60–69	13	4.46	2.29	8.46	6.60
70–79	32	4.40	4.99	6.62	8.70
80–89	9	3.22	2.43	5.22	6.02
90–96	2	3.00	2.83	4.00	2.83
All informants	93	4.58	4.70	7.28	8.95

The age of informants ranged from 20 to 96 years with 70–79 years being the most frequent age class (32 informants) (Table 1). No significant correlation was observed either with the mean number of reported species (Spearman Test: R = −0.14) and the mean number of reported uses (Spearman Test: R = −0.11). Informants aged from 30 to 59 seem to be the most knowledgeable ones, with a mean number of mentioned species and uses of 5.18 (±0.51) and 7.88 (±1.63), respectively. At the same time, the 11 informants aged ≥ 80 presented the lower rates of knowledge (mean number of species: 3.11 ± 2.68; mean number of uses: 4.61 ± 4.42).

The same pattern was observed by Bruschi et al. (2019) [8] in an Alpine area close to Valle Imagna. We can interpret this pattern as a result of the cultural interest in medicinal plants by younger generations within the green wave and healthy lifestyle system. Part of the knowledge gathered from the younger informants came from several different sources other than the memories from their parents and grandparents. Some of them had recently rediscovered the wild plant species of their own territory through broad consultation of books, TV programs, and the Internet (i.e., a poultice of leaves *Delphinium consolida* L. for the treatment of fractures, or the macerated oil of *Hedera helix* L. leaves used to improve legs circulation against cellulitis). Others combined family reminiscences with new understanding gained through schools and university courses, upon deciding to return to their home territory with innovative agropastoral activities. An explanation of the low rate of knowledge owned by the older informants could be found in the history that distinguished the XIX century Valle Imagna. During the post World War II era, this valley was in fact characterized by sheer poverty, while its inhabitants were barely surviving through their rural lives with what the territory would offer [24]. From then on, many of them started looking for a better life by moving to neighbouring countries, mainly Switzerland and France, but also Germany; most of them never to return. Some of the people that came back years, sometimes even decades later, had lost most of their memories regarding that past farm life, thus leaving behind a great part of their knowledge on the traditional uses of the spontaneous plants of Valle Imagna. We personally interviewed some of these older people and we could determine first-hand that while they hardly recalled them, the uses that they were able to mention were also the ones most rooted to the territory and to the ancient traditions of the valley. As a way of example, we cite an ointment of *Allium sativum* L. underground organs and pork fat against contusions, the use of leaves of *Agrimonia eupatoria* externally applied fresh to treat deep wounds, or the aerial parts of agrimony drunk in anti-inflammatory infusions for the digestive and urinary tract. Additionally, it is interesting to note that these occurrences may have partly thwarted the intergenerational knowledge transfer of these traditional uses in Valle Imagna.

2.2. Diachronic Analysis

Before focusing on the plant species, it is important to note that some of the ancient non-herbal remedies described in the booklet and the ones from 20th centuries referred by Maconi are interestingly similar. Prolapsed hemorrhoids, for example, were usually treated with herbal or non-herbal ointments in order to slide them back and were then kept in position with a balled up linen cloth [17]. Maconi reports that until the early decades of the 20th century, children often suffered from rectal prolapse, which was treated with olive oil (with the same purpose of the ancient ointments) and that then the children were forced to sit naked on a cold stone, to keep the rectum in place [19]. Additionally, some of the ritualistic and religious elements that used to characterize the remedies from 18th century could also be found in the descriptions given by Maconi and from some of the stories recounted by our informants, especially the older ones. Religious formulas, such as fragments of prayers or the sign of the cross while taking or applying the remedies, were in some cases considered an essential part of them and were thought to enhance their power.

Figure 4 shows the comparison between the historical and current taxa and uses. A strong erosion in plant-related knowledge can be observed over time. Eighteenth century manuscript reported the highest number of taxa (184) and uses (358). These numbers are lower in Maconi's book (125; 313) and in fieldwork (99; 224). The highest overlap was between the manuscript and the Maconi book for taxa (JI = 0.67) and between the current investigation and the Maconi book for uses (JI = 0.38) while the lowest one was between the manuscript and the current investigation (JI = 0.48 for taxa; JI = 0.16 for uses). Only 42 taxa (23% of the taxa reported in the manuscript) and 34 uses (9.5% of the uses reported in the manuscript) are currently known by the people living in the valley; in the 20th century investigation by Maconi, taxa and uses in common with those reported in the manuscript were 66 (35%) and 57 (16%), respectively.

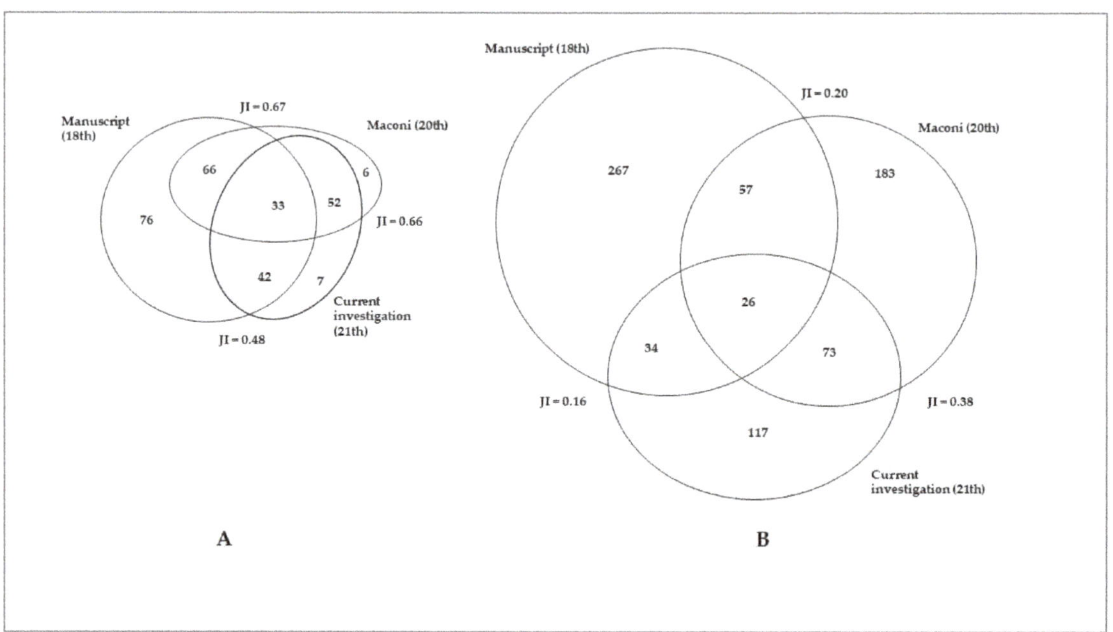

Figure 4. Venn diagrams comparing taxa (**A**) and uses (**B**) in the three data sources. JI = corrected Jaccard Index.

In order to explain the observed differences between the manuscript and the other considered sources, it has to be highlighted that the manuscript includes remedies taken from other sources, even from earlier times to 18th century, not always connected with the territory of Valle Imagna. For example, among these "external" sources identified by Milani and Fico (2021), *"Prospectus Phamaceuticus sub quo Antidotario Mediolanense"*, published in 1668 and updated until 1729; *"Pratica universale nella Medicina"*, dated 1693 and written by friar Felice Passera from Bergamo; and *"De Secreti del Reverendo Donno Alessio Piemontese"*, dated 1557 and written by Girolamo Ruscelli. Even the high number of exotic species reported in the manuscript (54, 14 of which neophytes) compared to Maconi (29; 8) and the fieldwork (17; 7) can further explain the observed differences. Since the 16th to 18th century, Bergamo was part of the *"Repubblica Serenissima di Venezia"*, one of the most important centers of trade in Europe at that time. For this reason, doctors and apothecaries could obtain a wide variety of spices and exotic ingredients.

The comparison between Maconi's book and the results of our fieldwork show that in a few decades, information on 73 taxa (58.4%) and on 240 uses (77%) was lost. A rapid loss of ethnomedicinal knowledge has been detected in many industrialized countries [8,25,26] and has been explained as a consequence of the cultural erosion caused by the influence of modern culture and education systems, the globalization of trade, and access to modern medicine [25].

Taking into consideration the categories of use, skin and digestive tract diseases are the ones with the highest number of taxa with a homogeneous distribution in the three datasets (Figure 5).

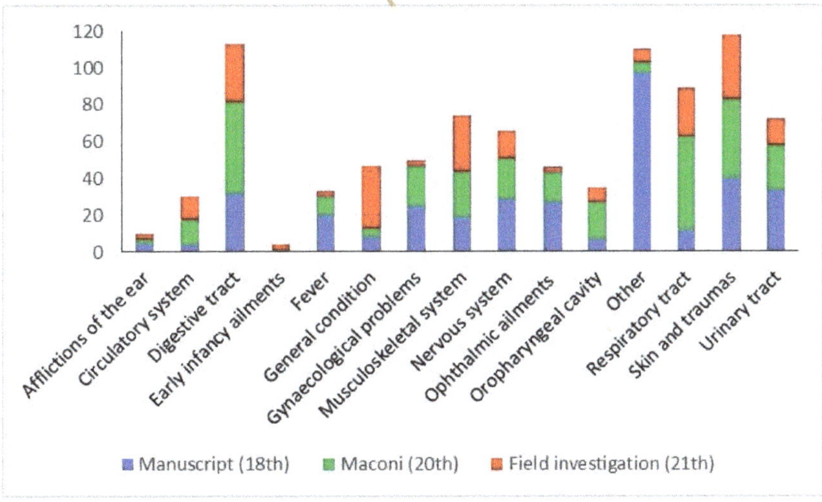

Figure 5. Distribution of the use categories in the three data sources.

Regarding the manuscript, "others" is the most represented category with 97 recorded taxa; it mostly includes diseases or symptoms which are not easily interpretable and classifiable in accordance with Western nosologies (i.e., *'ponta'*, *'brossole'*, *'tarlo'*, which cannot be translated). Taxa reported in all the three data sources were found in the following categories: skin diseases and traumas (10 taxa: *Chelidonium majus* L., *Citrus* x *limon*, *Malva sylvestris*, *Matricaria chamomilla* L., *Olea europaea* L., *Plantago major* L., *Plantago* ssp., *Salvia officinalis*, *Sambucus nigra*, *Urtica* spp.), respiratory tract diseases (5: *Linum usitatissimum* L., *M. sylvestris*, *Plantago* spp., *Rosa canina*, *Rosmarinus officinalis*), digestive tract diseases (2: *Artemisia absinthium*, *R. officinalis*), circulatory system disorders (1: *Urtica* spp.), urinary tract diseases (1: *M. syslvestris*), nervous system diseases (1: *M. chamomilla*), gynaecological disorders, obstetric, and puerperal problems (1: *M. sylvestris*), general condition (1: *S. officinalis*). A constant decrease in the number of the recorded taxa over time can be observed in the categories fever (20:10:3), gynaecological disorders, obstetric, and puerperal problems (25:22:3), nervous system diseases (29:22:15), ophthalmic ailments (27:16:3), and urinary tract diseases (34:24:14). Skin diseases (40:43:35) and digestive tract diseases (30:50:31) categories showed a more heterogeneous pattern but always with a reduction in the more recent years. Similar findings were observed by Söukand et al. (2022) [27] in Estonia and by Guarrera in Italy (2006) [28]. On the contrary, an increase can be detected for musculoskeletal disorders (19:25:30), circulatory system disorders (4:14:12), and general condition (8:5:34). As pointed out by Dal Cero et al. (2023) [6], the increasing importance of plants for treating circulatory system disorders and plants used as preventive measures, like plants administrated as blood and organs purifier or appetite stimulant/tonic, reflect the progress of medical knowledge and also epidemiological changes occurring in modern society [27].

Figure 6 shows the uses categories of the 10 most cited species during the interviews compared to those reported in the two historical sources. Only in 6% of the cases (*M. sylvestris* used to treat skin problems and urinary tract diseases; *T. cordata* used to treat respiratory tract diseases; *S. nigra* used to treat skin problems; *R. canina* used to treat respiratory tract diseases; *C. x limon* used to treat skin problems; *R. officinalis* used to treat respiratory tract diseases), the use was the same in the three sources. Fourteen percent of the uses were reported only in Maconi and in the fieldwork. Twenty-seven percent were new (i.e., were cited by the informants during the interviews but they were not reported in the two historical sources). In particular, new uses were recorded in the category general condition, a finding further confirming the increasing importance of preventive plants in the current herbal medicine. Although the use of *Taraxacum* as remedy to treat digestive, urinary, respiratory diseases, and as a blood purifier has been known since ancient times [29,30], no records were found in the two historical sources analysed in this study. This finding can partly be explained by possible identification problems of species included in the 18th century manuscript; on the other hand, no information about medicinal uses of Taraxacum spp. over the last 200 years were reported in Dal Cero et al. (2023) [6] for Central Europe and Dal Cero et al. (2014) [31] for Switzerland. In general, these data are consistent with what was reported by other diachronic studies. For example, when comparing the uses of *M. sylvestris*, *H. perforatum*, *S. nigra*, *R. canina*, and *A. eupatoria* to those reported in Dal Cero et al. (2023) for the same species, we found a Jaccard similarity index of 0.72.

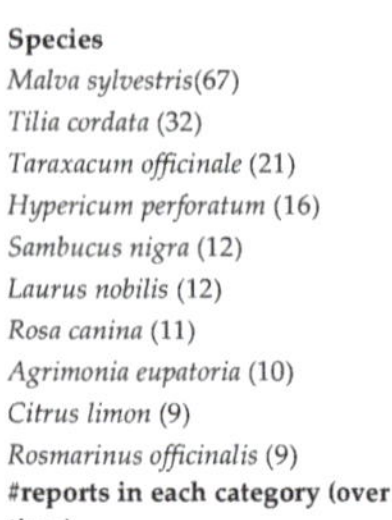
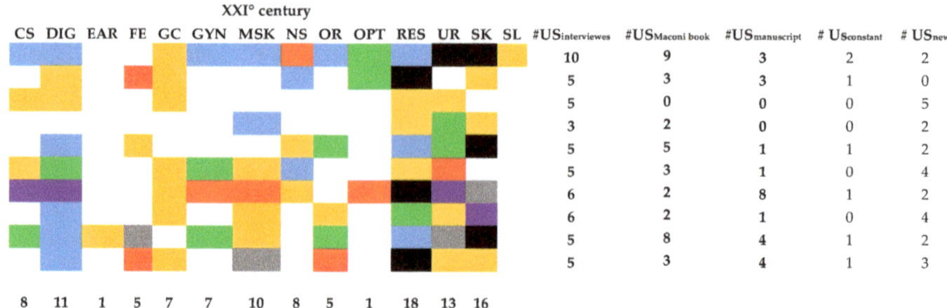

Species	CS	DIG	EAR	FE	GC	GYN	MSK	NS	OR	OPT	RES	UR	SK	SL	#US$_{interviewes}$	#US$_{Maconi\ book}$	#US$_{manuscript}$	# US$_{constant}$	# US$_{new}$
Malva sylvestris(67)															10	9	3	2	2
Tilia cordata (32)															5	3	3	1	0
Taraxacum officinale (21)															5	0	0	0	5
Hypericum perforatum (16)															3	2	0	0	2
Sambucus nigra (12)															5	5	1	1	2
Laurus nobilis (12)															5	3	1	0	4
Rosa canina (11)															6	2	8	1	2
Agrimonia eupatoria (10)															6	2	1	0	4
Citrus limon (9)															5	8	4	1	2
Rosmarinus officinalis (9)															5	3	4	1	3
#reports in each category (over time)	8	11	1	5	7	7	10	8	5	1	18	13	16						

XXI° century

Figure 6. Categories of use for the 10 most cited species in the field investigation: comparison between the three data sources. CS = Circulatory system disorders; DIG = Digestive tract disorders; EAR = Afflictions of the ear; FE = Fever; GC = General condition; GYN = Gynaecological disorders, obstetric, and puerperal problems; MSK = Musculoskeletal system disorders and traumas; NS = Nervous system disorders; OR = Oropharyngeal cavity affections; OPT = Ophthalmic ailments; RES = Respiratory tract infections; UR = Urinary tract disorders; SK = Skin diseases and traumas; SL = Slimming. In brackets: number of informants citing the species. Black: the use has been cited during the interviews and is reported in the two books; orange: the use has been cited during the interviews but is not present in the two books; blue: the use has been cited during the interviews and is reported in Maconi; violet: the use has been cited during the interviews and is reported in the manuscript; green: the use is reported only in Maconi; red: the use is reported only in the manuscript; grey: the use is reported both in Maconi and the manuscript. #UScostant = number of uses surviving through the time; #USnew = number of uses recently appeared.

2.3. Comparison among Similar Preparations in 18th and 21st Century in Valle Imagna

Out of the 42 species shared between the manuscript and our field investigation, 12 were used in remedies that were at least comparable, sometimes even almost identical. All information regarding the comparison can be found in Table 2.

Table 2. Comparison among similar preparations found in the manuscript [17] and during the field work in Valle Imagna.

Species	Use Described in the Manuscript [17]	Use Described in Valle Imagna	Activity
Amaryllidaceae			
Allium sativum L. Garlic	Wine decoction with garlic, drunk against hip pain. Eaten raw against gout.	Ointment made of smashed garlic and pork fat to be externally applied against contusions and pain.	Anti-inflammatory for the treatment of musculoskeletal problems.
Asteraceae			
Artemisia absinthium L. Absinth	Pills or aqueous preparation with absinth salts (obtained from the ashes of *A. absinthium*) as diuretic and antipyretic. Powdered absinth mixed with honey. Kept inside the mouth as an anti-inflammatory for the tongue.	Infusion of the leaves or the entire above ground part drunk as an antipyretic, anti-inflammatory, and diuretic.	Anti-inflammatory, antipyretic, and diuretic.
Matricaria chamomilla L. Chamomile	Chamomile oil mixed with other ingredients, clysters against hip pain and sciatica.	Macerated chamomile oil, applied externally against muscular pain and inflammation.	Anti-inflammatory for the treatment of musculoskeletal problems.
Brassicaceae			
Brassica oleracea L. Cabbage	Cabbage leaves dried in the oven, powdered, and mixed with pork fat. The ointment is applied against hip pain.	Fresh leaves, smashed and applied on contusions and joint pain and inflammations.	Anti-inflammatory for the treatment of musculoskeletal problems.
Lamiaceae			
Salvia officinalis L. Sage	Sage leaves, rosemary, and pomegranate boiled in wine. Mouthwashes against painful and loosening teeth.	Sage leaves rubbed on teeth and gums as an anti-inflammatory. Sage infusion used as an anti-inflammatory and disinfectant mouthwash for teeth and gums.	Antibacterial Anti-inflammatory

Table 2. *Cont.*

Species	Use Described in the Manuscript [17]	Use Described in Valle Imagna	Activity
Linaceae			
Linum usitatissimum L. Flax	Flax and fenugreek seeds boiled in water. The seeds are squeezed, and the mucilaginous water is mixed with butter. The ointment is applied on the chest against children cough.	Flax seeds boiled in water until a preparation similar to porridge is obtained. The mucilaginous poultice is applied warm on the chest against cough.	Antitussive Expectorant
Oleaceae			
Olea europaea L. Olive	An ointment of olive oil and bee wax or of olive oil and tallow applied on fissured hands and feet and on burns.	An ointment of olive oil and bee wax applied on burns and as soothing agent on inflamed skin.	Anti-inflammatory Soothing Wound healing
Papaveraceae			
Chelidonium majus L. Greater celandine	Latex of celandine with latex of parsnip applied on warts.	Latex of celandine applied on warts.	Antiviral Caustic
Plantaginaceae			
Plantago spp. *Plantago major* L. *Plantago lanceolata* L. Plantain	Poultice of plantain mixed with butter (and other ingredients) for two different ointments applied on inflamed nipples and on wounds.	Leaves applied externally on burns and as a wound healing and anti-inflammatory agent.	Anti-inflammatory Disinfectant Wound healing
Rosaceae			
Agrimonia eupatoria L. Agrimony	Complex remedy applied on deep leg ulcers. Above ground parts of agrimony and dried roses are boiled in wine, which is then used to disinfect the ulcers. A mix of powdered herbal and mineral ingredients are then applied on the wounds. Finally, the sediment of cooked agrimony and rose is smeared over the powder, and all is set in place with a gauze.	Fresh leaves are applied externally on deep wounds and set in place with a gauze. Compresses of the infusion of leaves or of above ground parts of agrimony are applied on deep wounds.	Anti-inflammatory Disinfectant Wound healing
Rosa canina L. Dog rose	Rose water, obtained from the petals, mixed with other ingredients and drunk in order to 'refresh the kidneys'.	Infusions of the false fruits drunk as diuretic.	Diuretic
Vitaceae			
Vitis vinifera L. *Vitis* spp. Grapevine	Mouth washes with wine in which sage, rosemary, and pomegranate were boiled.	Mouth washes with grappa.	Anti-inflammatory Disinfectant (Alcohol?)

More specifically, for 5 species (*Chelidonium majus*, *Linum usitatissimum*, *Plantago* spp., *Salvia officinalis*, and *Vitis vinifera*) the 18th century remedies and the traditional uses in Valle Imagna were similar not only in terms of part of the plant used, but also of methods of preparation and administration, and the same uses were reported by Maconi in his book.

For *V. vinifera*, byproducts of the species were used, namely wine or grappa. Wine was one of the mediums in 18th century in which plants were boiled to obtain disinfectant mouthwashes in case of gingivitis, while nowadays in Valle Imagna another byproduct of *V. vinifera*, grappa, could be used pure for the same purpose. Admittedly, in this specific case, the disinfectant action is probably sought after in the alcoholic nature of the preparations, as well as in the plant species infused. As a matter of fact, the author described a preparation of leaves of sage and rosemary boiled in wine with pomegranate used as mouthwashes against painful and loosening teeth. Our informants reported the use of sage leaves rubbed directly on teeth and gums as an anti-inflammatory remedy and of a sage infusion used as anti-inflammatory and disinfectant mouthwashes for teeth and gums. It is interesting to highlight that Maconi recounted in his book that, until the first half of 20th century, mouthwashes obtained by boiling sage in wine were a typical treatment for gum and tooth ache in the valley [19].

A similar consistency with the data reported by Maconi could also be found in the cases of *C. majus*, *L. usitatissimum*, and *Plantago* spp. As a matter of fact, the orange latex obtained by snapping the stems of celandine was mentioned as a useful treatment for warts in all three of the consulted sources. In a similar fashion, a warm poultice of boiled flaxseed was externally applied on the chest to treat productive cough and was reported by both Maconi and our informants with the same vernacular name (*'linusa'*), while in 18th century the mucilaginous water obtained by boiling the flaxseeds was added to fresh butter and then applied on the chest, specifically in children. As for *Plantago* spp., our informants, in accordance with what was written by Maconi, reported the use of leaves of plantain applied externally on burns and as a wound healing and anti-inflammatory remedy. For similar purposes, the author of the manuscript described the preparation of two different ointments, both obtained by mixing a poultice of plantain leaves with butter and other ingredients.

At last, an interesting case is represented by *Agrimonia eupatoria*. In the manuscript, the author referred to a complex remedy to treat leg ulcers. Specifically, they wrote: "*Alle Piaghe: D'ogni sorte nelle Gambe, sebbene la Gamba fosse scoperta e mangiata fino all'osso*" (To [treat] any type of leg ulcers, even if the leg was open raw and consumed to the bone) [17]. This remedy was prepared by boiling above ground parts of agrimony and dried roses in wine, which was then dabbed on the leg to disinfect the ulcers. A mix of powdered herbal and mineral ingredients was then applied on the wounds. Finally, the sediment of boiled agrimony and rose was smeared over the powder, and all was set in place with a gauze. While recounting the use of agrimony in Valle Imagna, our informants spoke about the external application of fresh leaves of the plant on deep wounds, which were considered 'untreatable with any other method'. The leaf was then set in place with a gauze that was periodically changed. Moreover, they also described the use of compresses of the leaves or infusion of above ground parts, for the treatment of deep wounds as well. It is interesting to note that the use of agrimony was mentioned almost exclusively by elderly people (75 years or older) and that the plant species was only referred to by its vernacular name, which is '*Erba del Vinil*', as not even the Italian common name was known to them. Furthermore, the primary data of our field investigation revealed that *A. eupatoria* was used in this fashion mainly until the 1960s–1970s, coinciding with the great migration of people from rural areas to the big cities and foreign countries. Finally, extensive bibliographic research conducted on ethnobotanical published works on neighboring areas has shown that in Lombardy, *A. eupatoria* was either used to treat different pathologies (Vitalini et al., 2009–infusion drunk for laryngitis [32]; Vitalini et al., 2015–compresses of the infusion for mild dermatitis and infusion drunk as an astringent agent, [33]), or not mentioned at all [8,34–37].

3. Materials and Methods

3.1. Area of Investigation

Valle Imagna is located in the western area of Orobic Prealps, in the province of Bergamo, Lombardy, Northern Italy (Figure 7). The highest peak of the Valley is mount Resegone, with its 1.875 m a.s.l. Although some of its municipalities do not exceed 500 m a.s.l., the whole territory is labelled as mountainous. The main river of the area, called Imagna, gives its name to the Valley. The mild weather, with its abundant rainfalls and limited temperature excursion, allows for the presence of rich plant biodiversity. The woodlands are characterized almost exclusively by deciduous trees. Specifically, at the higher altitudes, beech woods are so extensive that several toponyms derive from the beech Italian common name, *'faggio'* and the vernacular name *'fó'*. Concerning herbaceous plant species, among the most abundant we cite *Achillea millefolium* L., *Agrimonia eupatoria* L., *Arctium lappa* L., *Artemisia* L. spp, *Carum carvi* L., *Chelidonium majus* L., *Cichorium intybus* L., *Equisetum arvense* L., *Polypodium vulgare* L., *Filipendula ulmaria* (L.) Maxim., *Foeniculum vulgare* Mill., *Gentiana* L. spp, *Hedera helix* L., *Hypericum perforatum* L., *Malva sylvestris* L., *Ruta graveolens* L., *Salvia* L. spp, *Urtica dioica* L., and *Vaccinium myrtillus* L. [19]

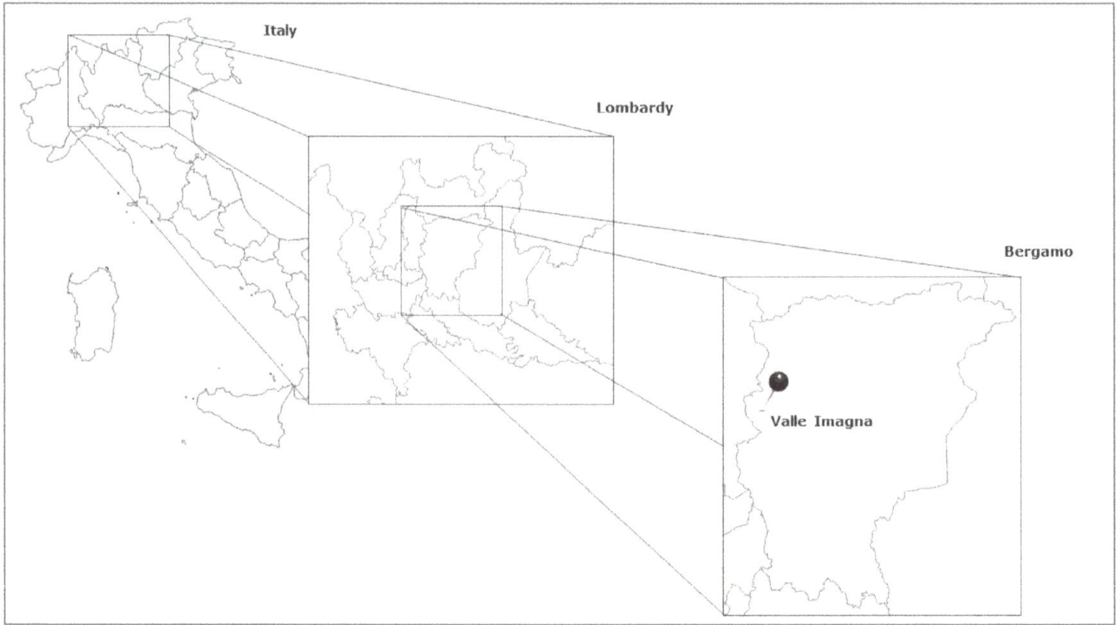

Figure 7. Valle Imagna is located in the province of Bergamo, Lombardy region, Northern Italy. Map modified from the original free maps at https://d-maps.com/carte.php?num_car=284951&lang=it (accessed on 14 July 2023), https://d-maps.com/carte.php?num_car=22383&lang=it (accessed on 14 July 2023), and https://d-maps.com/carte.php?num_car=213806&lang=it, (accessed on 14 July 2023).

For our investigation, we focused on upper Valle Imagna, specifically on the municipalities of Berbenno (675 m a.s.l.), Brumano (911 m a.s.l.), Corna Imagna (736 m a.s.l.), Costa Valle Imagna (1.014 m a.s.l.), Fuipiano Valle Imagna (1.055 m a.s.l.), Rota d'Imagna (690 m a.s.l.), and Sant'Omobono Terme (427 m a.s.l.).

3.2. Ethnobotanical Survey, Data Archiving and Processing

In preparation of the field work, preliminary bibliographic research on the territory and the spontaneous flora of Valle Imagna was performed through the consultation of local botany textbooks [20–23]. A list of autochthonous species of the Valley was produced. The scientific names and family categorization follow Pignatti et al., 2017 [38]. The list was then enriched by a photographic archive of the plants, in order to facilitate plant identification by the informants during the field work.

Subsequently, open- and semi-structured interviews were conducted. All the information gathered during the interviews was archived in Microsoft Word™ files (Microsoft, Redmond, WA, USA): an 'Informant Sheet' for data concerning the informants and a 'Species Sheet' for data concerning the plant species cited and their traditional uses throughout the Valley. Specifically, the "Species Sheet" consisted of a 7-column table arranged as follows: Species (common and vernacular name), Field of use, Detailed use, Preparation form, Administration form, Part of the plant, Other information. Each "Informant Sheet" was then matched to the corresponding "Species Sheet" through a one-to-one identification code. Identification of the species was performed by Professor Gelsomina Fico, Professor Claudia Giuliani, and Dr. Paola Sira Colombo following Pignatti et al., 2017 [38]. Plant species nomenclature was according to Pignatti et al., 2017 for the ones found on Italian territory and to http://www.worldfloraonline.org/ (accessed on 14 July 2023) [39] for the other ones.

3.3. Data Analysis

All data was filed away in a database, organized in an Excel™ spreadsheet (Microsoft, Redmond, WA, USA) where each row represents a 'citation', defined as 'a single use reported for a single species by a single informant'. Every citation was considered as 'distinct', when differing from one another in at least one of the following fields: species, informant id code, category of use, part of the plant, preparation, and administration form. Data was processed and analysed by means of Pivot tables.

Spearman's correlation test was carried out to detect a relationship between the age of informants and their plant-related knowledge. Mann–Whitney was performed to check for statistical significance between sex of informants and their ethnobotanical knowledge. To measure the similarity of the different data sources in terms of reported taxa and uses, we performed the Jaccard Similarity Index (corrected for maximum possible values, as reported in Pitman et al. (2005), in order to account for artefacts due to the difference in diversity between the considered data sources) [40].

For the comparisons among the different sources, we defined 'use' as 'category of use', namely the affected organ groups, such as 'digestive tract disorders', 'urinary tract disorders', etc. The followed criterion for the classification of the different categories of use was based on previous published works [34,35,41–43].

3.4. Bibliographic Research

For the comparison between the 18th and 20th century remedies and the 21st century ones, we consulted Milani and Fico, 2021 [17] and Maconi, 2006 [19], respectively. Comparing information in diachronic studies is complicated by the strong heterogeneity existing among the different sources [16]. Different authors employ different methodological approaches to record, systemize and present data, apply different categories and terminologies to describe different plant uses, often without any annotation, and use different plant nomenclatures. We used a rigorous approach to achieve data harmonization, punctually identifying, removing, and aligning all the study differences; this could have led to a loss of information but also provided a comparable view of data from different sources and revealed valid inferences from the analysis of pooled data.

Extensive bibliographic research in the ethnobotanical, phytochemical, and concerning the biological activities of the 99 identified species mentioned during the interviews was carried out through search engines and online databases, such as PubMed, MEDLINE, Google Scholar, and the bibliographic research online tool known as J.A.N.E. Concerning the ethnobotanical research, the scientific or English common name of the species was combined with the keywords 'ethnobotany', 'ethnopharmacology', 'traditional medicine', or 'folk medicine'. Special attention was paid to Italian ethnobotanical works published and, specifically, the ones concerning the neighboring areas to Valle Imagna. As for phytochemistry and biological activities, we paired the scientific or English common name with specific keywords concerning the category of use cited in Valle Imagna (i.e., *Artemisia absinthium* AND digestive system, *Agrimonia eupatoria* AND skin) and then the specific pathology or activity (i.e., *Artemisia absinthium* AND carminative, *Agrimonia eupatoria* AND wound healing, etc.). We focused our attention particularly on systematic reviews and meta-analyses, if possible, or on single in vitro, in vivo, and clinical trials studies, and we applied no time filters. Table S1 of Supplementary Materials was then produced [32–37,44–245].

4. Conclusions

Our previous analysis of a little 18th century manuscript found in Valle Imagna was the starting point of the case study reported in this paper. The library where the manuscript was retrieved was part of a private 17th century house, stocked with historical documents related to the valley, and the presence in the text of vernacular words from the province of Bergamo could be indication of the provenance of the author. As for the purpose of such a piece, the personal and professional notes that the anonymous author added to various remedies would suggest that the work was considered more than a mere copy from other sources. Even the size of the booklet hinted at its practicality to be carried around and produced when a consultation was needed, maybe even by a medical practitioner of the territory.

The first comparison with published books concerning 20th centuries folk medicine in the valley prompted us to deepen our understanding of the traditional knowledge still surviving nowadays in Valle Imagna, specifically on the uses of plant species. Our survey revealed that, especially after the 1960s and 1970s great migrations from the valley, this knowledge was almost completely lost to the passage of time and the fading of memories. However, a few older people of the valley could still recall some of the traditional uses of spontaneous plant species and being able to retrieve this information and partly stop the loss of memories was certainly an important step.

The diachronic comparison among the historical sources involved in this analysis confirmed a general pattern of decline in the information concerning both number of taxa and their uses across the centuries, from the 1700s, represented by the manuscript, through the early 1900s, described by Maconi, to the first two decades of the 2000s, highlighted by our field investigation. Another clear example of the pending loss of knowledge concerns the uses of the first 10 frequently mentioned species during the fieldwork, compared to the ones of the historical sources.

However, among the similar uses, the 18th and 21st century preparations regarding *Agrimonia eupatoria* undoubtedly piqued our interest, due not only to the shared purpose of use and desired outcome of the remedies, but also to some similarities concerning the part of the plant used and their administration form. Moreover, although some of the other species are commonly used throughout Lombardy, the bibliographic investigation concerning the ethnobotanical studies conducted in neighboring areas underlined the specific and identifiable use of agrimony in Valle Imagna.

Further investigation is certainly mandatory but ultimately, our analysis of the sources allowed us to likely sketch a *fil rouge* across centuries of traditional knowledge of Valle Imagna, with unfortunately only slivers of the ancient remedies of the 18th century surviving the passage of time.

Supplementary Materials: The following supporting information can be downloaded at: https://www.mdpi.com/article/10.3390/plants12142748/s1, Table S1: Traditional medicinal uses in 21st century Valle Imagna and related bibliographic research. For the complete reference list, please see the main text.

Author Contributions: Conceptualization, G.F.; Methodology, G.F. and P.B.; Validation, G.F., P.B. and C.G.; Formal analysis, F.M. and M.B.; Investigation, F.M., M.B., L.B., C.G., L.C. and P.S.C.; Resources,

C.G. and G.F.; Writing—original draft preparation, F.M. and P.B.; Writing—review and editing, F.M., P.B., M.B., L.C., P.S.C., C.G. and G.F.; Visualization, all the authors; Supervision, G.F. All authors have read and agreed to the published version of the manuscript.

Funding: This research received no external funding.

Institutional Review Board Statement: The study protocol received the acknowledgment by the Ethics Committee of the University of Milan (Protocol 25.05.2023).

Informed Consent Statement: The interviews followed the Ethical Guidelines of the International Society of Ethnobiology, 2006. All the people interviewed were informed about the objectives of our investigation and gave their oral consent to share the information.

Data Availability Statement: This paper contains all data concerning the medicinal uses detected during the field work, as well as the one concerning the comparison among sources.

Acknowledgments: We would like to heartfully thank the population of upper Valle Imagna, for their memories, trust, and generosity. We wish we could name each and every one of them. Our sincerest gratitude goes to Giorgio Locatelli and Antonio Carminati, respectively, president and director of the Centro Studi Valle Imagna, for their help, enthusiasm, and for their ever-present kindness. We thank them also for the concession of the use of images concerning the manuscript for Figure 1a,b. We also acknowledge Fuipiano municipal administration for their precious support. Finally, we would like to thank Leonardo Molino for revising the English text.

Conflicts of Interest: The authors declare no conflict of interest.

References

1. da Silva, T.C.; Medeiros, P.M.; Balcazár, A.L.; de Sousa Araújo, T.A.; Pirondo, A.; Medeiros, M.F.T. Historical ethnobotany: An overview of selected studies. *Ethnobiol. Conserv.* **2014**, *3*, 4. [CrossRef]
2. Fontefrancesco, M.F.; Pieroni, A. Renegotiating situativity: Transformations of local herbal knowledge in a Western Alpine valley during the past 40 years. *J. Ethnobiol. Ethnomed.* **2020**, *16*, 58. [CrossRef]
3. Mattalia, G.; Graetz, F.; Harms, M.; Segor, A.; Tomarelli, A.; Kieser, V.; Zerbe, S.; Pieroni, A. Temporal Changes in the Use of Wild Medicinal Plants in Tyrol, South Trentino. *Plants* **2023**, *12*, 2372. [CrossRef] [PubMed]
4. Leonti, M.; Casu, L.; Sanna, F.; Bonsignore, L. A comparison of medicinal plant use in Sardinia and Sicily-De Materia Medica revisited? *J. Ethnopharmacol.* **2009**, *121*, 255–267. [CrossRef] [PubMed]
5. Leonti, M.; Cabras, S.; Weckerle, C.S.; Solinas, M.N.; Casu, L. The causal dependence of present plant knowledge on herbals-Contemporary medicinal plant use in Campania (Italy) compared to Matthioli (1568). *J. Ethnopharmacol.* **2010**, *130*, 379–391. [CrossRef] [PubMed]
6. Dal Cero, M.; Saller, R.; Leonti, M.; Weckerle, C.S. Trends of Medicinal Plant Use over the Last 2000 Years in Central Europe. *Plants* **2023**, *12*, 135. [CrossRef]
7. Stepp, J.R.; Moerman, D.E. The importance of weeds in ethnopharmacology. *J. Ethnopharmacol.* **2001**, *75*, 19–23. [CrossRef]
8. Bruschi, P.; Sugni, M.; Moretti, A.; Signorini, M.A.; Fico, G. Children's versus adult's knowledge of medicinal plants: An ethnobotanical study in Tremezzina (Como, Lombardy, Italy). *Brazilian J. Pharmacogn.* **2019**, *29*, 644–655. [CrossRef]
9. Aziz, M.A.; Volpato, G.; Fontefrancesco, M.F.; Pieroni, A. Perceptions and Revitalization of Local Ecological Knowledge in Four Schools in Yasin Valley, North Pakistan. *Mt. Res. Dev.* **2022**, *42*, R1–R9. [CrossRef]
10. Arjona-García, C.; Blancas, J.; Beltrán-Rodríguez, L.; López Binnqüist, C.; Colín Bahena, H.; Moreno-Calles, A.I.; Sierra-Huelsz, J.A.; López-Medellín, X. How does urbanization affect perceptions and traditional knowledge of medicinal plants? *J. Ethnobiol. Ethnomed.* **2021**, *17*, 48. [CrossRef]
11. Applequist, W.L.; Brinckmann, J.A.; Cunningham, A.B.; Hart, R.E.; Heinrich, M.; Katerere, D.R.; Van Andel, T. Scientists warning on climate change and medicinal plants. *Planta Med.* **2020**, *86*, 10–18. [CrossRef] [PubMed]
12. Shrestha, U.B.; Lamsal, P.; Ghimire, S.K.; Shrestha, B.B.; Dhakal, S.; Shrestha, S.; Atreya, K. Climate change-induced distributional change of medicinal and aromatic plants in the Nepal Himalaya. *Ecol. Evol.* **2022**, *12*, e9204. [CrossRef] [PubMed]
13. Mattalia, G.; Sõukand, R.; Corvo, P.; Pieroni, A. "We Became Rich and We Lost Everything": Ethnobotany of Remote Mountain Villages of Abruzzo and Molise, Central Italy. *Hum. Ecol.* **2021**, *49*, 217–224. [CrossRef]
14. Mattalia, G.; Stryamets, N.; Balázsi, Á.; Molnár, G.; Gliga, A.; Pieroni, A.; Sõukand, R.; Reyes-García, V. Hutsuls' Perceptions of Forests and Uses of Forest Resource in Ukrainian and Romanian Bukovina. *Int. For. Rev.* **2022**, *24*, 393–410. [CrossRef]
15. Leonti, M. The future is written: Impact of scripts on the cognition, selection, knowledge and transmission of medicinal plant use and its implications for ethnobotany and ethnopharmacology. *J. Ethnopharmacol.* **2011**, *134*, 542–555. [CrossRef]
16. Prakofjewa, J.; Kalle, R.; Belichenko, O.; Kolosova, V.; Sõukand, R. Re-written narrative: Transformation of the image of Ivan-chaj in Eastern Europe. *Heliyon* **2020**, *6*, e04632. [CrossRef]
17. Milani, F.; Fico, G. *Raccolta di Diversi Rimedj a Varj Mali. Studio Etnobotanico di un Manoscritto Lombardo del Diciottesimo Secolo*; Centro Sudi Valle Imagna: Bergamo, Italy, 2021; ISBN 978-88-6417-106-7.

18. Colonna, R.; Piscitelli, A.; Iadevaia, V. Una breve storia della farmacologia occidentale. *G. Ital. di Farmacol. Clin.* **2019**, *33*, 86–106. [CrossRef]

19. Maconi, G. *La Medicina Popolare in Valle Imagna. Componenti Magiche, Religiose ed Empiriche Tradizionali tra L'Ottocento e il Novecento*; Centro Studi Valle Imagna: Bergamo, Italy, 2006.

20. Mangili, C. GEV Valle Imagna. In *Piccola Flora della Valle Imagna*; Comunità Montana Valle Imagna: Bergamo, Italy, 2016.

21. Flora Alpina Bergamasca. Parco delle Orobie Bergamasche. In *Fiori Delle Orobie 2—Gli Alberi*; EQUA Editrice: Bergamo, Italy, 2015.

22. Flora Alpina Bergamasca. Parco delle Orobie Bergamasche. In *Fiori delle Orobie 1—Collina e Bassa Montagna*; EQUA Editrice: Bergamo, Italy, 2014.

23. Flora Alpina Bergamasca. Parco delle Orobie Bergamasche. In *Fiori delle Orobie 3—Media e Alta Montagna*; EQUA Editrice: Bergamo, Italy, 2016.

24. Sgalippa, G.; Silva, M. *Gente e Terra d'Imagna. Atti del Convegno di Sant'Omobono Imagna, 15 Aprile—13 maggio 1993*; Il Pomerio Srl di Lodi: Lodi, Italy, 1996.

25. Vandebroek, I.; Balick, M.J. Globalization and loss of plant knowledge: Challenging the paradigm. *PLoS ONE* **2012**, *7*, e37643. [CrossRef]

26. Petelka, J.; Plagg, B.; Säumel, I.; Zerbe, S. Traditional medicinal plants in South Tyrol (northern Italy, southern Alps): Biodiversity and use. *J. Ethnobiol. Ethnomed.* **2020**, *16*, 74. [CrossRef]

27. Sõukand, R.; Kalle, R.; Pieroni, A. Homogenisation of Biocultural Diversity: Plant Ethnomedicine and Its Diachronic Change in Setomaa and Võromaa, Estonia, in the Last Century. *Biology* **2022**, *11*, 192. [CrossRef]

28. Guarrera, P.M. *Usi e Tradizioni della Flora Italiana. Medicina Popolare ed Etnobotanica*; Aracne: Rome, Italy, 2006.

29. Schütz, K.; Carle, R.; Schieber, A. Taraxacum-A review on its phytochemical and pharmacological profile. *J. Ethnopharmacol.* **2006**, *107*, 313–323. [CrossRef] [PubMed]

30. Lis, B.; Olas, B. Pro-health activity of dandelion (*Taraxacum officinale* L.) and its food products—History and present. *J. Funct. Foods* **2019**, *59*, 40–48. [CrossRef]

31. Dal Cero, M.; Saller, R.; Weckerle, C.S. The use of the local flora in Switzerland: A comparison of past and recent medicinal plant knowledge. *J. Ethnopharmacol.* **2014**, *151*, 253–264. [CrossRef]

32. Vitalini, S.; Tomè, F.; Fico, G. Traditional uses of medicinal plants in Valvestino (Italy). *J. Ethnopharmacol.* **2009**, *121*, 106–116. [CrossRef]

33. Vitalini, S.; Puricelli, C.; Mikerezi, I.; Iriti, M. Plants, people and traditions: Ethnobotanical survey in the Lombard Stelvio National Park and neighbouring areas (Central Alps, Italy). *J. Ethnopharmacol.* **2015**, *173*, 435–458. [CrossRef]

34. Bottoni, M.; Milani, F.; Colombo, L.; Nallio, K.; Colombo, P.S.; Giuliani, C.; Bruschi, P.; Fico, G. Using Medicinal Plants in Valmalenco (Italian Alps): From Tradition to Scientific Approaches. *Molecules* **2020**, *25*, 4144. [CrossRef]

35. Bottoni, M.; Colombo, L.; Gianoli, C.; Milani, F.; Colombo, P.S.; Bruschi, P.; Giuliani, C.; Fico, G. Alpine ethnobotanical knowledge in Sondalo (SO, Lombardy, Italy). *Ethnobot. Res. Appl.* **2022**, *24*, 1–63. [CrossRef]

36. Dei Cas, L.; Pugni, F.; Fico, G. Tradition of use on medicinal species in Valfurva (Sondrio, Italy). *J. Ethnopharmacol.* **2015**, *163*, 113–134. [CrossRef]

37. Vitalini, S.; Iriti, M.; Puricelli, C.; Ciuchi, D.; Segale, A.; Fico, G. Traditional knowledge on medicinal and food plants used in Val San Giacomo (Sondrio, Italy)—An alpine ethnobotanical study. *J. Ethnopharmacol.* **2013**, *145*, 517–529. [CrossRef]

38. Pignatti, S.; Guorino, R.; La Rosa, M. *Flora d'Italia*, 2nd ed.; Edagricole-New Business Media: Bologna, Italy, 2018.

39. World Flora Online. Available online: http://www.worldfloraonline.org/ (accessed on 29 June 2023).

40. Pitman, N.C.A.; Cerón, C.E.; Reyes, C.I.; Thurber, M.; Arellano, J. Catastrophic natural origin of a species-poor tree community in the world's richest forest. *J. Trop. Ecol.* **2005**, *21*, 559–568. [CrossRef]

41. Cook, F.E. *Economic Botanic Data Collection Standard*; Prendergast, H., Ed.; Royal Botanic Gardens, Kew: London, UK, 1995; ISBN 0947643710.

42. González, J.A.; García-Barriuso, M.; Amich, F. Ethnobotanical study of medicinal plants traditionally used in the Arribes del Duero, western Spain. *J. Ethnopharmacol.* **2010**, *131*, 343–355. [CrossRef]

43. Staub, P.O.; Geck, M.S.; Weckerle, C.S.; Casu, L.; Leonti, M. Classifying diseases and remedies in ethnomedicine and ethnopharmacology. *J. Ethnopharmacol.* **2015**, *174*, 514–519. [CrossRef]

44. Lucchetti, L.; Zitti, S.; Taffetani, F. Ethnobotanical uses in the Ancona district (Marche region, Central Italy). *J. Ethnobiol. Ethnomed.* **2019**, *15*, 9. [CrossRef]

45. Guarrera, P.M. Traditional phytotherapy in Central Italy (Marche, Abruzzo, and Latium). *Fitoterapia* **2005**, *76*, 1–25. [CrossRef]

46. Cornara, L.; La Rocca, A.; Terrizzano, L.; Dente, F.; Mariotti, M.G. Ethnobotanical and phytomedical knowledge in the North-Western Ligurian Alps. *J. Ethnopharmacol.* **2014**, *155*, 463–484. [CrossRef]

47. Pagano, C.; Marinozzi, M.; Baiocchi, C.; Beccari, T.; Calarco, P.; Ceccarini, M.R.; Chielli, M.; Orabona, C.; Orecchini, E.; Ortenzi, R.; et al. Bioadhesive polymeric films based on red onion skins extract for wound treatment: An innovative and eco-friendly formulation. *Molecules* **2020**, *25*, 318. [CrossRef]

48. Dorsch, W.; Ring, J. Suppression of Immediate and Late Anti-IgE-Induced Skin Reactions by Topically Applied Alcohol/Onion Extract. *Allergy* **1984**, *39*, 43–49. [CrossRef] [PubMed]

49. Batiha, G.E.S.; Beshbishy, A.M.; El-Mleeh, A.; Abdel-Daim, M.M.; Devkota, H.P. Traditional uses, bioactive chemical constituents, and pharmacological and toxicological activities of *Glycyrrhiza glabra* L. (fabaceae). *Biomolecules* **2020**, *10*, 352. [CrossRef]

50. Shang, A.; Cao, S.Y.; Xu, X.Y.; Gan, R.Y.; Tang, G.Y.; Corke, H.; Mavumengwana, V.; Li, H. Bin Bioactive compounds and biological functions of garlic (*Allium sativum* L.). *Foods* **2019**, *8*, 246. [CrossRef]

51. Sarker, S.; Nahar, L. Natural Medicine:The Genus Angelica. *Curr. Med. Chem.* **2004**, *11*, 1479–1500. [CrossRef] [PubMed]

52. Facino, R.M.; Carini, M.; Stefani, R.; Aldini, G.; Saibene, L. Anti-Elastase and Anti-Hyaluronidase Activities of Saponins and Sapogenins from *Hedera helix, Aesculus hippocastanum,* and *Ruscus aculeatus*: Factors Contributing to their Efficacy in the Treatment of Venous Insufficiency. *Arch. Pharm.* **1995**, *328*, 720–724. [CrossRef]

53. Vázquez-Castilla, S.; De la Puerta, R.; García Giménez, M.D.; Fernández-Arche, M.A.; Guillén-Bejarano, R. Bioactive constituents from "triguero" asparagus improve the plasma lipid profile and liver antioxidant status in hypercholesterolemic rats. *Int. J. Mol. Sci.* **2013**, *14*, 21227–21239. [CrossRef]

54. World Health Organization (Ed.) *WHO Monographs on Selected Medicinal Plants*; WHO Library Cataloguing in Publication Data: Geneva, Switzerland, 2002; Volume 2.

55. Applequist, W.L.; Moerman, D.E. Yarrow (*Achillea millefolium* L.): A Neglected Panacea? A Review of Ethnobotany, Bioactivity, and Biomedical Research1. *Econ. Bot.* **2011**, *65*, 209–225. [CrossRef]

56. Ali, S.I.; Gopalakrishnan, B.; Venkatesalu, V. Pharmacognosy, Phytochemistry and Pharmacological Properties of *Achillea millefolium* L.: A Review. *Phyther. Res.* **2017**, *31*, 1140–1161. [CrossRef] [PubMed]

57. Saeidnia, S.; Gohari, A.R.; Mokhber-Dezfuli, N.; Kiuchi, F. A review on phytochemistry and medicinal properties of the genus Achillea. *DARU, J. Pharm. Sci.* **2011**, *19*, 173–186.

58. Mattalia, G.; Quave, C.L.; Pieroni, A. Traditional uses of wild food and medicinal plants among Brigasc, Kyé, and Provençal communities on the Western Italian Alps. *Genet Resour Crop Evol.* **2013**, *60*, 587–603. [CrossRef]

59. Pieroni, A.; Giusti, M.E. Alpine ethnobotany in Italy: Traditional knowledge of gastronomic and medicinal plants among the Occitans of the upper Varaita valley, Piedmont. *J. Ethnobiol. Ethnomed.* **2009**, *5*, 32. [CrossRef] [PubMed]

60. Bellia, G.; Pieroni, A. Isolated, but Transnational: The Glocal Nature of Waldensian Ethnobotany, Western Alps, NW Italy. *J. Ethnobiol. Ethnomed.* **2015**, *11*. [CrossRef]

61. Kriplani, P.; Guarve, K.; Baghael, U.S. *Arnica montana* L.—A plant of healing: Review. *J. Pharm. Pharmacol.* **2017**, *69*, 925–945. [CrossRef]

62. Lyss, G.; Schmidt, T.J.; Merfort, I.; Pahl, H.L. Helenalin, an anti-inflammatory sesquiterpene lactone from *Arnica*, selectively inhibits transcription factor NF-κB. *Biol. Chem.* **1997**, *378*, 951–961. [CrossRef] [PubMed]

63. Sharma, S.; Arif, M.; Nirala, R.K.; Gupta, R.; Thakur, S.C. Cumulative therapeutic effects of phytochemicals in *Arnica montana* flower extract alleviated collagen-induced arthritis: Inhibition of both pro-inflammatory mediators and oxidative stress. *J. Sci. Food Agric.* **2015**, *96*, 1500–1510. [CrossRef] [PubMed]

64. World Health Organization (Ed.) *WHO Monographs on Selected Medicinal Plants*; WHO Library Cataloguing in Publication Data: Geneva, Switzerland, 2007; Volume 3.

65. Hall, I.H.; Lee, K.H.; Starenes, C.O.; Sumida, Y.; Wu, R.Y.; Waddell, T.G.; Cochran, J.W.; Gerhart, K.G. Anti-inflammatory activity of sesquiterpene lactones and related compounds. *J. Pharm. Sci.* **1979**, *68*, 537–542. [CrossRef] [PubMed]

66. Hall, I.H.; Starenes, C.O.; Lee, K.H.; Waddell, T.G. Mode of action of sesquiterpene lactones as anti-inflammatory agents. *J. Pharm. Sci.* **1980**, *69*, 537–543. [CrossRef]

67. Capasso, F.; Grandolini, G.; Izzo, A.A. *Fitoterapia. Impiego Razionale delle Droghe Vegetali*; Springer: Berlin/Heidelberg, Germany, 2006.

68. McMullen, M.K.; Whitehouse, J.M.; Whitton, P.A.; Towell, A. Bitter tastants alter gastric-phase postprandial haemodynamics. *J. Ethnopharmacol.* **2014**, *154*, 719–727. [CrossRef] [PubMed]

69. McMullen, M.K.; Whitehouse, J.M.; Towell, A. Bitters: Time for a new paradigm. *Evidence-Based Complement. Altern. Med.* **2015**, *2015*, 670504. [CrossRef] [PubMed]

70. Batiha, G.E.; Olatunde, A.; El-mleeh, A.; Hetta, H.F.; Al-rejaie, S.; Alghamdi, S.; Zahoor, M.; Beshbishy, A.M. Pharmacokinetics of Wormwood (*Artemisia absinthium*). *Antibiotics* **2020**, *9*, 353. [CrossRef]

71. Brockhoff, A.; Behrens, M.; Massarotti, A.; Appending, G.; Meyerhof, W. Broad tuning of the human bitter taste receptor hTAS2R46 to various sesquiterpene lactones, clerodane and labdane diterpenoids, strychnine, and denatonium. *J. Agric. Food Chem.* **2007**, *55*, 6236–6243. [CrossRef]

72. Gras, A.; Parada, M.; Vallès, J.; Garnatje, T. The Role of Traditional Plant Knowledge in the Fight Against Infectious Diseases: A Meta-Analytic Study in the Catalan Linguistic Area. *Front. Pharmacol.* **2021**, *12*, 744616. [CrossRef]

73. Khattak, S.G.; Gilani, S.N.; Ikram, M. Antipyretic studies on some indigenous Pakistani medicinal plants. *J. Ethnopharmacol.* **1985**, *14*, 45–51. [CrossRef]

74. Rakhshandeh, H.; Heidari, A.; Pourbagher-Shahri, A.M.; Rashidi, R.; Forouzanfar, F. Hypnotic Effect of *A. absinthium* Hydroalcoholic Extract in Pentobarbital-Treated Mice. *Neurol. Res. Int.* **2021**, *2021*, 5521019. [CrossRef]

75. Amat, N.; Upur, H.; Blažeković, B. In vivo hepatoprotective activity of the aqueous extract of *Artemisia absinthium* L. against chemically and immunologically induced liver injuries in mice. *J. Ethnopharmacol.* **2010**, *131*, 478–484. [CrossRef]

76. World Health Organization (Ed.) *WHO Monographs on Selected Medicinal Plants*; WHO Library Cataloguing in Publication Data: Geneva, Switzerland, 1999; Volume 1.

77. Dinda, M.; Mazumdar, S.; Das, S.; Ganguly, D.; Dasgupta, U.B.; Dutta, A.; Jana, K.; Karmakar, P. The Water Fraction of *Calendula officinalis* Hydroethanol Extract Stimulates In Vitro and In Vivo Proliferation of Dermal Fibroblasts in Wound Healing. *Phyther. Res.* **2016**, *30*, 1696–1707. [CrossRef]

78. Fonseca, Y.M.; Catini, C.D.; Vicentini, F.T.M.C.; Nomizo, A.; Gerlach, R.F.; Fonseca, M.J.V. Protective effect of *Calendula officinalis* extract against UVB-induced oxidative stress in skin: Evaluation of reduced glutathione levels and matrix metalloproteinase secretion. *J. Ethnopharmacol.* **2010**, *127*, 596–601. [CrossRef]

79. Jahdi, F.; Khabbaz, A.; Kashian, M.; Taghizadeh, M.; Haghani, H. The impact of *Calendula* ointment on cesarean wound healing: A randomized controlled clinical trial. *J. Fam. Med. Prim. Care* **2018**, *7*, 893. [CrossRef]

80. Nicolaus, C.; Junghanns, S.; Hartmann, A.; Murillo, R.; Ganzera, M.; Merfort, I. In vitro studies to evaluate the wound healing properties of *Calendula officinalis* extracts. *J. Ethnopharmacol.* **2017**, *196*, 94–103. [CrossRef]

81. Givol, O.; Kornhaber, R.; Visentin, D.; Cleary, M.; Haik, J.; Harats, M. A systematic review of *Calendula officinalis* extract for wound healing. *Wound Repair Regen.* **2019**, *27*, 548–561. [CrossRef]

82. Okada, N.; Kobayashi, S.; Moriyama, K.; Miyataka, K.; Abe, S.; Sato, C.; Kawazoe, K. *Helianthus tuberosus* (Jerusalem artichoke) tubers improve glucose tolerance and hepatic lipid profile in rats fed a high-fat diet. *Asian Pac. J. Trop. Med.* **2017**, *10*, 439–443. [CrossRef] [PubMed]

83. McKay, D.L.; Blumberg, J.B. A Review of the bioactivity and potential health benefits of chamomile tea (*Matricaria recutita* L.). *Phyther. Res.* **2006**, *20*, 519–530. [CrossRef]

84. Miraj, S.; Alesaeidi, S. A systematic review study of therapeutic effects of *Matricaria recutita* chamomile (chamomile). *Electron. Physician* **2016**, *8*, 3024–3031. [CrossRef] [PubMed]

85. Keefe, J.R.; Mao, J.J.; Soeller, I.; Li, Q.S.; Amsterdam, J.D. Short-term open-label chamomile (*Matricaria chamomilla* L.) therapy of moderate to severe generalized anxiety disorder. *Phytomedicine* **2016**, *23*, 1699–1705. [CrossRef] [PubMed]

86. El Mihyaoui, A.; Esteves da Silva, J.C.G.; Charfi, S.; Candela Castillo, M.E.; Lamarti, A.; Arnao, M.B. Chamomile (*Matricaria chamomilla* L.): A Review of Ethnomedicinal Use, Phytochemistry and Pharmacological Uses. *Life* **2022**, *12*, 479. [CrossRef]

87. Marmouzi, I.; Bouyahya, A.; Ezzat, S.M.; El Jemli, M.; Kharbach, M. The food plant *Silybum marianum* (L.) Gaertn.: Phytochemistry, Ethnopharmacology and clinical evidence. *J. Ethnopharmacol.* **2021**, *265*, 113303. [CrossRef] [PubMed]

88. Li, Y.; Chen, Y.; Sun-Waterhouse, D. The potential of dandelion in the fight against gastrointestinal diseases: A review. *J. Ethnopharmacol.* **2022**, *293*, 115272. [CrossRef]

89. González-Castejón, M.; Visioli, F.; Rodriguez-Casado, A. Diverse biological activities of dandelion. *Nutr. Rev.* **2012**, *70*, 534–547. [CrossRef]

90. Guarrera, P.M. Food medicine and minor nourishment in the folk traditions of Central Italy (Marche, Abruzzo and Latium). *Fitoterapia* **2003**, *74*, 515–544. [CrossRef] [PubMed]

91. Liu, C.; Wu, H.; Wang, L.; Luo, H.; Lu, Y.; Zhang, Q.; Tang, L.; Wang, Z. *Farfarae Flos*: A review of botany, traditional uses, phytochemistry, pharmacology, and toxicology. *J. Ethnopharmacol.* **2020**, *260*, 113038. [CrossRef] [PubMed]

92. Wu, Q.Z.; Zhao, D.X.; Xiang, J.; Zhang, M.; Zhang, C.F.; Xu, X.H. Antitussive, expectorant, and anti-inflammatory activities of four caffeoylquinic acids isolated from *Tussilago farfara*. *Pharm. Biol.* **2016**, *54*, 1117–1124. [CrossRef]

93. Welna, M.; Szymczycha-Madeja, A.; Pohl, P. Simplified ICP OES-based method for determination of 12 elements in commercial bottled birch saps: Validation and bioaccessibility study. *Molecules* **2020**, *25*, 1256. [CrossRef]

94. Cameron, M.; Chrubasik, S. Topical herbal therapies for treating osteoarthritis. *Cochrane Database Syst. Rev.* **2013**, *5*, CD010538. [CrossRef]

95. Frost, R.; MacPherson, H.; O'Meara, S. A critical scoping review of external uses of comfrey (*Symphytum* spp.). *Complement. Ther. Med.* **2013**, *21*, 724–745. [CrossRef]

96. Rakhecha, B.; Agnihotri, P.; Dakal, T.C.; Saquib, M.; Monu; Biswas, S. Anti-inflammatory activity of nicotine isolated from *Brassica oleracea* in rheumatoid arthritis. *Biosci. Rep.* **2022**, *42*, BSR20211392. [CrossRef]

97. Kazimierski, M.; Regula, J.; Molska, M. Cornelian cherry (*Cornus mas* L.)—characteristics, nutritional and pro-health properties. *Acta Sci. Pol. Technol. Aliment.* **2019**, *18*, 5–12. [CrossRef]

98. Carneiro, D.M.; Freire, R.C.; Honório, T.C.D.D.; Zoghaib, I.; Cardoso, F.F.D.S.; Tresvenzol, L.M.F.; de Paula, J.R.; Sousa, A.L.L.; Jardim, P.C.B.V.; Cunha, L.C.; et al. Randomized, Double-Blind Clinical Trial to Assess the Acute Diuretic Effect of *Equisetum arvense* (Field Horsetail) in Healthy Volunteers. *Evidence-Based Complement. Altern. Med.* **2014**, *2014*, 760683. [CrossRef] [PubMed]

99. Al-Snafi, A.E. The pharmacology of *Equisetum arvense*—A review. *IOSR J. Pharm.* **2017**, *7*, 31–42. [CrossRef]

100. Asgarpanah, J. Phytochemistry and pharmacological properties of *Equisetum arvense* L. *J. Med. Plants Res.* **2012**, *6*, 3689–3693. [CrossRef]

101. World Health Organization (Ed.) *WHO Monographs on Selected Medicinal Plants*; WHO Library Cataloguing in Publication Data: Geneva, Switzerland, 2009; Volume 4.

102. Deng, H.W.; Tian, Y.; Zhou, X.J.; Zhang, X.M.; Meng, J. Effect of Bilberry Extract on Development of Form-Deprivation Myopia in the Guinea Pig. *J. Ocul. Pharmacol. Ther.* **2016**, *32*, 196–202. [CrossRef] [PubMed]

103. Gizzi, C.; Belcaro, G.; Gizzi, G.; Feragalli, B.; Dugall, M.; Luzzi, R.; Cornelli, U. Bilberry extracts are not created equal: The role of non anthocyanin fraction. Discovering the "dark side of the force" in a preliminary study. *Eur. Rev. Med. Pharmacol. Sci.* **2016**, *20*, 2418–2424.

104. Ogawa, K.; Tsuruma, K.; Tanaka, J.; Kakino, M.; Kobayashi, S.; Shimazawa, M.; Hara, H. The Protective Effects of Bilberry and Lingonberry Extracts against UV Light-Induced Retinal Photoreceptor Cell Damage in Vitro. *J. Agric. Food Chem.* **2013**, *61*, 10345–10353. [CrossRef]

105. Ogawa, K.; Kuse, Y.; Tsuruma, K.; Kobayashi, S.; Shimazawa, M.; Hara, H. Protective effects of bilberry and lingonberry extracts against blue light-emitting diode light-induced retinal photoreceptor cell damage in vitro. *BMC Complement. Altern. Med.* **2014**, *14*, 120. [CrossRef]

106. Ozawa, Y.; Kawashima, M.; Inoue, S.; Inagaki, E.; Suzuki, A.; Ooe, E.; Kobayashi, S.; Tsubota, K. Bilberry extract supplementation for preventing eye fatigue in video display terminal workers. *J. Nutr. Health Aging* **2015**, *19*, 548–554. [CrossRef]

107. Wang, Y.; Zhao, L.; Lu, F.; Yang, X.; Deng, Q.; Ji, B.; Huang, F.; Kitts, D.D. Retinoprotective effects of bilberry anthocyanins via antioxidant, anti-inflammatory, and anti-apoptotic mechanisms in a visible light-induced retinal degeneration model in pigmented rabbits. *Molecules* **2015**, *20*, 22395–22410. [CrossRef]

108. Canter, P.H.; Ernst, E. Anthocyanosides of *Vaccinium myrtillus* (bilberry) for night vision—A systematic review of placebo-controlled trials. *Surv. Ophthalmol.* **2004**, *49*, 38–50. [CrossRef]

109. Szajdek, A.; Borowska, E.J. Bioactive Compounds and Health-Promoting Properties of Berry Fruits: A Review. *Plant Foods Hum. Nutr.* **2008**, *63*, 147–156. [CrossRef]

110. Tunaru, S.; Althoff, T.F.; Nüsing, R.M.; Diener, M.; Offermanns, S. Castor oil induces laxation and uterus contraction via ricinoleic acid activating prostaglandin EP3 receptors. *Proc. Natl. Acad. Sci. USA* **2012**, *109*, 9179–9184. [CrossRef]

111. Chávez-Santoscoy, R.A.; Gutiérrez-Uribe, J.A.; Serna-Saldívar, S.O. Effect of Flavonoids and Saponins Extracted from Black Bean (*Phaseolus vulgaris* L.) Seed Coats as Cholesterol Micelle Disruptors. *Plant Foods Hum. Nutr.* **2013**, *68*, 416–423. [CrossRef] [PubMed]

112. Wölfle, U.; Seelinger, G.; Schempp, C.M. Topical application of St John's wort (*Hypericum perforatum*). *Planta Med.* **2014**, *80*, 109–120.

113. Uslusoy, F.; Nazıroğlu, M.; Övey, İ.S.; Sönmez, T.T. *Hypericum perforatum* L. supplementation protects sciatic nerve injury-induced apoptotic, inflammatory and oxidative damage to muscle, blood and brain in rats. *J. Pharm. Pharmacol.* **2019**, *71*, 83–92. [CrossRef]

114. Khiljee, S.; Rehman, N.U.; Khiljee, T.; Loebenberg, R.; Ahmad, R.S. Formulation and clinical evaluation of topical dosage forms of Indian Penny Wort, walnut and turmeric in eczema. *Pak. J. Pharm. Sci.* **2015**, *28*, 2001–2007. [PubMed]

115. Cavanagh, H.M.A.; Wilkinson, J.M. Biological activities of lavender essential oil. *Phyther. Res.* **2002**, *16*, 301–308. [CrossRef] [PubMed]

116. Pérez-Recalde, M.; Ruiz Arias, I.E.; Hermida, É.B. Could essential oils enhance biopolymers performance for wound healing? A systematic review. *Phytomedicine* **2018**, *38*, 57–65. [CrossRef]

117. Rai, V.K.; Sinha, P.; Yadav, K.S.; Shukla, A.; Saxena, A.; Bawankule, D.U.; Tandon, S.; Khan, F.; Chanotiya, C.S.; Yadav, N.P. Anti-psoriatic effect of *Lavandula angustifolia* essential oil and its major components linalool and linalyl acetate. *J. Ethnopharmacol.* **2020**, *261*, 113127. [CrossRef]

118. Luo, J.; Jiang, W. A critical review on clinical evidence of the efficacy of lavender in sleep disorders. *Phyther. Res.* **2022**, *36*, 2342–2351. [CrossRef]

119. Nawrot, J.; Gornowicz-Porowska, J.; Budzianowski, J.; Nowak, G.; Schroeder, G.; Kurczewska, J. Medicinal Herbs in the Relief of Neurological, Cardiovascular, and Respiratory Symptoms after COVID-19 Infection a Literature Review. *Cells* **2022**, *11*, 1897. [CrossRef] [PubMed]

120. Takahashi, M.; Yamanaka, A.; Asanuma, C.; Asano, H.; Satou, T.; Koike, K. Anxiolytic-like effect of inhalation of essential oil from *Lavandula officinalis*: Investigation of changes in 5-HT turnover and involvement of olfactory stimulation. *Nat. Prod. Commun.* **2014**, *9*, 1023–1026. [CrossRef] [PubMed]

121. Brown, M.E.; Browning, J.C. A case of psoriasis replaced by allergic contact dermatitis in a 12-year-old boy. *Pediatr. Dermatol.* **2016**, *33*, e125–e126. [CrossRef]

122. Dobros, N.; Zawada, K.D.; Paradowska, K. Phytochemical Profiling, Antioxidant and Anti-Inflammatory Activity of Plants Belonging to the *Lavandula* Genus. *Molecules* **2023**, *28*, 256. [CrossRef] [PubMed]

123. Sasannejad, P.; Saeedi, M.; Shoeibi, A.; Gorji, A.; Abbasi, M.; Foroughipour, M. Lavender essential oil in the treatment of migraine headache: A placebo-controlled clinical trial. *Eur. Neurol.* **2012**, *67*, 288–291. [CrossRef]

124. Shamabadi, A.; Akhondzadeh, S. Efficacy and tolerability of *Lavandula angustifolia* in treating patients with the diagnosis of depression: A systematic review of randomized controlled trials. *J. Complement. Integr. Med.* **2021**, *20*, 81–91. [CrossRef]

125. Rauf, A.; Akram, M.; Semwal, P.; Mujawah, A.A.H.; Muhammad, N.; Riaz, Z.; Munir, N.; Piotrovsky, D.; Vdovina, I.; Bouyahya, A.; et al. Antispasmodic Potential of Medicinal Plants: A Comprehensive Review. *Oxid. Med. Cell. Longev.* **2021**, *2021*, 4889719. [CrossRef]

126. Naghdi, F.; Gholamnezhad, Z.; Boskabady, M.H.; Bakhshesh, M. Muscarinic receptors, nitric oxide formation and cyclooxygenase pathway involved in tracheal smooth muscle relaxant effect of hydro-ethanolic extract of *Lavandula angustifolia* flowers. *Biomed. Pharmacother.* **2018**, *102*, 1221–1228. [CrossRef]

127. Ueno-Iio, T.; Shibakura, M.; Yokota, K.; Aoe, M.; Hyoda, T.; Shinohata, R.; Kanehiro, A.; Tanimoto, M.; Kataoka, M. Lavender essential oil inhalation suppresses allergic airway inflammation and mucous cell hyperplasia in a murine model of asthma. *Life Sci.* **2014**, *108*, 109–115. [CrossRef] [PubMed]

128. Haybar, H.; Javid, A.Z.; Haghighizadeh, M.H.; Valizadeh, E.; Mohaghegh, S.M.; Mohammadzadeh, A. The effects of *Melissa officinalis* supplementation on depression, anxiety, stress, and sleep disorder in patients with chronic stable angina. *Clin. Nutr. ESPEN* **2018**, *26*, 47–52. [CrossRef]

129. Bardot, V.; Escalon, A.; Ripoche, I.; Denis, S.; Alric, M.; Chalancon, S.; Chalard, P.; Cotte, C.; Berthomier, L.; Leremboure, M.; et al. Benefits of the ipowder®extraction process applied to: *Melissa officinalis* L.: Improvement of antioxidant activity and in vitro gastro-intestinal release profile of rosmarinic acid. *Food Funct.* **2020**, *11*, 722–729. [CrossRef]

130. Aubert, P.; Guinobert, I.; Blondeau, C.; Bardot, V.; Ripoche, I.; Chalard, P.; Neunlist, M. Basal and spasmolytic effects of a hydroethanolic leaf extract of *Melissa officinalis* L. on intestinal motility: An ex vivo study. *J. Med. Food* **2019**, *22*, 653–662. [CrossRef]

131. Farzaei, M.H.; Bahramsoltani, R.; Ghobadi, A.; Farzaei, F.; Najafi, F. Pharmacological activity of *Mentha longifolia* and its phytoconstituents. *J. Tradit. Chin. Med.* **2017**, *37*, 710–720. [CrossRef]

132. Caro, D.C.; Rivera, D.E.; Ocampo, Y.; Franco, L.A.; Salas, R.D. Pharmacological Evaluation of *Mentha spicata* L. and *Plantago major* L., Medicinal Plants Used to Treat Anxiety and Insomnia in Colombian Caribbean Coast. *Evidence-Based Complement. Altern. Med.* **2018**, *2018*, 5921514. [CrossRef] [PubMed]

133. Ben-Arye, E.; Dudai, N.; Eini, A.; Torem, M.; Schiff, E.; Rakover, Y. Treatment of upper respiratory tract infections in primary care: A randomized study using aromatic herbs. *Evidence-Based Complement. Altern. Med.* **2011**, *2011*, 690346. [CrossRef]

134. Sun, Z.; Wang, H.; Wang, J.; Zhou, L.; Yang, P. Chemical Composition and Anti-Inflammatory, Cytotoxic and Antioxidant Activities of Essential Oil from Leaves of *Mentha piperita* Grown in China. *PLoS ONE* **2014**, *9*, e114767. [CrossRef]

135. Li, Y.X.; Liu, Y.B.; Ma, A.Q.; Bao, Y.; Wang, M.; Sun, Z.L. In vitro antiviral, anti-inflammatory, and antioxidant activities of the ethanol extract of *Mentha piperita* L. *Food Sci. Biotechnol.* **2017**, *26*, 1675–1683. [CrossRef]

136. Verma, S.M.; Arora, H.; Dubey, R. Anti—inflammatory and sedative—hypnotic activity of the methanolic extract of the leaves of *Mentha arvensis*. *Anc. Sci. Life* **2003**, *23*, 95–99.

137. Atta, A.H.; Alkofahi, A. Anti-nociceptive and anti-inflammatory effects of some Jordanian medicinal plant extracts. *J. Ethnopharmacol.* **1998**, *60*, 117–124. [CrossRef]

138. Mahendran, G.; Verma, S.K.; Rahman, L.U. The traditional uses, phytochemistry and pharmacology of spearmint (*Mentha spicata* L.): A review. *J. Ethnopharmacol.* **2021**, *278*, 114266. [CrossRef]

139. Chen, H.P.; Yang, K.; You, C.X.; Lei, N.; Sun, R.Q.; Geng, Z.F.; Ma, P.; Cai, Q.; Du, S.S.; Deng, Z.W. Chemical constituents and insecticidal activities of the essential oil of *Cinnamomum camphora* leaves against *Lasioderma serricorne*. *J. Chem.* **2014**, *2014*, 963729. [CrossRef]

140. Arumugam, G.; Swamy, M.K.; Sinniah, U.R. *Plectranthus amboinicus* (Lour.) Spreng: Botanical, Phytochemical, Pharmacological and Nutritional Significance. *Molecules* **2016**, *21*, 369. [CrossRef]

141. Lemos, I.C.S.; De Araújo Delmondes, G.; Ferreira Dos Santos, A.D.; Santos, E.S.; De Oliveira, D.R.; De Figueiredo, P.R.L.; De Araújo Alves, D.; Barbosa, R.; De Menezes, I.R.A.; Coutinho, H.D.; et al. Ethnobiological survey of plants and animals used for the treatment of acute respiratory infections in children of a traditional community in the municipality of barbalha, Ceará, Brazil. *African J. Tradit. Complement. Altern. Med.* **2016**, *13*, 166–175. [CrossRef] [PubMed]

142. Ahmed, H.M.; Babakir-Mina, M. Investigation of rosemary herbal extracts (*Rosmarinus officinalis*) and their potential effects on immunity. *Phyther. Res.* **2020**, *34*, 1829–1837. [CrossRef]

143. Machado, D.G.; Bettio, L.E.B.; Cunha, M.P.; Capra, J.C.; Dalmarco, J.B.; Pizzolatti, M.G.; Rodrigues, A.L.S. Antidepressant-like effect of the extract of *Rosmarinus officinalis* in mice: Involvement of the monoaminergic system. *Prog. Neuro-Psychopharmacol. Biol. Psychiatry* **2009**, *33*, 642–650. [CrossRef]

144. Malvezzi, L.; Mendes, É.; Militao, L.; Lacalendola, L.; Artem, J.; Barbosa, E.; Gava, P. Rosemary (*Rosmarinus officinalis* L., syn *Salvia rosmarinus* Spenn.) and Its Topical Applications: A Review. *Plants* **2020**, *9*, 651.

145. Shin, H.B.; Choi, M.S.; Ryu, B.; Lee, N.R.; Kim, H.I.; Choi, H.E.; Chang, J.; Lee, K.T.; Jang, D.S.; Inn, K.S. Antiviral activity of carnosic acid against respiratory syncytial virus. *Virol. J.* **2013**, *10*, 303. [CrossRef]

146. Bieski, I.G.C.; Leonti, M.; Arnason, J.T.; Ferrier, J.; Rapinski, M.; Violante, I.M.P.; Balogun, S.O.; Pereira, J.F.C.A.; Figueiredo, R.D.C.F.; Lopes, C.R.A.S.; et al. Ethnobotanical study of medicinal plants by population of Valley of Juruena Region, Legal Amazon, Mato Grosso, Brazil. *J. Ethnopharmacol.* **2015**, *173*, 383–423. [CrossRef] [PubMed]

147. Jamila, F.; Mostafa, E. Ethnobotanical survey of medicinal plants used by people in Oriental Morocco to manage various ailments. *J. Ethnopharmacol.* **2014**, *154*, 76–87. [CrossRef]

148. Noureddine, B.; Mostafa, E.; Mandal, S.C. Ethnobotanical, pharmacological, phytochemical, and clinical investigations on Moroccan medicinal plants traditionally used for the management of renal dysfunctions. *J. Ethnopharmacol.* **2022**, *292*, 115178. [CrossRef] [PubMed]

149. Haloui, M.; Louedec, L.; Michel, J.B.; Lyoussi, B. Experimental diuretic effects of *Rosmarinus officinalis* and *Centaurium erythraea*. *J. Ethnopharmacol.* **2000**, *71*, 465–472. [CrossRef] [PubMed]

150. Ghorbani, A.; Esmaeilizadeh, M. Pharmacological properties of *Salvia officinalis* and its components. *J. Tradit. Complement. Med.* **2017**, *7*, 433–440. [CrossRef]

151. Lemle, K.L. Salvia officinalis used in pharmaceutics. *IOP Conf. Ser. Mater. Sci. Eng.* **2018**, *294*, 012037. [CrossRef]

152. Kerkoub, N.; Panda, S.K.; Yang, M.-R.; Lu, J.-G.; Jiang, Z.-H.; Nasri, H.; Luyten, W. Bioassay-Guided Isolation of Anti-Candida Biofilm Compounds from Methanol Extracts of the Aerial Parts of *Salvia officinalis* (Annaba, Algeria). *Front. Pharmacol.* **2018**, *9*, 1418. [CrossRef] [PubMed]

153. Karpiński, T.M.; Ożarowski, M.; Seremak-Mrozikiewicz, A.; Wolski, H. Anti-Candida and Antibiofilm Activity of Selected Lamiaceae Essential Oils. *Front. Biosci.* **2023**, *28*, 28. [CrossRef]

154. Jakovljević, M.; Jokić, S.; Molnar, M.; Jašić, M.; Babić, J.; Jukić, H.; Banjari, I. Bioactive Profile of Various *Salvia officinalis* L. Preparations. *Plants* **2019**, *8*, 55. [CrossRef]

155. Ehrnhöfer-Ressler, M.M.; Fricke, K.; Pignitter, M.; Walker, J.M.; Walker, J.; Rychlik, M.; Somoza, V. Identification of 1,8-Cineole, Borneol, Camphor, and Thujone as Anti-inflammatory Compounds in a *Salvia officinalis* L. Infusion Using Human Gingival Fibroblasts. *J. Agric. Food Chem.* **2013**, *61*, 3451–3459. [CrossRef]

156. Karimzadeh, S.; Farahpour, M.R. Topical application of Salvia officinalis hydroethanolic leaf extract improves wound healing process. *Indian J. Exp. Biol.* **2017**, *55*, 98–106. [PubMed]

157. Ayrle, H.; Mevissen, M.; Kaske, M.; Nathues, H.; Gruetzner, N.; Melzig, M.; Walkenhorst, M. Medicinal plants—Prophylactic and therapeutic options for gastrointestinal and respiratory diseases in calves and piglets? A systematic review. *BMC Vet. Res.* **2016**, *12*, 89. [CrossRef] [PubMed]

158. Jarić, S.; Mitrović, M.; Pavlović, P. Review of Ethnobotanical, Phytochemical, and Pharmacological Study of *Thymus serpyllum* L. *Evidence-Based Complement. Altern. Med.* **2015**, *2015*, 101978. [CrossRef] [PubMed]

159. Salehi, B.; Prakash Mishra, A.; Shukla, I.; Sharifi-Rad, M.; Del Mar Contreras, M.; Segura-Carretero, A.; Fathi, H.; Nasri Nasrabasi, N.; Kobarfard, F.; Sharifi-Rad, J. Thymol, Thyme, and Other Plant Sources: Health and Potential Uses. *Phyther. Res.* **2018**, *32*, 1688–1706. [CrossRef]

160. Shakeri, F.; Ghorani, V.; Saadat, S.; Gholamnezhad, Z.; Boskabady, M.H. The stimulatory effects of medicinal plants on $\beta2$-adrenoceptors of tracheal smooth muscle. *Iran. J. Allergy Asthma Immunol.* **2019**, *18*, 12–26. [CrossRef] [PubMed]

161. Sakkas, H.; Papadopoulou, C. Antimicrobial Activity of Basil, Oregano, and Thyme Essential Oils. *J. Microbiol. Biotechnol.* **2017**, *27*, 429–438. [CrossRef] [PubMed]

162. Saleem, A.; Afzal, M.; Naveed, M.; Makhdoom, S.I.; Mazhar, M.; Aziz, T.; Khan, A.A.; Kamal, Z.; Shahzad, M.; Alharbi, M.; et al. HPLC, FTIR and GC-MS Analyses of *Thymus vulgaris* Phytochemicals Executing In Vitro and In Vivo Biological Activities and Effects on COX-1, COX-2 and Gastric Cancer Genes Computationally. *Molecules* **2022**, *27*, 8512. [CrossRef]

163. El Yaagoubi, M.; Mechqoq, H.; El Hamdaoui, A.; Jrv Mukku, V.; El Mousadik, A.; Msanda, F.; El Aouad, N. A review on Moroccan *Thymus* species: Traditional uses, essential oils chemical composition and biological effects. *J. Ethnopharmacol.* **2021**, *278*, 114205. [CrossRef]

164. Mustafa, B.; Hajdari, A.; Pieroni, A.; Pulaj, B.; Koro, X.; Quave, C.L. A cross-cultural comparison of folk plant uses among Albanians, Bosniaks, Gorani and Turks living in south Kosovo. *J. Ethnobiol. Ethnomed.* **2015**, *11*, 39. [CrossRef]

165. Silva, P.T.M.; Silva, M.A.F.; Silva, L.; Seca, A.M.L. Ethnobotanical knowledge in sete cidades, azores archipelago: First ethnomedicinal report. *Plants* **2019**, *8*, 256. [CrossRef]

166. Aziz, M.A.; Adnan, M.; Khan, A.H.; Shahat, A.A.; Al-Said, M.S.; Ullah, R. Traditional uses of medicinal plants practiced by the indigenous communities at Mohmand Agency, FATA, Pakistan. *J. Ethnobiol. Ethnomed.* **2018**, *14*, 2. [CrossRef]

167. Micucci, M.; Protti, M.; Aldini, R.; Frosini, M.; Corazza, I.; Marzetti, C.; Mattioli, L.B.; Tocci, G.; Chiarini, A.; Mercolini, L.; et al. *Thymus vulgaris* L. Essential oil solid formulation: Chemical profile and spasmolytic and antimicrobial effects. *Biomolecules* **2020**, *10*, 860. [CrossRef]

168. Rivera, D.; Alcaraz, F.; Obón, C. Wild and cultivated plants used as food and medicine by the Cimbrian ethnic minority in the Alps. *I Int. Symp. Med. Aromat. Nutraceutical Plants Mt. Areas* **2011**, *955*, 31–39. [CrossRef]

169. Alsakhawy, S.A.; Baghdadi, H.H.; El-Shenawy, M.A.; Sabra, S.A.; El-Hosseiny, L.S. Encapsulation of thymus vulgaris essential oil in caseinate/gelatin nanocomposite hydrogel: In vitro antibacterial activity and in vivo wound healing potential. *Int. J. Pharm.* **2022**, *628*, 122280. [CrossRef] [PubMed]

170. Al-Mijalli, S.H.; Mrabti, H.N.; Ouassou, H.; Flouchi, R.; Abdallah, E.M.; Sheikh, R.A.; Alshahrani, M.M.; Awadh, A.A.A.; Harhar, H.; Omari, N.E.; et al. Chemical Composition, Antioxidant, Anti-Diabetic, Anti-Acetylcholinesterase, Anti-Inflammatory, and Antimicrobial Properties of *Arbutus unedo* L. and *Laurus nobilis* L. Essential Oils. *Life* **2022**, *12*, 1876. [CrossRef] [PubMed]

171. Lee, E.H.; Shin, J.H.; Kim, S.S.; Lee, H.; Yang, S.R.; Seo, S.R. *Laurus nobilis* leaf extract controls inflammation by suppressing NLRP3 inflammasome activation. *J. Cell. Physiol.* **2019**, *234*, 6854–6864. [CrossRef]

172. Bouadid, I.; Amssayef, A.; Eddouks, M. Study of the Antihypertensive Effect of *Laurus nobilis* in Rats. *Cardiovasc. Hematol. Agents Med. Chem.* **2022**, *21*, 42–54. [CrossRef]

173. Benso, B.; Rosalen, P.L.; Alencar, S.M.; Murata, R.M. *Malva sylvestris* Inhibits Inflammatory Response in Oral Human Cells. An In Vitro Infection Model. *PLoS ONE* **2015**, *10*, e0140331. [CrossRef]

174. Braga, A.S.; Pires, J.G.; Magalhães, A.C. Effect of a mouthrinse containing *Malva sylvestris* on the viability and activity of microcosm biofilm and on enamel demineralization compared to known antimicrobials mouthrinses. *Biofouling* **2018**, *34*, 252–261. [CrossRef]

175. Gasparetto, J.C.; Martins, C.A.F.; Hayashi, S.S.; Otuky, M.F.; Pontarolo, R. Ethnobotanical and scientific aspects of *Malva sylvestris* L.: A millennial herbal medicine: Scientific evidences of *Malva sylvestris*. *J. Pharm. Pharmacol.* **2012**, *64*, 172–189. [CrossRef]

176. Martins, C.; Campos, M.; Irioda, A.; Stremel, D.; Trindade, A.; Pontarolo, R. Anti-Inflammatory Effect of *Malva sylvestris*, *Sida cordifolia*, and *Pelargonium graveolens* Is Related to Inhibition of Prostanoid Production. *Molecules* **2017**, *22*, 1883. [CrossRef]

177. Jeambey, Z.; Johns, T.; Talhouk, S.; Batal, M. Perceived health and medicinal properties of six species of wild edible plants in north-east Lebanon. *Public Health Nutr.* **2009**, *12*, 1902–1911. [CrossRef] [PubMed]

178. Prudente, A.S.; Loddi, A.M.V.; Duarte, M.R.; Santos, A.R.S.; Pochapski, M.T.; Pizzolatti, M.G.; Hayashi, S.S.; Campos, F.R.; Pontarolo, R.; Santos, F.A.; et al. Pre-clinical Anti-Inflammatory Aspects of a Cuisine and Medicinal Millennial Herb: *Malva sylvestris* L. *Food Chem. Toxicol.* **2013**, *58*, 324–331. [CrossRef] [PubMed]

179. Fahimi, S.; Abdollahi, M.; Mortazavi, S.A.; Hajimehdipoor, H.; Abdolghaffari, A.H.; Rezvanfar, M.A. Wound Healing Activity of a Traditionally Used Poly Herbal Product in a Burn Wound Model in Rats. *Iran. Red Crescent Med. J.* **2015**, *17*, e19960. [CrossRef] [PubMed]

180. Pirbalouti, A.G.; Shahrzad, A.; Abed, K.; Hamedi, B. Wound healing activity of Malva sylvestris and *Punica granatum* in alloxan-induced diabetic rats. *Acta Pol. Pharm.—Drug Res.* **2010**, *67*, 511–516.

181. Prudente, A.S.; Sponchiado, G.; Mendes, D.A.G.B.; Soley, B.S.; Cabrini, D.A.; Otuki, M.F. Pre-clinical efficacy assessment of *Malva sylvestris* on chronic skin inflammation. *Biomed. Pharmacother.* **2017**, *93*, 852–860. [CrossRef]

182. Mohamadi Yarijani, Z.; Najafi, H.; Shackebaei, D.; Madani, S.H.; Modarresi, M.; Jassemi, S.V. Amelioration of renal and hepatic function, oxidative stress, inflammation and histopathologic damages by *Malva sylvestris* extract in gentamicin induced renal toxicity. *Biomed. Pharmacother.* **2019**, *112*, 108635. [CrossRef]

183. Javid, A.; Motevalli Haghi, N.; Emami, S.A.; Ansari, A.; Zojaji, S.A.; Khoshkhui, M.; Ahanchian, H. Short-course administration of a traditional herbal mixture ameliorates asthma symptoms of the common cold in children. *Avicenna J. Phytomed.* **2019**, *9*, 126–133.

184. Bohlooli, S.; Mohebipoor, A.; Mohammadi, S.; Kouhnavard, M.; Pashapoor, S. Comparative study of fig tree efficacy in the treatment of common warts (*Verruca vulgaris*) vs. cryotherapy. *Int. J. Dermatol.* **2007**, *46*, 524–526. [CrossRef]

185. Juergens, U.R. Anti-inflammatory properties of the monoterpene 18-cineole: Current evidence for co-medication in inflammatory airway diseases. *Drug Res.* **2014**, *64*, 638–646. [CrossRef]

186. Zougagh, S.; Belghiti, A.; Rochd, T.; Zerdani, I.; Mouslim, J. Medicinal and Aromatic Plants Used in Traditional Treatment of the Oral Pathology: The Ethnobotanical Survey in the Economic Capital Casablanca, Morocco (North Africa). *Nat. Products Bioprospect.* **2019**, *9*, 35–48. [CrossRef] [PubMed]

187. Rosas-Piñón, Y.; Mejía, A.; Díaz-Ruiz, G.; Aguilar, M.I.; Sánchez-Nieto, S.; Rivero-Cruz, J.F. Ethnobotanical survey and antibacterial activity of plants used in the Altiplane region of Mexico for the treatment of oral cavity infections. *J. Ethnopharmacol.* **2012**, *141*, 860–865. [CrossRef]

188. Marchese, A.; Barbieri, R.; Coppo, E.; Orhan, I.E.; Daglia, M.; Nabavi, S.F.; Izadi, M.; Abdollahi, M.; Nabavi, S.M.; Ajami, M. Antimicrobial activity of eugenol and essential oils containing eugenol: A mechanistic viewpoint. *Crit. Rev. Microbiol.* **2017**, *43*, 668–689. [CrossRef]

189. Elmaksoud, H.A.A.; Motawea, M.H.; Desoky, A.A.; Elharrif, M.G.; Ibrahimi, A. Hydroxytyrosol alleviate intestinal inflammation, oxidative stress and apoptosis resulted in ulcerative colitis. *Biomed. Pharmacother.* **2021**, *142*, 112073. [CrossRef]

190. Fernández-Aparicio, Á.; Correa-Rodríguez, M.; Castellano, J.M.; Schmidt-RioValle, J.; Perona, J.S.; González-Jiménez, E. Potential Molecular Targets of Oleanolic Acid in Insulin Resistance and Underlying Oxidative Stress: A Systematic Review. *Antioxidants* **2022**, *11*, 1517. [CrossRef] [PubMed]

191. Lin, T.K.; Zhong, L.; Santiago, J.L. Anti-inflammatory and skin barrier repair effects of topical application of some plant oils. *Int. J. Mol. Sci.* **2018**, *19*, 70. [CrossRef] [PubMed]

192. Melguizo-rodríguez, L.; de Luna-Bertos, E.; Ramos-torrecillas, J.; Illescas-montesa, R.; Costela-ruiz, V.J.; García-martínez, O. Potential effects of phenolic compounds that can be found in olive oil on wound healing. *Foods* **2021**, *10*, 1642. [CrossRef] [PubMed]

193. Foxlee, R.; Johansson, A.C.; Wejfalk, J.; Dooley, L.; Del Mar, C.B. Topical analgesia for acute otitis media. *Cochrane Database Syst. Rev.* **2006**, *2011*, CD005657. [CrossRef]

194. Nawrot, J.; Wilk-jędrusik, M.; Nawrot, S.; Nawrot, K.; Wilk, B.; Dawid-Pać, R.; Urbańska, M.; Micek, I.; Nowak, G.; Gornowicz-porowska, J. Milky sap of greater celandine (*Chelidonium majus* L.) and anti-viral properties. *Int. J. Environ. Res. Public Health* **2020**, *17*, 1540. [CrossRef]

195. Nawrot, R. Defense-related Proteins from *Chelidonium majus* L. as Important Components of its Latex. *Curr. Protein Pept. Sci.* **2017**, *18*, 864–880. [CrossRef]

196. Rautio, M.; Sipponen, A.; Peltola, R.; Lohi, J.; Jokinen, J.J.; Papp, A.; Carlson, P.; Sipponen, P. Antibacterial effects of home-made resin salve from Norway spruce (*Picea abies*). *Apmis* **2007**, *115*, 335–340. [CrossRef] [PubMed]

197. Jokinen, J.J.; Sipponen, A. Refined Spruce Resin to Treat Chronic Wounds: Rebirth of an Old Folkloristic Therapy. *Adv. Wound Care* **2016**, *5*, 198–207. [CrossRef]

198. Goels, T.; Eichenauer, E.; Tahir, A.; Glasl, S.; Prochaska, P.; Hoeller, F.; Heiß, E.H. Exudates of *Picea abies*, *Pinus nigra*, and *Larix decidua*: Chromatographic Comparison and Pro-Migratory Effects on Keratinocytes In Vitro. *Plants.* **2023**, *11*, 599. [CrossRef]

199. Mitić, Z.S.; Jovanović, B.; Jovanović, S.Č.; Mihajilov-Krstev, T.; Stojanović-Radić, Z.Z.; Cvetković, V.J.; Mitrović, T.L.; Marin, P.D.; Zlatković, B.K.; Stojanović, G.S. Comparative study of the essential oils of four *Pinus* species: Chemical composition, antimicrobial and insect larvicidal activity. *Ind. Crops Prod.* **2018**, *111*, 55–62. [CrossRef]

200. Ciuman, R.R. Phytotherapeutic and naturopathic adjuvant therapies in otorhinolaryngology. *Eur. Arch. Oto-Rhino-Laryngol.* **2012**, *269*, 389–397. [CrossRef]

201. Majkić, T.; Bekvalac, K.; Beara, I. Plantain (*Plantago* L.) species as modulators of prostaglandin E2 and thromboxane A2 production in inflammation. *J. Ethnopharmacol.* **2020**, *262*. [CrossRef]

202. Kováč, I.; uľurkáč, J.; Hollý, M.; Jakubčová, K.; Peržeľová, V.; Mučaji, P.; Švajdlenka, E.; Sabol, F.; Legáth, J.; Belák, J.; et al. *Plantago lanceolata* L. water extract induces transition of fibroblasts into myofibroblasts and increases tensile strength of healing skin wounds. *J. Pharm. Pharmacol.* **2015**, *67*, 117–125. [CrossRef] [PubMed]

203. Kurt, B.; Bilge, N.; Sözmen, M.; Aydın, U.; Önyay, T. Özaydın Effects of *Plantago lanceolata* L. extract on full-thickness excisional wound healing in a mouse model. *Biotech. Histochem.* **2018**, *93*, 249–257. [CrossRef] [PubMed]

204. Miraj, S. A review study of pharmacological properties of *Plantago major* L. *Der Pharma Chem.* **2016**, *8*, 21–25.

205. Adom, M.B.; Taher, M.; Mutalabisin, M.F.; Amri, M.S.; Abdul Kudos, M.B.; Wan Sulaiman, M.W.A.; Sengupta, P.; Susanti, D. Chemical constituents and medical benefits of *Plantago major*. *Biomed. Pharmacother.* **2017**, *96*, 348–360. [CrossRef]

206. Najafian, Y.; Hamedi, S.S.; Kaboli Farshchi, M.; Feyzabadi, Z. *Plantago major* in Traditional Persian Medicine and modern phytotherapy: A narrative review. *Electron. Physician* **2018**, *10*, 6390–6399. [CrossRef]

207. Bhangale, J.; Acharya, S. Antiarthritic activity of *Cynodon dactylon* (L.) Pers. *Indian J. Exp. Biol.* **2014**, *52*, 215–222. [PubMed]

208. Sadki, C.; Hacht, B.; Souliman, A.; Atmani, F. Acute diuretic activity of aqueous Erica multiflora flowers and *Cynodon dactylon* rhizomes extracts in rats. *J. Ethnopharmacol.* **2010**, *128*, 352–356. [CrossRef]

209. Golshan, A.; Hayatdavoudi, P.; Hadjzadeh, M.A.-R.; Khajavi Rad, A.; Mohamadian Roshan, N.; Abbasnezhad, A.; Mousavi, S.M.; Pakdel, R.; Zarei, B.; Aghaee, A. Kidney stone formation and antioxidant effects of *Cynodon dactylon* decoction in male Wistar rats. *Avicenna J. Phytomed.* **2017**, *7*, 180–190.

210. Zhang, Y.; Liu, J.; Guan, L.; Fan, D.; Xia, F.; Wang, A.; Bao, Y.; Xu, Y. By-Products of *Zea mays* L.: A Promising Source of Medicinal Properties with Phytochemistry and Pharmacological Activities: A Comprehensive Review. *Chem. Biodivers.* **2023**, *20*, e202200940. [CrossRef] [PubMed]

211. Zhou, Y.X.; Xin, H.L.; Rahman, K.; Wang, S.J.; Peng, C.; Zhang, H. *Portulaca oleracea* L.: A review of phytochemistry and pharmacological effects. *BioMed Res. Int.* **2015**, *2015*, 925631. [CrossRef]

212. Vasilenko, T.; Kovac, I.; Slezak, M.; Durkac, J.; Perzelova, V.; Coma, M.; Kanuchova, M.; Urban, L.; Szabo, P.; Dvorankova, B.; et al. *Agrimonia eupatoria* L. Aqueous Extract Improves Skin Wound Healing: An In Vitro Study in Fibroblasts and Keratinocytes and In Vivo Study in Rats. *In Vivo (Brooklyn).* **2022**, *36*, 1236–1244. [CrossRef]

213. Mouro, C.; Dunne, C.P.; Gouveia, I.C. Designing New Antibacterial Wound Dressings: Development of a Dual Layer Cotton Material Coated with Poly(Vinyl Alcohol)_Chitosan Nanofibers Incorporating *Agrimonia eupatoria* L. Extract. *Molecules* **2021**, *26*, 83. [CrossRef]

214. Qian, H.; Wen-Jun, H.; Hao-Jie, H.; Qing-Wang, H.; Li, C.; Liang, D.; Jie-Jie, L.; Xiang, L.; Ya-Jing, Z.; Ying-Zhi, M.; et al. The four-herb chinese medicine ANBP enhances wound healing and inhibits scar formation via bidirectional regulation of transformation growth factor pathway. *PLoS ONE* **2014**, *9*, e112274. [CrossRef]

215. Malheiros, J.; Simões, D.M.; Figueirinha, A.; Cotrim, M.D.; Fonseca, D.A. *Agrimonia eupatoria* L.: An integrative perspective on ethnomedicinal use, phenolic composition and pharmacological activity. *J. Ethnopharmacol.* **2022**, *296*, 115498. [CrossRef]

216. Watkins, F.; Pendry, B.; Sanchez-Medina, A.; Corcoran, O. Antimicrobial assays of three native British plants used in Anglo-Saxon medicine for wound healing formulations in 10th century England. *J. Ethnopharmacol.* **2012**, *144*, 408–415. [CrossRef] [PubMed]

217. Santos, T.N.; Costa, G.; Ferreira, J.P.; Liberal, J.; Francisco, V.; Paranhos, A.; Cruz, M.T.; Castelo-Branco, M.; Figueiredo, I.V.; Batista, M.T. Antioxidant, Anti-Inflammatory, and Analgesic Activities of *Agrimonia eupatoria* L. Infusion. *Evidence-based Complement. Altern. Med.* **2017**, *2017*, 8309894. [CrossRef]

218. Martino, E.; Collina, S.; Rossi, D.; Bazzoni, D.; Gaggeri, R.; Bracco, F.; Azzolina, O. Influence of the extraction mode on the yield of hyperoside, vitexin and vitexin-2-O-rhamnoside from *Crataegus monogyna* Jacq. (Hawthorn). *Phytochem. Anal.* **2008**, *19*, 534–540. [CrossRef]

219. Hooman, N.; Mojab, F.; Nickavar, B.; Pouryousefi-Kermani, P. Diuretic effect of powdered *Cerasus avium* (cherry) tails on healthy volunteers. *Pak. J. Pharm. Sci.* **2009**, *22*, 381–383. [PubMed]

220. Negrean, O.R.; Farcas, A.C.; Pop, O.L.; Socaci, S.A. Blackthorn—A Valuable Source of Phenolic Antioxidants with Potential Health Benefits. *Molecules* **2023**, *28*, 3456. [CrossRef]

221. Tiboni, M.; Coppari, S.; Casettari, L.; Guescini, M.; Colomba, M.; Fraternale, D.; Gorassini, A.; Verardo, G.; Ramakrishna, S.; Guidi, L.; et al. *Prunus spinosa* extract loaded in biomimetic nanoparticles evokes in vitro anti-inflammatory and wound healing activities. *Nanomaterials* **2021**, *11*, 36. [CrossRef]

222. Coppari, S.; Colomba, M.; Fraternale, D.; Brinkmann, V.; Romeo, M.; Rocchi, M.B.L.; Di Giacomo, B.; Mari, M.; Guidi, L.; Ramakrishna, S.; et al. Antioxidant and anti-inflammaging ability of prune (*Prunus spinosa* L.) extract result in improved wound healing efficacy. *Antioxidants* **2021**, *10*, 374. [CrossRef]

223. Ayati, Z.; Amiri, M.S.; Ramezani, M.; Delshad, E.; Sahebkar, A.; Emami, S.A. Phytochemistry, Traditional Uses and Pharmacological Profile of Rose Hip: A Review. *Curr. Pharm. Des.* **2018**, *24*, 4101–4124. [CrossRef] [PubMed]

224. van Wietmarschen, H.; van Steenbergen, N.; van der Werf, E.; Baars, E. Effectiveness of herbal medicines to prevent and control symptoms of urinary tract infections and to reduce antibiotic use: A literature review. *Integr. Med. Res.* **2022**, *11*, 100892. [CrossRef]

225. Wang, Y.; Zhao, Y.; Liu, X.; Li, J.; Zhang, J.; Liu, D. Chemical constituents and pharmacological activities of medicinal plants from *Rosa* genus. *Chinese Herb. Med.* **2022**, *14*, 187–209. [CrossRef] [PubMed]

226. Malanga, G.A.; Yan, N.; Stark, J. Mechanisms and efficacy of heat and cold therapies for musculoskeletal injury. *Postgrad. Med.* **2015**, *127*, 57–65. [CrossRef]

227. Mokhtar, M.; Youcefi, F.; Keddari, S.; Saimi, Y.; Otsmane Elhaou, S.; Cacciola, F. Phenolic Content and in Vitro Antioxidant, Anti-Inflammatory and antimicrobial Evaluation of Algerian *Ruta graveolens* L. *Chem. Biodivers.* **2022**, *19*, e202200545. [CrossRef]

228. Ratheesh, M.; Sindhu, G.; Helen, A. Anti-inflammatory effect of quinoline alkaloid skimmianine isolated from *Ruta graveolens* L. *Inflamm. Res.* **2013**, *62*, 367–376. [CrossRef]

229. Lans, C. Do recent research studies validate the medicinal plants used in British Columbia, Canada for pet diseases and wild animals taken into temporary care? *J. Ethnopharmacol.* **2019**, *236*, 366–392. [CrossRef]

230. Kenny, O.M.; McCarthy, C.M.; Brunton, N.P.; Hossain, M.B.; Rai, D.K.; Collins, S.G.; Jones, P.W.; Maguire, A.R.; O'Brien, N.M. Anti-inflammatory properties of potato glycoalkaloids in stimulated Jurkat and Raw 264.7 mouse macrophages. *Life Sci.* **2013**, *92*, 775–782. [CrossRef] [PubMed]

231. Visvanathan, R.; Jayathilake, C.; Chaminda Jayawardana, B.; Liyanage, R. Health-beneficial properties of potato and compounds of interest: Health-beneficial properties of potato. *J. Sci. Food Agric.* **2016**, *96*, 4850–4860. [CrossRef] [PubMed]

232. Basilicata, M.G.; Pepe, G.; Rapa, S.F.; Merciai, F.; Ostacolo, C.; Manfra, M.; Di Sarno, V.; Autore, G.; De Vita, D.; Marzocco, S.; et al. Anti-Inflammatory and Antioxidant Properties of Dehydrated Potato-Derived Bioactive Compounds in Intestinal Cells. *Int. J. Mol. Sci.* **2019**, *20*, 6087. [CrossRef]

233. Joshi, B.C.; Mukhija, M.; Kalia, A.N. Pharmacognostical review of *Urtica dioica* L. *Int. J. Green Pharm.* **2014**, *8*, 201–209. [CrossRef]

234. Asgarpanah, J.; Mohajerani, R. Phytochemistry and pharmacologic properties of *Urtica dioica* L. *J. Med. Plants Res.* **2012**, *6*, 5714–5719.

235. Bahramsoltani, R.; Rostamiasrabadi, P.; Shahpiri, Z.; Marques, A.M.; Rahimi, R.; Farzaei, M.H. *Aloysia citrodora* Paláu (Lemon verbena): A review of phytochemistry and pharmacology. *J. Ethnopharmacol.* **2018**, *222*, 34–51. [CrossRef]

236. Picon, P.D.; Picon, R.V.; Costa, A.F.; Sander, G.B.; Amaral, K.M.; Aboy, A.L.; Henriques, A.T. Randomized clinical trial of a phytotherapic compound containing *Pimpinella anisum*, *Foeniculum vulgare*, *Sambucus nigra*, and *Cassia augustifolia* for chronic constipation. *BMC Complement. Altern. Med.* **2010**, *10*, 17. [CrossRef]

237. Porter, R.S.; Bode, R.F. A Review of the Antiviral Properties of Black Elder (*Sambucus nigra* L.) Products: Antiviral Properties of Black Elder (*Sambucus nigra* L.). *Phyther. Res.* **2017**, *31*, 533–554. [CrossRef] [PubMed]

238. Chen, C.; Zuckerman, D.M.; Brantley, S.; Sharpe, M.; Childress, K.; Hoiczyk, E.; Pendleton, A.R. *Sambucus nigra* extracts inhibit infectious bronchitis virus at an early point during replication. *BMC Vet. Res.* **2014**, *10*, 12. [CrossRef] [PubMed]

239. Hawkins, J.; Baker, C.; Cherry, L.; Dunne, E. Black elderberry (*Sambucus nigra*) supplementation effectively treats upper respiratory symptoms: A meta-analysis of randomized, controlled clinical trials. *Complement. Ther. Med.* **2019**, *42*, 361–365. [CrossRef] [PubMed]

240. Harnett, J.; Oakes, K.; Carè, J.; Leach, M.; Brown, D.; Holger, C.; Pinder, T.-A.; Steel, A.; Anheyer, D. The effects of *Sambucus nigra* berry on acute respiratory viral infections: A rapid review of clinical studies. *Adv. Integr. Med.* **2020**, *7*, 240–246. [CrossRef]

241. Skowrońska, W.; Granica, S.; Czerwińska, M.E.; Osińska, E.; Bazylko, A. *Sambucus nigra* L. leaves inhibit TNF-α secretion by LPS-stimulated human neutrophils and strongly scavenge reactive oxygen species. *J. Ethnopharmacol.* **2022**, *290*, 115116. [CrossRef] [PubMed]

242. Poles, J.; Karhu, E.; Mcgill, M.; Mcdaniel, H.R.; Lewis, J.E. The effects of twenty-one nutrients and phytonutrients on cognitive function: A narrative review. *J. Clin. Transl. Res.* **2021**, *7*, 333–376. [CrossRef]

243. Mao, Q.Q.; Xu, X.Y.; Cao, S.Y.; Gan, R.Y.; Corke, H.; Beta, T.; Li, H. Bin Bioactive compounds and bioactivities of ginger (*Zingiber officinale* Roscoe). *Foods* **2019**, *8*, 185. [CrossRef]

244. Karampour, N.S.; Arzi, A.; Rezaie, A.; Pashmforoosh, M.; Kordi, F. Gastroprotective effect of zingerone on ethanol-induced gastric ulcers in rats. *Medicina* **2019**, *55*, 64. [CrossRef]

245. Nikkhah Bodagh, M.; Maleki, I.; Hekmatdoost, A. Ginger in gastrointestinal disorders: A systematic review of clinical trials. *Food Sci. Nutr.* **2019**, *7*, 96–108. [CrossRef]

Article

Diverse in Local, Overlapping in Official Medical Botany: Critical Analysis of Medicinal Plant Records from the Historic Regions of Livonia and Courland in Northeast Europe, 1829–1895

Julia Prakofjewa [1], Martin Anegg [1], Raivo Kalle [2,*], Andra Simanova [3,4], Baiba Prūse [1], Andrea Pieroni [2,5] and Renata Sõukand [1]

1   Department of Environmental Sciences, Informatics and Statistics, Ca' Foscari University of Venice, Via Torino 155, Mestre, 30172 Venezia, Italy; yuliya.prakofyeva@unive.it (J.P.); martin.anegg@gmx.at (M.A.); baiba.pruse@unive.it (B.P.); renata.soukand@unive.it (R.S.)
2   University of Gastronomic Sciences, Piazza Vittorio Emanuele II 9, 12042 Pollenzo, Italy; a.pieroni@unisg.it
3   Department of Latvian and Baltic Studies, University of Latvia, Visvalža Str. 4a, LV-1050 Riga, Latvia; andra.simanova@inbox.lv
4   Institute for Environmental Solutions, "Lidlauks", Priekuļi Parish, LV-4126 Cēsis, Latvia
5   Medical Analysis Department, Tishk International University, 100 Meter Street & Mosul Road, Erbil 44001, Iraq
*   Correspondence: raivo.kalle@mail.ee

Citation: Prakofjewa, J.; Anegg, M.; Kalle, R.; Simanova, A.; Prūse, B.; Pieroni, A.; Sõukand, R. Diverse in Local, Overlapping in Official Medical Botany: Critical Analysis of Medicinal Plant Records from the Historic Regions of Livonia and Courland in Northeast Europe, 1829–1895. *Plants* **2022**, *11*, 1065. https://doi.org/10.3390/plants11081065

Academic Editor: Ana Maria Carvalho

Received: 8 March 2022
Accepted: 10 April 2022
Published: 13 April 2022

Publisher's Note: MDPI stays neutral with regard to jurisdictional claims in published maps and institutional affiliations.

Abstract: Works on historical ethnobotany can help shed light on past plant uses and humankind's relationships with the environment. We analyzed medicinal plant uses from the historical regions of Livonia and Courland in Northeast Europe based on three studies published within the 19th century by medical doctors researching local ethnomedicine. The sources were manually searched, and information extracted and entered into a database. In total, there were 603 detailed reports of medicinal plant use, which refer to 219 taxa belonging to 69 families and one unidentified local taxon. Dominant families were Asteraceae (14%), Solanaceae (7%), Rosaceae (6%), and Apiaceae (5%). The majority of use reports were attributed to the treatment of four disease categories: digestive (24%), skin (22%), respiratory (11%), and general (11%). The small overlapping portion (14 taxa mentioned by all three authors and another 27 taxa named by two authors) contained a high proportion of taxa (46%) mentioned in Dioscorides, which were widespread during that period in scholarly practice. Despite the shared flora, geographical vicinity, and culturally similar backgrounds, the medicinal use of plants in historical Courland and Livonia showed high biocultural diversity and reliance on wild taxa. We encourage researchers to study and re-evaluate the historical ethnobotanical literature and provide some suggestions on how to do this effectively.

Keywords: historical ethnobotany; local ecological knowledge; old herbals; scholarly medicine; Livonia; Courland

1. Introduction

Historical ethnobotanical research has recently become an area of growing importance for researchers. The analysis of such data provides the grounds for a better understanding of how the Local Ecological Knowledge (LEK) of societies evolved over time, how these societies have used (local) plants, and how they have interacted with the environment and its components [1–4]. Nevertheless, comparing historical and contemporary data is not as easy as it may appear because of changes in the social, cultural, and socio-economic conditions of the studied societies [5]. However, investigators should continue the current trend of systemizing historical data collected by representatives of diverse scientific fields and extract the local ecological knowledge for future analysis [6,7].

Historical texts on plant use date back to around 3100 BC. One text from the common era is *"De Materia Medica"* [8], written by Greek physician Pedanius Dioscorides (AD 40–90), which inspired the medicine of that period and influenced many herbals published in the second half of the second millennium and especially earlier in the Middle Ages. During that period, herbal texts and recipe books were used as a standard means of creating knowledge about the medicinal usages of plants available. In the 19th century, inspired by Swedish botanist Carl von Linné (1707–1778), the gathering of local ethnomedicinal knowledge became popular, especially in the Nordic hemisphere. Such collections were also sometimes analyzed and published, inspiring future ethnobotanical research [6,9–11]. Such books, in addition to doctors and pharmacies, were important resources for the literate population for acquiring knowledge of simpler and more affordable medicines, and they contained, among local uses, traces of Dioscorides's work. Leonti et al. [12–14] clearly demonstrated that in Italy the influence of the doctor and naturalist Pietro Andrea Mattioli's (1501–1577) work is still apparent in studies from the last few decades: up to 20% can be traced back to Mattioli, and thus also, Dioscorides [14].

Although historical materials may be of crucial significance for understanding the evolution of local ecological knowledge, special attention needs to be given to the background and conditions in which the works were complied. Sõukand et al. [15] presented an example of a specific taxon (*Epilobium angustifolium*) that through confusion created by the existence of multiple concurrent names for the same species, incorrect translations, and illustrations supporting the transfer of usages from other species, led to a chain of misunderstandings and misinterpretations. In addition, not only the nomenclature of plants, but also disease names, change over time and specific knowledge is needed for interpretation. Therefore, special care needs to be taken in analyzing historical publications.

1.1. Background

While digitizing the German-language literature containing Local Ecological Knowledge of the 19th century for a database [16], as a practical part of the Master's thesis of the second author [17], we selected three scientifically sound and valuable ethnobotanical works published in German on neighboring Livonia and Courland [18–20], which we consider to be the first summary studies on folk herbal medicine in that region.

Dr. Johann Wilhelm Ludwig von Luce's (1756–1842) book *"Heilmittel der Ehsten auf der Insel Oesel"* (*Remedies of the Estonians of Oesel Island*), which was issued in 1829 in Pernau (Pärnu) [18], presents an original study on the local ethnomedicine of the island (currently Saaremaa). Having worked on the island for 38 years, first as a pastor and later as a doctor, Luce presented on 159 pages his own experiments as well as the local ethnomedical knowledge. Although his earlier book (in 1823) *"Topographische Nachrichten von der Insel Oesel, in medicinischer und ökonomischer Hinsicht"* (*Topographical News from the Island of Oesel, in Medical and Economic Terms*) [21] already contained the plant uses described in his earlier publication, we decided to focus our analysis only on the later publication as it is more voluminous.

Luce came to Saaremaa from Germany in 1781 as a pastor. After the tragic death of his wife, he returned to Germany to study medicine in 1789–1792. In 1801, he defended his doctor's exam in St. Petersburg and after that became a practicing doctor in Saaremaa. His knowledge of botany was profound and his contribution to the scientific discipline is acknowledged by the presence of his own botanical author abbreviation. He wrote his book in order to share his experiences with local ecological knowledge, making it available to a wider audience [18,22]. He divided his book into chapters covering mineral, herbal (including pharmaceuticals herbs), animal, instrumental (also bloodletting, steam baths, and massage), and fantastic (magical rituals) components of respective remedies. All the reports were gathered almost exclusively on Oesel, although it has often been cited as reflecting data covering the whole of Estonia.

Jewish doctor Emil Aronson's (1863–1942) article *"Ueber die Volksheilmittel der Letten"* (*On the Folk Remedies of the Latvians*), which was issued in 1891, is a 19-page-long contribution to the 19th volume of the journal "*Magazin lettisch-literarischen Gesellschaft*" (*Magazine of the Latvian Literature Society*) [19]. He explained the need for his work by citing the lack of Latvian data outlined in the doctoral dissertation of Dr. Wassily Demitsch (Василий Федорович Демич) (1858–1930), "*Literärische Studien über die wichtigsten russischen Volksheilmittel aus dem Pflanzenreiche*" (*Literary Studies about the Most Important Russian Folk Remedies from the Plant Kingdom*), which the latter defended at the University of Dorpat (Tartu) in 1888 [23]. Aronson studied medicine at the same university. In addition, he followed the structure of Luce's book: sorted his own article by type of medicine, categorizing them into mineral, herbal, animal constituents, and applications. The data presented by Aronson originated almost exclusively from Libau (currently Liepāja), where his own doctor's office was located, although he presented the data as filling the gap for Latvia in general. Aronson sometimes compared his results with those of Luce and Demitsch or referred to them for additional information.

In the foreword to his work, Aronson acknowledged the importance of documenting lay uses without any prejudice or contempt. Aronson wanted to show that, although most usages are superstition- or curiosity-driven, some could still be useful for contemporary medicinal science, and thus, there is the need to identify the good ones. He considered local medicines affordable and obtainable by everyone, while their effectiveness was supported by local beliefs and culture. Aronson did not provide any local names for plants or diseases. In 1893, he relocated to Dallas, USA, where he became a pioneer in public health.

Latvian medical student Jēkabs Alksnis's (1870–1957) article "*Materialien zur lettischen Volksmedizin*" (*Materials on Latvian Folk Medicine*) [20] was issued in 1894 in the fourth yearbook of the University of Dorpat (Tartu), "*Historische Studien aus dem pharmakologischen Institute der kaiserlichen Universität Dorpat*" (*Historical Studies from the Pharmacological Institute of the Imperial University of Dorpat*). Alksnis studied in Tartu from 1890 to 1895. In the preface of the article, he mentions that he started this work at the request of Professor Eduard Rudolf Kobert (1854–1918), who also instructed and advised him. Alksnis apologizes for leaving a lot of material out of this work because of the length limit of the article, which was 117 pages long. It was mainly a summary translation into German of information from Latvian and Russian sources. For example, he translated with the permission of folklorist and poet Fricis Brīvzemnieks [Fricis Treilands] (1846–1907), a chapter on incantations from a book issued in 1881 in the Russian language "*Труды Этнографического отдела, Материалы по этнографии латышского племени*" (*Proceedings of the Ethnographic Department, Materials on the Ethnography of the Latvian Tribe*). In addition, he used many newspapers, such as the supplement newspaper "*Dienas Lapa*", which published Latvian ethnographic writings. In the same newspaper, he published a call in 1892 for the collection of folk medicine and provided recommendations on how to collect data correctly [24]. Alksnis added many of his own experiences and mentions the names of other doctors, one sent him dried plant samples (Dr. P. Kalniņš) and another helped him to describe folk diseases (Dr. Raphael). The plants sent by Dr. P. Kalniņš were identified later by a botanist named Dr. Johannes Christoph Klinge. Alksnis also used folk medicine material from the Riga Latvian Society, which was sent to the society or collected by members of the society themselves.

It was not until 1898, however, that Alksnis published previously unpublished materials held by the Riga Latvian Society [25]. However, we have not analyzed this data in the present work, since at that time, general research on Latvian folk medicine became more active, and many similar articles began to appear in Latvian. One of these was written by the first Latvian botanist, Jānis Ilsters (1851–1889), who highlighted the use of Latvian folk medicines and folk plant names. His book contains descriptions of plants, yet lacks information about the source, clearly presenting facts from other countries [26]. In addition, he published several appeals to the public to help in reporting plant names and their application in folk medicine [27], but the data he collected were published posthumously in 1891 [28]. However, Ilster's article is not included in the literature cited at the end of Alksnis's article. Also, Riga pharmacist Ernests Birzmanis (Birsmanis) (1860–1900) published a call in 1897 in the newspaper "*Latweeschu Awises*" for the collection of folk medicine to gather information about treatments using plants. However, unlike Alksnis, he also paid attention to folk plant names [29]. Furthermore, he published one of the first Latvian-language books on Latvian medicinal plants [30], in which he indicated the most common folk plant names. He, too, studied at the University of Dorpat (Tartu), where he graduated in 1892 with a Master's degree in pharmacy.

Alksnis outlines the scientific goals of his work as translating existing knowledge from Russian, educating Latvian doctors on local ethnomedicine and promoting the rationalization of drug administration, reflecting the scientific approach in the way his article is organized: background information on a disease is complimented by details on healing practices, drug preparations, and components. He provides local names only for the diseases, and not for the plants. Alksnis's article can be seen as representative, as it covered the entire area inhabited by Latvians in Livonia and Courland. He later worked as a surgeon and was a professor of medicine at the University of Latvia from 1924 to 1944, emigrating to Germany during World War II and after the war to the UK. Today, the work of Alksnis holds great value for Latvian folk medicine history [31].

1.2. Aims of the Work

Dr. Wassily Demitsch's doctoral dissertation was the first scientific study of plants used in folk medicine to summarize areas of Russia. However, as he stated, he reported only a small part of his work, describing only the most popular plants (he had over 65 species on the list). He says that there was a great deal of overlap with ancient Greco-Roman plant uses [23]. Other authors have suggested that there is a high degree of overlap with Hippocrates and Dioscorides in Russian territories [32]. At the same time, Demitsch stressed that doctors with an academic education did not care about or evaluate folk medicine and only relied on active ingredient-based (very expensive) treatments. However, people have prejudices and beliefs that prevent new therapies from being accepted by them. Demitsch notes that it is the cultural background of the community that could help the doctor to better explain to the people what is rational and what is not rational. A few years later, Leopold Glück (1854–1907), a Polish physician and public figure of Jewish descent, also emphasized the need to study the non-rational methods of folk medicine [33]. Thus, at that time, the contribution of general practitioner doctors to the study of folk medicine was significant.

What was also decisive for our choice was the fact that Luce, Aronson, and Alksnis shared a feature in common: they were doctors or medical students. As they lived during the same century and within close proximity to each other, the analysis of their work provides grounds for a detailed comparison. This allows shedding light on the biocultural diversity related to ethnomedicine and provide comparative data for field studies from the region.

The aims of this study are (a) to update and reconcile with current knowledge the identification of plants and diseases, identifying potential mistakes in the initial sources; (b) to compare the local plant uses described by the three authors; and (c) to evaluate the diversity of the sources. We expect to find high biocultural diversity in the three published sources.

2. Results
2.1. Disease and Plant Identification

The majority of all medical conditions described in the books could be assigned to specific modern disease categories. There were, however, some exceptions; for example, we assigned *artheibisches Fieber* mentioned by Luce solely to the general fever category, while *rose* mentioned by Alksnis can generally be identified as a skin disease, as there are various types of *rose* according to Luce. Likewise, the symptom *sich verhoben haben/sich verrissen haben/Verreissung*, attributed to "working too hard or with the wrong posture" [18,20], is a common condition described in Estonian folklore [34]. Not having more details to rely on, we interpreted it as indifferent back pain (musculoskeletal disease category).

Plant identification was sometimes a challenge, even though most of the time the authors provided the Latin name of the plant. One such example for Luce is *Cnicus serratuloides*, which is *Cirsium serratuloides* according to Plants of the World Online (https://powo.science.kew.org/, accessed on 7 March 2022). However, it was absent from the local floras, as the study area is outside of its natural range. The Estonian local name did not provide further clarity, and therefore, plant identification was kept at the genus level: *Cirsium* sp. Another example is *Ononis repens*, which was identified instead of *Ononis spinosa*, as given by Luce: *O. spinosa* does not grow in Saaremaa and according to a book of Estonian folk plant names [35], the term *luuderohi* is associated with *O. repens* in the Püha parish in Saaremaa. There are also mistakes outside the same genus, for example: *Hippocrepis comosa* was re-identified as *Argentina anserina*, as this taxon does not grow in Saaremaa and it is a clear misidentification by Luce, since the two taxa are visually very similar, while the name *hoolmerohi* was widespread in Kihelkonna parish in Saaremaa for *Argentina anserina* according to the book of Estonian folk plant names [35]. A difficult case of identification was a plant identified by Luce as *Equisetum fragile*, as such a name does not exist; however, its German name is *Engelsüß*, local name *rinna rohhi*, and it was used to treat a cough referred to as *Polypodium vulgare*.

The book by Alksnis also contained cases of difficult identification, like *Lappa* and *Lappa tournefortii*. While *Lappa* could refer to *Arctium lappa*, the most common taxon in the region is *Arctium tomentosum*, and the two are not differentiated on the popular level. Thus, we assigned both records to *Arctium* spp. *Thymus chamaedrys* is a wild thyme species in Western Europe, but it does not grow wild in modern-day Latvia, and thus it was re-identified as the local wild thyme *Thymus serpyllum*.

The only instance in which Alksnis confesses to having failed to identify a plant is described as follows: The Latvian people have a disease which they call the "suffocating" or the "choking" (speedejs un schnaudsejs). It is very bad: the sick roll on their beds and tear their hair out in despair "pinched". From this description it follows that we are dealing with colic here. A herb called "speedeja sahle" (i.e., herb against the colic) is said to be very effective against this condition, its Latin name I have not yet been able to determine. [20] (pp. 191–192). We list the plant in our table as an unidentified taxon, but do not consider it in other analyses, unless explicitly named.

Two other taxa in Alksnis's records needed special attention. *Sedum vulgare* is absent in the studied local floras [36–38] and has a single record in Estonia from 1864, the other records only come from Central Europe. Similarly, *Aconitum lycoctonum* has only a few records [37,39]. However, it is not similar enough to the local widespread taxon *Aconitum napellus*, which has blue flowers. Therefore, we identified the two taxa as *Sedum* sp. and *Aconitum* sp., respectively.

Alksnis in particular, but the other two authors as well, sometimes provided only the German name of some household cultivars like *Linde* (lime tree—*Tilia* spp.), *Kohl* (cabbage—*Brassica oleracea* L.), *Pflaumensaft* (plume juice), and *Turmkraut* (tower mustard—*Turritis glabra* L.). Cabbage was identified on the basis of the way in which it was prepared (fermented), as *Brassica oleraceae* L. was the only possible species that was prepared in that way at the time.

We identified some taxa solely to the genus level, which were also not differentiated on the popular level; for example, there are two taxa of *Tilia* growing in the area, of these *Tilia cordata* was most likely used, yet *T. latifolia* is also common, especially in cultivation. Of the four possible *Betula* species growing in the region, the ones most likely used were *B. pendula* and *B. pubescens*.

2.2. Overview of the Reported Taxa and Comparison between the Three Authors

In total, there were 603 detailed reports of medicinal plant use, which refer to 219 taxa belonging to 69 families and one unidentified local taxon (Table 1). The dominant families were Asteraceae (14%), Solonaceae (7%), Rosaceae (6%), and Apiaceae (5%) (Figure 1). The majority of Detailed Use Report (DUR) was attributed to the treatment of four disease categories: digestive (24%), skin (22%), respiratory (11%), and general (11%).

Table 1. Plants named by all three authors and general disease categories.

Family	Latin Name	Name in the Source	Local Name	Origin of Plant	Luce	Aronson	Alksnis
Acoraceae	*Acorus calamus* L.	*Acorus Calamus*		C, W [F]			DIGE, GEUN, MUSC
Amaranthaceae	*Atriplex* sp.	*Atriplex*		W, C			SKIN
	Beta vulgaris L.	*Beta vulgaris*		C			DIGE, RESP
Amaryllidaceae	*Allium cepa* L.	*Allium cepa*	Sibbulas	C	GEUN	SKIN	GENI, PSYC, RESP, SKIN
	Allium sativum L.	*Allium sativum*	Küislauk	C	DIGE		EAR
	Allium schoenoprasum L.	*Allium schoenoprasum*		W			RESP
	Alloideae sp.	*Allioideae*		W			NEUR
Apiaceae	*Angelica archangelica* L.	radix *Angelicae*		C [F], P(?)			DIGE
	Carum carvi L.	*Carum carvi*	Köömled	W, C, P [F]	DIGE, PCFP		DIGE, ENDO, RESP
	Cicuta virosa L.	*Cicuta virosa*		W [F]			GEUN, NEUR, SKIN
	Daucus carota L.	*Daucus Carota*		C			DIGE
	Ferula assa-foetida L.	*Ferula asa foetida, Scorodosma foetidum*	Tiwistriik	P	CULT	DIGE	PSYC
	Laserpitium latifolium L.	*Laserpitium latifolium*		W			DIGE
	Levisticum officinale W.D.J.Koch	*Levisticum officinale*	Liibstocki rohhi	C [F]	SKIN	GEUN	CARD GEUN MUSC NEUR
	Petroselinum crispum (Mill.) Fuss	*Petroselinum crispum*		C			DIGE, GEUN, SKIN, UROL
	Peucedanum ostruthium (L.) W.D.J.Koch	*Peucedanum ostruthium,* Radix Imperatoriae		P			DIGE
	Pimpinella sp.	*Pimpinella* L.		W	GEUN		CARD
Araliaceae	*Hedera helix* L.	*Hedera helix*	Ragga mailase rohhi, lude rohhi	W [F]	MUSC, SKIN		
Asparagaceae	*Polygonatum odoratum* (Mill.) Druce	*Convallaria polygonatum*		W [F]			MUSC
Asphodelaceae	*Aloe* sp.	*Aloe*		C			DIGE
Asteraceae	*Achillea millefolium* L.	*Achillea millefolium*	Raudrohhi	W, C [F]	SKIN	RESP	BLIM, RESP, SKIN
	Anthemis arvensis L.	*Anthemis arvensis*		W			SKIN
	Arctium spp.	*Lappa* [AL], *Lappa tournefortii* [AL], *Arctium Lappa* L. [D]		W [F]			NEUR, SKIN

Table 1. *Cont.*

Family	Latin Name	Name in the Source	Local Name	Origin of Plant	Luce	Aronson	Alksnis
	Arnica montana L.	*Arnica montana* [L, AL], not stated [AR]	Ärratöstmise-haiguse rohhi	P, (W [F])	MUSC	GEUN, MUSC	DIGE, GEUN, MUSC, RESP, SKIN
	Artemisia abrotanum L.	*Artemisia abrotanum*		C [F]			GENI, SKIN
	Artemisia absinthium L.	*Artemisia absinthium* [L], *Artemisia Absynthium* [AL], *Artemisia Absynthium* [AR]	Koi rohhi	C, (W [F])	DIGE, GEUN, SKIN	DIGE, GEUN	DIGE, GEUN, PSYC
	Artemisia cina Berg ex Poljakov	*Flores cinae*		P		DIGE	DIGE
	Artemisia sieberi Besser	*Artemisia sieberi*	Ussi rohhi	P	DIGE		
	Artemisia vulgaris L.	*Artemisia vulgaris*		W [F]			NEUR
	Calendula officinalis L.	*Calendula officinalis*	Koltsed aja öied	C	GENI, RESP, SKIN		DIGE
	Cirsium vulgare (Savi) Ten.	*Cirsium lanceolatum*		W			RESP
	Cota tinctoria (L.) J.Gay	*Anthemis tinctoria*		W [F]			DIGE
	Centaurea cyanus L.	*Centaurea cyanus*		W, C [F]			EYE, PSYC, RESP, UROL
	Helichrysum arenarium (L.) Moench	*Helichrysum arenarium*		W			SKIN
	Jacobaea vulgaris Gaertn.	*Jacobaea vulgaris*	Rist hoolmete rohhi	W	GENI		
	Leucanthemum vulgare (Vaill.) Lam.	*Chrysanthemum Leucanthemum*		W			DIGE, SKIN
	Matricaria chamomilla L.	*Matricaria Chamomilla* [L], not stated [AL]	Kummelid	P, C [F]	GEUN	EYE, PCFP, PSYC	CARD, DIGE, GEUN, PCFP
	Solidago virgaurea L.	*Solidago virgaurea*	Hoolmete rohhi	W [F]	DIGE, SKIN		
	Tanacetum vulgare L.	*Tanacetum vulgare*	Reinware rohhi, solika rohhi	W, C [F]	DIGE		DIGE
	Taraxacum officinale F.H.Wigg. (coll.)	*Taraxacum campylodes* [L], *Leontodon Taraxacum* [AL, AR]	Sea öied, sea pima rohhi, sea nuppud, woi rosid	W [F]	SKIN	SKIN	SKIN
	Tussilago farfara L.	*Tussilago farfara*	Paiso lehhed	W [F]	SKIN		GEUN
Balsaminaceae	*Impatiens nolitangere* L.	*Impatiens tangere noli*		W [F]			SKIN
	Alnus glutinosa (L.) Gaertn.	*Alnus glutinosa*		W			DIGE, SKIN
Betulaceae	*Betula* spp.	*Betula pubescens* [L], *Betula alba* [AR]	Kasse pu	W, C [F]	BLIM, DIGE	DIGE	DIGE, MUSC, PFCP, SKIN

Table 1. *Cont.*

Family	Latin Name	Name in the Source	Local Name	Origin of Plant	Luce	Aronson	Alksnis
Boraginaceae	*Myosotis* sp.	*Myosotis*		W			PSYC
	Symphytum officinale L.	*Symphytum officinale*		W [F]			SKIN
Brassicaceae	*Armoracia rusticana* P.Gaertn., B.Mey. & Scherb.	*Cochlearia armoracia*		C [F]			DIGE
	Berteroa incana (L.) DC.	*Berteroa incana*		W			PSYC
	Brassica oleraceae L.	Saures Kohlblatt, sauerer Kohl		C			NEUR
	Cardamine pratensis L.	*Cardamine*		W			CARD
	Cochlearia officinalis L.	*Cochlearia officinalis*		P			CARD
	Raphanus raphanistrum subsp. *sativus* (L.) Domin	*Raphanus niger*		W[F]			RESP, MUSC
Campanulaceae	*Campanula trachelium* L.	*Campanula trachelium*		W			GEUN
Cannabaceae	*Cannabis sativa* L.	*Cannabis sativa*		C, (W) [F]			DIGE, RESP
	Humulus lupulus L.	*Humulus lupulus*	Hummalad	W, C [F]	DIGE		
Caprifoliaceae	*Succisa pratensis* Moench	*Succisa pratensis* [L], *Scabiosa succisa* [AL]	Tölbi jurega pibe lehhed, peetri pibe lehhed	W [F]	DIGE, GEUN		DIGE
	Valeriana officinalis L.	*Valeriana officinalis*	Paldrian, üllekäija rohhi	W [L], W [F]	DIGE, PFCP	DIGE, GENI, GEUN, NEUR	CARD, DIGE, PSYC, RESP
Caryophyllaceae	*Dianthus deltoides* L.	*Dianthus deltoides*		W			DIGE
	Herniaria glabra L.	*Herniaria glabra*	Söötreia rohhi	W	SKIN		
	Saponaria officinalis L.	*Saponaria officinalis*		W, C [F]			PSYC, SKIN
	Silene vulgaris (Moench) Garcke	*Silene vulgaris* [L], *Silene inflata* [AL]	Pöie rohhi	W	UROL		MUSC
	Stellaria media (L.) Vill.	*Stellaria media*		W			GEUN
Celastraceae	*Parnassia palustris* L.	*Parnassia palustris*		W			CARD, GEUN
Convolvulaceae	*Convolvulus arvensis* L.	*Convolvulus arvensis* [L], *Convolvulus* [AL]	Jooksja rohhi, kurre katlad, lippo rohhud, lippo warrekad	W	ENDO		SKIN
Crassulaceae	*Sedum acre* L.	*Sedum acre*		W [F]			GEUN, MUSC
	Sempervivum globiferum L.	*Sempervivum soboliferum*		W			EAR
Cucurbitaceae	*Cucumis sativus* L.	*Cucumis sativus* L.		C			DIGE
Cupressaceae	*Juniperus communis* L.	*Juniperus*		W, (C [F])			CARD, DIGE, EAR, RESP, SKIN
	Juniperus sabina L.	*Sabina*		W, (C) [F]			PCFP

Table 1. *Cont.*

Family	Latin Name	Name in the Source	Local Name	Origin of Plant	Luce	Aronson	Alksnis
Cyperaceae	*Carex arenaria* L.	*Carex arenaria*		W ^{F}			GENI, MUSC
	Carex flava L.	*Carex flava*		W ^{F}			RESP
Equisetaceae	*Equisetum hyemale* L.	*Equisetum hyemale*		W ^{F}			CARD, GENI
	Equisetum sp.	*Equisetum*		W ^{F}			MUSC
	Equisetum sylvaticum L.	*Equisetum sylvaticum*	Rammi rohhi	W	DIGE		
Ericaceae	*Arctostaphylos uvaursi* (L.) Spreng.	*Arctostaphylos uvaursi*		W ^{F}			DIGE
	Chimaphila umbellata (L.) Nutt.	*Chimaphila umbellata*		W			MUSC
	Empetrum nigrum L.	*Empetrum nigrum*		W, (C ^{F})			DIGE, SKIN
	Rhododendron tomentosum Harmaja	*Ledum palustre*	Käelud	W	SKIN		CARD, GEUN, MUSC, RESP, SKIN
	Pyrola rotundifolia L.	*Pyrola rotundifolia*	Lambakörwad, lutöbbi rohhi	W ^{F}	ENDO		
	Vaccinium oxycoccos L.	*Vaccinium oxycoccus*		W ^{F}			NEUR
	Vaccinium myrtillus L.	*Vaccinium myrtillus*		W ^{F}			DIGE, RESP
	Vaccinium vitisidaea L.	*Vaccinium vitisidaea*		W ^{F}			GEUN, MUSC, SKIN
Euphorbiaceae	*Euphorbia helioscopia* L.	*Euphorbia helioscopia*		W			DIGE
Fabaceae	*Cassia fistula* L.	*Cassia fistula*		P			RESP
	Glycyrrhiza glabra L.	*Glycyrrhiza glabra*	Kolne pu	P	GENI, MUSC		
	Ononis repens L.	*Ononis spinosa*	Lude rohhi	W ^{F}	ENDO, MUSC		
	Senna alexandrina Mill.	*Foliae sennae*		P			DIGE
	Trifolium aureum Pollich	*Trifolium agrarium*		W ^{F}			DIGE, GENI
Fagaceae	*Quercus infectoria* G.Olivier	*Quercus infectoria*		P			DIGE
	Quercus robur L.	*Quercus robur*	Tamme pu	W, C ^{F}	SKIN	DIGE	DIGE, GEUN
Gentianaceae	*Centaurium erythraea* Rafn.	*Erythraea centaurium*		W ^{F}			DIGE, MUSC
	Gentiana sp.	*Gentiana*		W			DIGE
	Gentianella amarella (L.) Harry Sm.	*Gentiana amarella*		W			DIGE, PSYC
Geraniaceae	*Erodium cicutarium* (L.) L'Hér.	*Erodium cicutarium*		W			DIGE
	Geranium pusillum L.	*Geranium pusillum*		W			RESP, SKIN
	Geranium robertianum L.	*Geranium robertianum*	Rülli küined, russekud, punnase rosi rohi	W ^{F}	SKIN		
	Geranium sp.	*Geranium*		C			EAR
	Geranium sylvaticum L.	*Geranium sylvaticum*		W			GENI

Table 1. *Cont.*

Family	Latin Name	Name in the Source	Local Name	Origin of Plant	Luce	Aronson	Alksnis
Grossulariaceae	*Ribes rubrum* L.	*Ribes rubrum*		W, C [F]			RESP
Hypericaceae	*Hypericum perforatum* L.	*Hypericum perforatum*	Emmaste rohhi, raeste punned	W [F]	DIGE, GENI, RESP, SKIN		GENI, GEUN
Iridaceae	*Crocus* sp.	*Crocus*		C			DIGE
	Gladiolus sp.	*Gladiolus*		C			DIGE
Lamiaceae	*Glechoma hederacea* L.	*Glechoma hederacea*	Rosi rohhi, kassi naered	W [F]	SKIN		DIGE, RESP
	Lamium album L.	*Lamium album*		W			GENI
	Mentha × *piperita* L.	*Mentha piperita*		C, (W) [F]			CARD, DIGE, MUSC, NEUR, RESP
	Mentha spicata L.	*Mentha crispa*		C, (W) [F]		NEUR	DIGE, RESP
	Origanum vulgare L.	*Origanum vulgare*	Naeste punned	W [F]	GENI		DIGE
	Prunella vulgaris L.	*Prunella vulgaris*		W			RESP
	Salvia rosmarinus Spenn.	Rosmarinöl		C, P			NEUR
	Salvia glutinosa L.	*Salvia glutinosa*		C			GENI
	Thymus serpyllum L.	*Thymus chamaedrys* [AL] *Thymus serpyllum* [L]	Rabanduse rohhi	W, (C) [F]	SKIN		RESP
Lauraceae	*Cinnamomum camphora* (L.) J.Presl	*Cinnamomum camphora* [L], *Laurus Camphora* [AR]	Kampwer	P	ENDO	DIGE, EAR, GEUN	
	Laurus nobilis L.	*Laurus nobilis*	Loorberid	P			GEUN, SKIN
Linaceae	*Linum catharticum* L.	*Linum catharticum*		W [F]			PSYC
	Linum usitatissimum L.	*Linum usitatissimum*		C			EYE, GEUN, SKIN
Loganiaceae	*Strychnos nuxvomica* L.	*Strychnos nuxvomica* [L, AR], *Nux vomica* [AL]	Rebbase rohhi	P	DIGE	DIGE, GEUN	DIGE, NEUR, SKIN
Lycopodiaceae	*Lycopodium clavatum* L.	*Lycopodium clavatum*	Nöia rohhi, terwise rohhi	W [F]	DIGE, SKIN		
	Huperzia selago (L.) Bernh. ex Schrank & Mart.	*Lycopodium selago*		W [F]			DIGE, SKIN
Malvaceae	*Tilia* sp.	Lindenblüthentee		W, C [F]			DIGE, GEUN, RESP, SKIN
Melanthiaceae	*Paris quadrifolia* L.	*Paris quadrifolia*	Hora marjad, ussilak	W [F]	GEUN		
Menyanthaceae	*Menyanthes trifoliata* L.	*Menyanthes trifoliata*		W [F]			CARD, DIGE, GEUN, NEUR, RESP
Nymphaeaceae	*Nuphar lutea* (L.) Sm.	*Nuphar lutea*	Koltsed kuppo lehhed	W, (C) [F]	CARD		
	Nymphaea alba L.	*Nymphaea alba*	Wallged kuppo lehhed	W [F]	CARD		
	Nymphaea sp.	*Nymphaea*		W			GEUN

Table 1. *Cont.*

Family	Latin Name	Name in the Source	Local Name	Origin of Plant	Luce	Aronson	Alksnis
Oleaceae	*Fraxinus excelsior* L.	*Fraxinus excelsior*		W, C [F]			MUSC, NEUR
	Syringa vulgaris L.	*Syringa*		W [F]			RESP
Orchidaceae	*Dactylorhiza maculata* (L.) Soó	*Orchis maculata*		(W) [F]			GENI, PCFP
	Epipactis palustris (L.) Crantz	*Epipactis palustris*		W			ENDO, PSYC
Orobanchaceae	*Pedicularis* sp.	*Pedicularis*		W			SKIN
Oxalidaceae	*Oxalis acetosella* L.	*Oxalis acetosella*		W [F]			GEUN
Papaveraceae	*Chelidonium majus* L.	*Chelidonium majus*	Werre urma rohhi	W, C [F]	DIGE, EYE, SKIN		SKIN
	Papaver somniferum L.	*Papaver somniferum*		W, C			PSYC
Pinaceae	*Picea abies* (L.) H.Karst.	Fichtenrinde		W			DIGE
	Pinus sylvestris L.	*Pinus sylvestris*	Manna pu	W, (C) [F]	DIGE, ENDO, SKIN		
Piperaceae	*Piper nigrum* L.	*Piper nigrum* [L], Pfeffer [AL]	Walge ja must pippar	P	DIGE		DIGE, EAR, RESP, SKIN
Plantaginaceae	*Linaria vulgaris* Mill.	*Linaria vulgaris*		W [F]			SKIN
	Plantago major L.	*Plantago major*	Tee lehhed	W [F]	SKIN		DIGE, SKIN, UROL
	Veronica agrestis L.	*Veronica agrestis*		W			PSYC
	Veronica arvensis L.	*Veronica arvensis*		W			PSYC
	Veronica beccabunga L.	*Veronica beccabunga*		W [F]			GEUN, MUSC
	Veronica longifolia L.	*Veronica longifolia*		W			SKIN
	Veronica officinalis L.	*Veronica officinalis*	Jooksja rohhi, jaani rohhi, mailase rohhi	W [F]	CULT, ENDO, GEUN, SKIN		
Poaceae	*Alopecurus pratensis* L.	Roggengras		W, C [F]			GEUN
	Avena sativa L.	Haferkörner, Haferstroh		(W, C) [F]			DIGE, RESP
	Briza media L.	*Briza media*		W			DIGE, GEUN
	Calamagrostis sp.	*Calamagrostis*		W			GENI
	Hordeum vulgare L.	Gerstenkörner, Gerstengrütze		C			EYE, SKIN
	Secale cereale L.	Roggenblüthe, Roggenähren, Roggenmehl, Roggenbrod		C			DIGE, RESP, SKIN
	Triticum sp.	Weizenmehl		(W, (C)) [F]			SKIN
Polygalaceae	*Persicaria maculosa* Gray	*Polygonum persicaria*		W [F]			SKIN
	Polygala amara L.	*Polygala amara*		W [F]			PSYC
	Polygala sp.	*Polygala*		(W, C) [F]			GEUN, PSYC

Table 1. *Cont.*

Family	Latin Name	Name in the Source	Local Name	Origin of Plant	Luce	Aronson	Alksnis
	Polygala vulgaris L.	*Polygala vulgaris*		W [F]			GENI
Polygonaceae	*Rumex crispus* L.	*Rumex crispus*		(W) [F]			DIGE, SKIN
	Rumex obtusifolius L.	*Rumex obtusifolius*	Hobbosehoblikad	W	DIGE, SKIN		
Polypodiaceae	*Dryopteris filixmas* (L.) Schott	*Aspidium filix mas*		W			SKIN
	Polypodium vulgare L.	*Equisetum fragile*, Engelsüß	Rinna rohhi	W [F]	RESP		
Primulaceae	*Lysimachia vulgaris* L.	*Lysimachia vulgaris*		W, (C [F])			DIGE
Ranunculaceae	*Aconitum napellus* L.	*Aconitum napellus*		W			SKIN
	Actaea spicata L.	*Actaea spicata*	Akkitse haiguse rohi	W [F]	GEUN, PSYC		PSYC
	Anemone nemorosa L.	*Anemone nemorosa*	Külma ellased	W [F]	EYE, SKIN		
	Caltha palustris L.	*Caltha palustris*	Warsa kabjad, kuller kuppud	W [F]	DIGE		
	Consolida regalis Gray	*Delphinium consolida*		W, (C [F])			DIGE, RESP
	Ranunculus ficaria L.	*Ranunculus ficaria*		W, (C) [F]			CARD
	Ranunculus acris L.	*Ranunculus acris*	Tullikad, sobia rohhi, jooksja rohhi, pöld ingwerid, tullililled	W [F]	CARD, ENDO, GEUN, MUSC, SKIN		SKIN
Rhamnaceae	*Frangula alnus* Mill.	*Rhamnus frangula*		W, (C) [F]			DIGE, MUSC, SKIN, UROL
	Rhamnus cathartica L.	*Rhamnus cathartica*	Paaks pu	W, (C) [F]	DIGE, SKIN		RESP
Rosaceae	*Argentina anserina* (L.) Rydb.	*Hippocrepis comosa*	Hoolmete rohhi	W	DIGE		
	Comarum palustre L.	*Comarum palustre*		W [F]			MUSC
	Filipendula ulmaria (L.) Maxim.	*Filipendula ulmaria* [L], *Spiraea ulmaria* [AL]	Wormid, naeste rohhi	W [F]	PCFP		DIGE, EYE, NEUR, SKIN
	Filipendula vulgaris Moench	*Spiraea vulgaris*		W, C [F]			DIGE
	Fragaria vesca L.	*Fragaria vesca*		W, (C) [F]			DIGE, RESP
	Geum urbanum L.	*Geum urbanum*		W [F]			GENI
	Malus sp.	Apfelbaumblätter, Sauere Aepfel		W/C			GEUN, SKIN
	Malus sylvestris (L.) Mill.	wilder Apfelbaum		W, C [F]			SKIN
	Potentilla erecta (L.) Raeusch.	*Tormentilla*		W [F]			DIGE, MUSC
	Prunus cerasus L.	*Prunus cerasus*		W, C [F]		PCFP	PCFP
	Prunus domestica L.	Pflaumensaft		C			PCFP

Table 1. *Cont.*

Family	Latin Name	Name in the Source	Local Name	Origin of Plant	Luce	Aronson	Alksnis
	Prunus padus L.	*Prunus padus*		W, C [F]		GEUN, SKIN	CARD, NEUR, UROL
	Rubus caesius L.	*Rubus caesius*		W [F]			CARD
	Rubus chamaemorus L.	*Rubus chamaemorus*	Murrakad, (kabbarad, kaas marjad)	W [F]	CARD		
	Rubus idaeus L.	*Rubus idaeus*		W, C [F]			RESP, SKIN
	Rubus saxatilis L.	*Rubus saxatilis*		W [F]			MUSC
	Sorbus aucuparia L.	*Sorbus aucuparia*		W, C [F]			GEUN, MUSC
Rubiaceae	*Galium odoratum* (L.) Scop.	*Asperula*		(W) [F]			DIGE
	Galium boreale L.	*Galium boreale*	Maddarad	W [F]	GENI		
Sapindaceae	*Aesculus hippocastanum* L.	*Aesculus hippocastanum*		C [F]			MUSC
Saxifragaceae	*Chrysosplenium alternifolium* L.	*Chrysosplenium alternifolium*		W			DIGE, SKIN
Scrophulariaceae	*Scrophularia nodosa* L.	*Scrophularia*		W			CARD
	Verbascum thapsus L.	*Verbascum thapsus*	Ühheksa mehhe wäggi	W [F]	RESP, SKIN	RESP	GEUN, MUSC, SKIN
Solanaceae	*Capsicum annuum* L.	*Capsicum annuum*	Türgi pippar	P	GEUN		DIGE, GEUN, MUSC
	Datura stramonium L.	*Datura stramonium*		C, (W) [F]			NEUR, RESP
	Hyoscyamus niger L.	*Hyoscyamus niger*	Hüllo koera rohhi, hüllo koera hänna rohhi	W [F]	DIGE		DIGE, MUSC, NEUR, PSYC, SKIN
	Nicotiana rustica L.	*Nicotiana rustica* [L AL], *Nicotiana tabac. Rustica* [AR]	Tubbaka lehhed	C	DIGE, SKIN	EYE, SKIN	DIGE, GEUN, RESP, SKIN
	Solanum americanum Mill.	*Solanum nigrum*		W [F]			PSYC
	Solanum dulcamara L.	*Solanum dulcamara*	Solika rohhi	W, (C) [F]	DIGE		DIGE, GEUN
	Solanum tuberosum L.	*Solanum tuberosum*		C			GEUN, NEUR, SKIN
Taxaceae	*Taxus baccata* L.	*Taxus baccata*	Juhha pu	W, (C) [F]	SKIN		
Thymelaeaceae	*Daphne mezereum* L.	*Daphne mezereum*		W, C [F]			DIGE, GEUN
Urticaceae	*Urtica urens* L.	*Urtica urens*		W [F]			MUSC, NEUR, RESP, SKIN, UROL
Viburnaceae	*Sambucus ebulus* L.	*Sambucus ebulus·*		W, C [F]			SKIN
	Sambucus nigra L.	*Sambucus niger*		W, C [F]			CARD, SKIN
	Viburnum opulus L.	*Viburnum opulus*		W, C [F]			SKIN
Violaceae	*Viola tricolor* L.	*Viola tricolor*	Mailase rohhi	W	SKIN		RESP
	Viola arvensis Murray	*Viola arvensis*		W			PSYC

Table 1. *Cont.*

Family	Latin Name	Name in the Source	Local Name	Origin of Plant	Luce	Aronson	Alksnis
Zingiberaceae	*Aframomum melegueta* K.Schum.	*Grana paradisi*		P			DIGE
	Alpinia galanga (L.) Willd.	*Alpinia galanga*	Jalgendi jured	P	PCFP		
	Alpinia officinarum Hance	*Radix Galangae*		P			DIGE
	Curcuma zedoaria (Christm.) Roscoe	*Curcuma zedoaria*		P			DIGE
Zygophyllaceae	*Guaiacum officinale* L.	*Tinct. Guajaci*	Plussas drape	P			NEUR
Unidentified	Unidentified	Speedeja sahle	speedeja sahle				GEUN

Name in original if listed by more than one source: (L) Luce, (AL) Alksnis, and (AR) Aronson. Origin of plant: if different in Friebe [40] (F): wild (W), cultivated (C), or purchased (P). Abbreviations of disease categories: BLIM—Blood, Blood Forming Organs and Immune Mechanism, CARD—Cardiovascular, CULT—Culture Bound Syndrome, DIGE—Digestive, EAR—Ear, ENDO—Endocrine/Metabolic and Nutritional, EYE—Eye, GENI—Female Genital, GEUN—General and Unspecified diseases, MUSC—Musculoskeletal, NEUR—Neurological, PCFP—Pregnancy, Childbearing, Family Planning, PSYC—Psychological, RESP—Respiratory, SKIN—Skin.

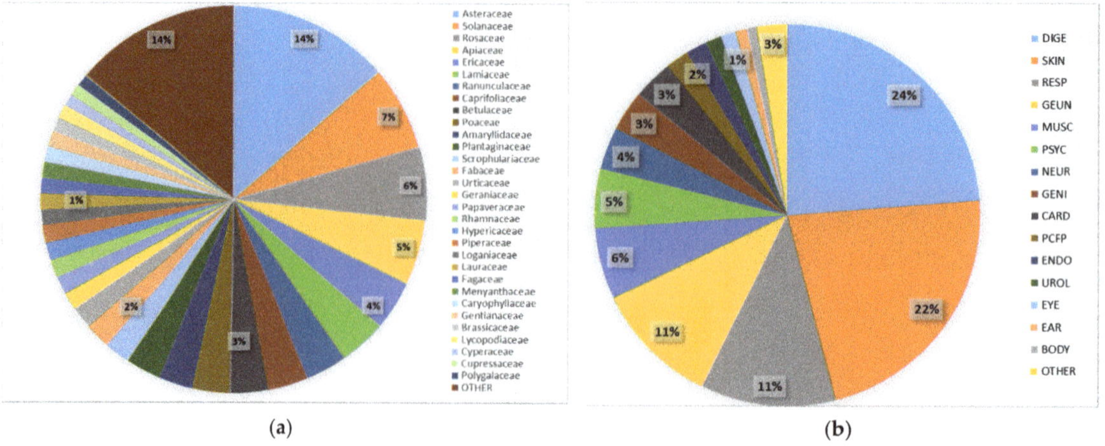

(a)   (b)

Figure 1. Proportional distribution of DUR among (**a**) plant families and (**b**) general disease categories. For abbreviations of the diseases see Table 1 below.

The proportion of single DUR is high in the three studies: 50% for Alksnis and Aronson and 60% for Luce. The records with more than two DURs constitute 16% in the works of Luce and Aronson and 28% in Alksnis. Examples of multifunctionality include *Artemisia absinthium*, for which Luce reported five DUR, Alksnis reported six DUR, and Aronson reported two DUR; *Valeriana officinalis*, for which Luce reported two DUR, Alksnis reported eight DUR, and Aronson reported six DUR; and *Taraxacum officinale* (one DUR in all works).

Of all the taxa described by the authors, 22 were clearly purchased (12 of them described only by Alksnis). There is a high number of taxa described as wild by Friebe [40], but according to current classifications they are regarded as cultivated.

Alksnis's article contained the greatest number of taxa, while he also had the highest number of taxa identified on the genus level (Figure 2, Table 2). Thirteen taxa were present in all three works: *Achillea millefolium*, *Allium cepa*, *Arnica montana*, *Artemisia absinthium*, *Betula*, *Levisticum officinale*, *Matricaria chamomilla*, *Nicotiana rustica*, *Strychnos nux-vomica*, *Taraxacum officinale*, *Valeriana officinalis*, *Verbascum thapsus*, and *Quercus robur*. Although

Alksnis clearly used Aronson's work (often referring to him), he did not include three taxa (*Cinnamomum camphora* shared with Luce and *Prunus cerasus* and *Ferula assa-foetida* mentioned solely by Aronson—of them only *Prunus cerasus* grows locally).

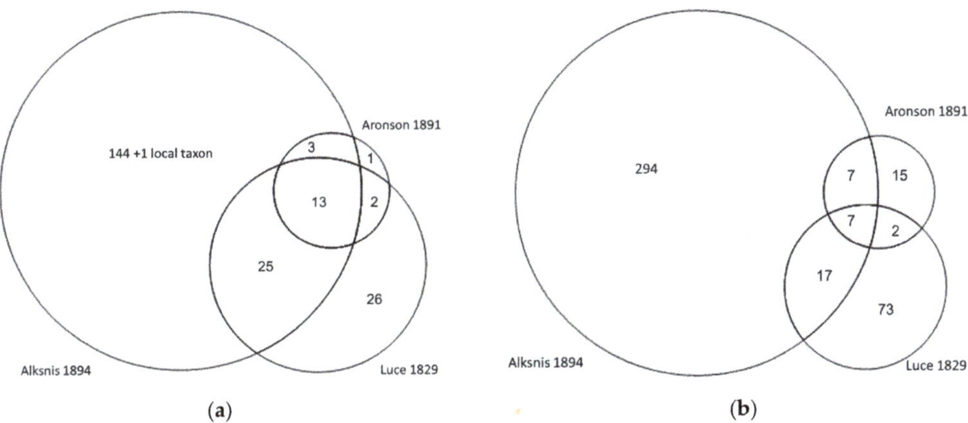

Figure 2. Proportional Venn diagrams showing the number of overlapping species (**a**) and Use Instances (UI) (**b**) among the three authors.

Table 2. Comparison of the three authors.

	Luce 1829	**Aronson 1891**	**Alksnis 1894**
Taxa	66	19	186
DUR	123	35	445
UI	99	31	325
Most-diversely used species (DUR)	*Ranunculus acris* (9), *Hypericum perforatum* (7), *Artemisia absinthium* (5)	*Valeriana officinalis* (6), *Cinnamomum camphora* (4), *Matricaria chamomilla* (3)	*Allium cepa* (10), *Urtica urens* (10), *Betula* spp. (9)
Most-mentioned etic disease categories (DUR)	SKIN (38), DIGE (33), GEUN (10), ENDO (9)	GEUN (9), DIGE (9), SKIN (5), RESP (2)	DIGE (104), SKIN (94), RESP (64), GEUN (46)

While for all three authors the Asteraceae family had the most mentioned taxa and DUR, the other botanical families differ. In Luce's work, the second and third most important families are Ranunculaceae and Apiaceae, while in Alksnis, they are Solanaceae and Rosaceae and in Aronson they are Caprifoliaceae and Lauraceae. In terms of species, Luce recorded the highest diversity of general categories in which the plant is used for *Ranunculus acris* and *Hypericum perforatum*; Alksnis for *Arnica montana, Juniperus communis, Urtica urens,* and *Hyoscyamus niger*; and Aronson for *Veronica officinalis* and *Matricaria chamomilla*. The number of DUR per plant varies from author to author.

Skin and digestive diseases are the most mentioned medicinal use categories for all three authors, and while general and unspecified diseases are the third most mentioned category for Luce and Aronson, they are number four for Alksnis. For Alksnis, respiratory diseases represent the third most-mentioned category, whereas the other two authors reported far fewer DUR in this category (Table 2).

The Jaccard Similarity Indexes show the highest overlap between Luce and Aronson on both the taxa and use instance (UI) level (Table 3). While there are some similar uses mentioned by all three, or at least two, of the authors, there are many more different applications. *Artemisia absinthium*, being the most diversely used plant, shows some overlaps (dysentery and abdominalgia in Luce and Alksnis, fever in Alksnis and Aroson) as well as divergences in use (ulcers and malaria in Luce, and internal diseases and actual neurosis in Alksnis). *Tanacetum vulgare* is a rather rare example of complete agreement in use: worm infestation in both Luce and Alksnis.

Table 3. Jaccard Indexes (JI) showing the proportional overlap between the authors.

Taxa/UI	Luce 1829	Aronson 1891	Alksnis 1894
Luce 1829	X	0.07438	0.06
Aronson 1891	0.214286	X	0.040936
Alksnis 1894	0.193878	0.084656	X

An interesting example is that of *Rhododendron tomentosum*, used to treat lice in Luce, to which Alksnis added uses against pulmonary tuberculosis, bone pain, and general deteriorating health. Luce reported the use of *Ranunculus acris* against gout, dropsy, vesicating, amaurosis, hip pain, rheumatism, and fever, while Alksnis reported its use to treat cold and burn wounds. Likewise, Luce mentioned the use of *Viola tricolor* against skin diseases, whereas Alksnis noted it use against whooping cough. While Luce recorded the use of *Silene vulgaris* against urological diseases, Alksnis mentioned its use to treat joint rheumatism (musculoskeletal category).

3. Discussion

3.1. Why Such Diversity?

The Jaccard Similarity Index is remarkably low: for comparison, the lowest JI from the region recorded in recent years was over 0.54 for all taxa. As the highest similarity was evident between the authors providing the fewest plants and uses, the size of the collection and the region covered play a significant role: while Luce collected on a small island, Aronson covered part of the mainland, which was about three times larger. Considering that the distance between the two regions was around 100 km over the sea and the time between collection dates was about 60 years (two generations), the difference is still considerable. Notably, both authors relied on long-standing personal experiences from the region in which they worked and their own discoveries. Another aspect to consider concerns the plant identifications made by the authors.

While the work of Aronson received little attention from detractors, Luce's work was criticized by some of his contemporaries. Schmidt [39] complained that Luce's works were not reliable, while Lehmann [41] accused him of listing "dubious species" that were not taken into account by later botanists. Regardless of such observations, the majority of later authors cite Luce (often referring to the whole of Estonia).

Aronson and Alksnis published during the same decade, yet the methodology of Alksnis was very different. Remarkably, Alksnis used Luce's and Aronson's works, but he did not copy them, instead he seems to have used them as a reference for a similar use. As Alksnis covered all of Livonia and Courland, he did not obtain the data by practicing there. Moreover, Alksnis's age (at the time of publishing his article he was just 24 years old) did not allow for him to accumulate much practical experience. In addition to his own collection, Alksnis relied on the data mediated by the newspaper and this aspect was heavily criticized by his contemporary colleague, pharmacist Ernests Birzmanis [42]. Birzmanis pointed out the lack of original (Latvian) names in the descriptions of medicinal remedies, questioning the ability to substantiate the accuracy of the Latin and German translations. In the preference to his work, Birzmanis emphasized that with the given report he did not want to diminish the value of Alksnis's published book, but rather with his notes help subsequent authors in the field to correct mistakes in their respective books of interest. Birzmanis noticed several errors in the translations for both taxa and illnesses; however, we cannot automatically consider those adequate. For example, while Alksnis mentions the use of *Artemisia vulgaris* bulbs for treating nerve diseases, Birzmanis emphasizes that "viboksne" ("vibotne"), which corresponds to *Artemisia vulgaris*, does not have bulbs and thus the translation of the Latvian name must have been incorrect. Additionally, Alksnis mentions "kruklis"—for which his book offers three different taxa having this name, while Birzmanis presents *Rhamnus frangula* as the corresponding one. Those two mistakes are probably the most serious additions to the text, yet as the exact source of the information cannot be verified, we decided to keep it unchanged. Birzmanis also criticizes German translations; however, here it is impossible to confirm them. For example, "pirts slota" should have been translated as "die Quast" not "der Basen". As for illnesses, Birzmanis notes "meris" should have been recorded as "die Pest" not "die Seuche" and adds that such translation errors can be found in several places. Birzmanis also adds that in many places the medicinal applications are quite disgusting and thus he doubts these uses are common, likely being utilized by only one or two strange individuals; and if this is in fact the case, he believes they cannot be considered part of Latvian folk medicine, thus trying to idealize Latvian medicine. For example, the use of "cukas zults ar lapu tabaku" (pig bile with tobacco leaves) to treat swellings corresponds to a single event and thus does not warrant inclusion in Latvian folk medicine. However, Birzmanis adds that for the materials, Alksnis himself collected or borrowed from Fricis Brīvzemnieks's book—such silly things are not found. Birzmanis points out that for the materials referred to by Dr. Raphael, the information seems miraculous and unbelievable [42].

Relying on current knowledge, we have observed that the authors, especially Luce, made several potential identification errors leading to the recording of taxa that do not grow in the region. Therefore, regardless of all our efforts to minimize errors and misinterpretations, we cannot guarantee that all the initial identification mistakes have been eliminated. This needs to be taken into consideration while working with the data.

Works by contemporaries of Luce support the idea of the diversity of ethnomedicine at that time. The closest in time to Luce was the unpublished manuscript of pastor and amateur botanist Johann Heinrich Rosenplänter (1782–1846), along with his field notes and loose-leaf herbarium vouchers, which have been thoroughly analyzed recently, showing just a few overlapping plants and no overlap on the use level with Luce's work originating from the same decade (1820–1830) [43].

Another comparable work is that of Dr. Mihkel Ostrov (1863–1940), who not only studied medicine at the University of Dorpat at almost the same time as Alksnis, but was also interested in folk medicine. Ostrov collected medicinal plant knowledge from across Estonia through an appeal in the newspaper. Unlike Alknis, Ostrov also placed great emphasis on the popular names of medicinal plants. Comparing Ostrov's data provided in the manuscript with the article in Alksnis, we found that nearly two-thirds of the used plants overlapped, whereas the uses differed in majority of cases [44]. However, the study of Ostrov was not complete, so the comparison is not fully informative.

Some more similarities can be seen on the genus level with the results obtained by Sile et al. [45], but as their methodology and the actual time of collection of the folklore which is the basis for their study is not stated, a more detailed comparison is not feasible. Moreover, as Alksnis and several other authors also later published in Latvian, the later folklore might have been influenced by literary sources and cannot be considered a good base for comparison.

Therefore, we can assume, regardless of any possibly remaining mistakes in identification, that plant use in the 19th century in the region was highly diverse and place-specific. Despite the criticism of Birzmanis towards presenting singular uses, we should not underestimate the importance of even singular uses of local wild taxa, which represent part of the local biocultural diversity. In this framework, for understanding the patterns of formation of LEK, we need to differentiate between locally developed and global, introduced knowledge, which may have already become local. Therefore, source criticism, taking into account the high possibility of error even in the interpreted text or data, as well as in the possible tracking of the origin of the data, is necessary [12,46].

The selected works are situated in very similar local environments and the diseases treated in the 19th century were quite similar. However, comparing the numbers of plants and uses each author provided, we also need to keep in mind the size of the region the authors covered, which can contribute to the perceived differences in the taxa use. Nevertheless, as we see high divergence in uses regardless of the limits set by flora and diseases, we can assert that an interchange of LEK within the 19th century only occurred for a limited number of species, then related to official medicine. This is very different from the current situation in the region, where overlap is very high and clear signs of homogenization caused by official medicine of the Soviet era can be seen [47].

3.2. Comparison with Dioscorides

Overlap with the taxa mentioned in Dioscorides's *Materia medica* differs among the authors, being highest for Aronson (39%), followed by Luce (38%) and then Alksnis (26%) (Figure 3). However, the highest percentage of overlap (47.5%) is found among the taxa that overlap between the authors (either all three or two of them), while the remaining taxa represent 25% for Luce and 21% for Alksnis. In all cases, the percentage is higher than the 20% proposed by Leonti et al. [14].

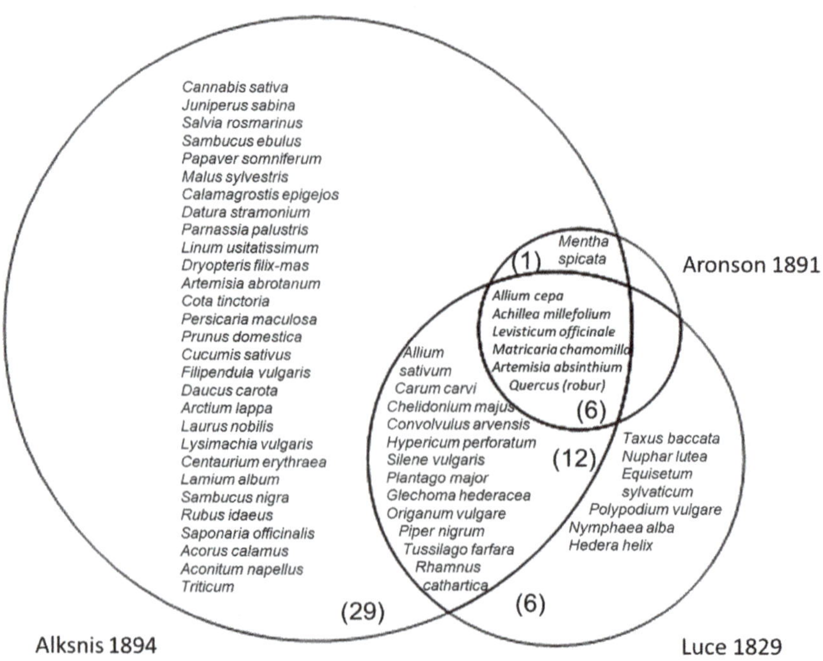

Figure 3. Taxa represented in the analyzed books and Dioscorides.

Looking at the species combined with their medicinal use, commonly mentioned taxa include *Artemisia absinthium* (digestive and general and unspecified), *Carum carvi* (digestive), *Plantago major* L. (skin), *Chelidonium majus* (skin), and *Achillea millefolium* (skin). Frequently mentioned usages of species are mostly from the medicinal use categories with the most DUR, such as skin, digestive, respiratory, general, and unspecified diseases and symptoms.

Such overlap suggests a potential, although not direct, influence of Dioscorides on local ethnomedicine. However, we need to take into account that both Luce and Aronson had a lot of practical experience and developed remedies on their own, thus not only influencing local ethnomedicine, but also perhaps including their own knowledge without explicitly acknowledging it, creating a "feedback loop" as described by Leonti [12]. This can, in fact, cause an overestimation of the impact of Dioscorides.

Another important point is that the influence of herbals and books on plants and folk medicine should be evaluated along with the literacy level of a society. In Livonia, the study area of this paper, the "reading revolution" did not take place until the middle of the 19th century [48]. Poorly literate people were not able to contribute to, understand, or have access to the increasing numbers of books and texts, but needed mediators able to transmit such knowledge. While we acknowledge that filtering from scientific publications to common use is a known phenomenon in society [49], we need to understand how such book-based knowledge transferred into practice.

We also cannot underestimate the potential self-discovery of some uses (even if they overlap with Dioscorides). A good example is provided by *Carum carvi*: peasants had an obligation to collect the seeds for the manors, and therefore, they were familiar with the plant and always had it available and used it for food [50]; as a result, many ad hoc uses could also have been tried on demand and entered into circulation after yielding positive results. Another example may be that of *Arnica monatna*. The first Latvian herbal book [42] says that *Arnica montana* grows in Courland, but this claim was later refuted. At that time, many similar yellow-flowered local species could have been mistakenly identified as *A. montana*, as happened among Estonians living in Livonia [51].

3.3. Aspects to Consider When Interpreting the Data

The increasing influence of pharmacies and doctors in the 19th century influenced people in their wild plant usage. Although this has an effect on people's medicinal plant usage, i.e., certain usages decrease or disappear as a result of switching to pharmaceutical products, this influence is, in retrospect, difficult to quantify, especially because other factors like people's beliefs in certain cultural treatments, as well as an aversion to pharmacies and doctors, also have to be considered here.

We recommend that future researchers wanting to interpret historical sources consider the following aspects (summarized in Figure 4):

(1) The greatest possible attention needs to be given to the background of the authors and the general context in which the work was written, the available supporting tools, and the existing knowledge of that time.

(2) It is important to understand the metalanguage of the source, e.g., one needs to study the source in the context of the time it was written and understand the natural, cultural, and societal settings of that time. This not only applies to historical archival sources, but also to earlier ethnobotanical literature.

(3) Keep in mind that current background knowledge (on nature, culture, society, literature, etc.) can influence the interpretation of historical data, potentially leading to misinterpretations as the researcher may involuntarily assume the contexts have some elements in common.

(4) Throughout all the work, and also after the interpretation is completed, vigilant source-criticism and self-criticism need to be present.

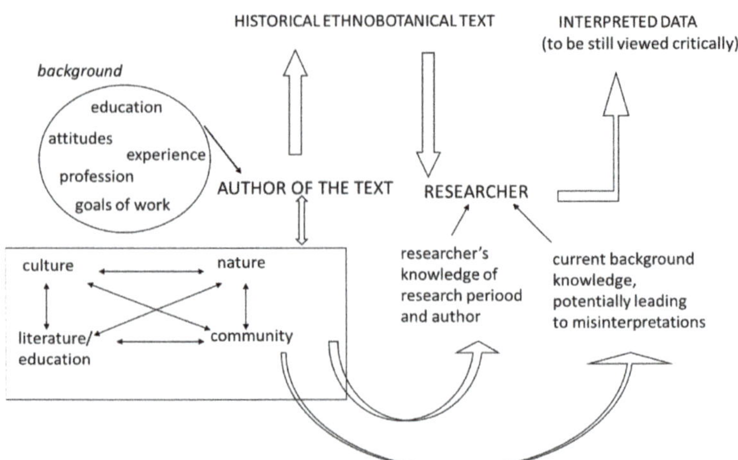

Figure 4. Aspects to consider when interpreting historical ethnobotanical works. Adapted from Sõukand [52].

4. Materials and Methods

An Excel database was created by manually selecting relevant information and entering it into the database. Every independent use in the sources was accounted for as a Detailed Use Report (DUR), where the informant *i* mentions a specific medicinal use, based on the use categories of the specific author, of the plant part (*p*, e.g., fruits, leaves, aerial parts, flowers, etc., if provided), considering also the form in which the plant part is used (*f*, e.g., fresh, dried, frozen, refrigerated) and specific way of preparation, if provided. Every DUR was entered on a separate row in the excel spreadsheet.

The historical medicinal data (originally mentioned disease or symptom recorded in German) was interpreted on the basis of the provided name and its correspondence to equivalents reported in historical and current literature [53,54]. To identify general disease categories, we relied on the symptoms or conditions associated with the disease and the ICPC-2 classification [55] was applied for comparative purposes. For comparison, we also calculated Use Instances (UI), where one UI corresponds to the specific plants used in an etic disease category according to the ICPC-2, regardless of the number of different emic diseases treated. Comparison between the three ethnobotanical sources was made using the Jaccard Similarity Index (JI), adopting the methodology of González-Tejero et al. [56]: JI = (C/(A+B − C)), where A represents the number of taxa/UI in sample A, B is the number of taxa/UI in sample B, and C is the number of taxa/UI common to A and B.

The plants and their names given in the books were checked for reliability using:

- Flora Europaea [57] to verify the plant identifications and with floras of that time [36–38] to confirm that the plants really grew in the region when the books were published;
- Vilbaste [35], Beiche [58], Genaust [59], and Hiller and Melzig [60] comparing the local and German names and descriptions;
- online biodiversity databases (https://elurikkus.ee/ and https://www.latvijasdaba.lv/, accessed on 29 October 2021) to confirm the presence and distribution of the taxa in the region;
- other herbal texts and books [61–63].

The current names provided follow Plants of the World Online [64], except for two taxa (*Taraxacum officinale* and *Ononis repens*), which are based on Flora Europaea [57].

Graphs and diagrams were created with Excel, while proportional Venn diagrams were created using the PAST Toolkit Venn diagram plotter software program (https://omics.pnl.gov/software/venn-diagram-plotter, accessed on 26 October 2021).

Region

The area of data gathering for Alksnis included present-day Latvia and southern Estonia, while Luce covered modern-day Saaremaa and Aronson, which are the surroundings of present-day Liepaja (Figure 5).

(a)   (b)

Figure 5. (a) Study region within Europe; (b) regions where authors worked (ArcGIS, historical borders according to H. Laakmann [65], edited by the second author M.A.).

Latvia's and Estonia's landscapes belong to the Eastern European hilly lowlands (the highest point being 318 m above sea level), and consist of approximately 50% (mainly) pine forest, alternating with meadows and swamps and mixed deciduous forest towards the south. A wide variety of wild berries and mushrooms also grow there [66], see also Lehmann [41] and Schmidt [39]. The climate in Estonia and Latvia is similar: bordered by the Baltic Sea, it is moist and humid with an annual precipitation between 600 and 700 mm. Podzol soils predominate [66].

Livonia and Courland underwent constant changes in rulership and repeated attempts at Germanization and Russification, heavily affecting the diverse local cultures. The first invasion started in the 13th century and German influence lasted with various interruptions of leadership until World War I [65,67,68]. Divided into two at the end of the 16th century, Livonia was conquered by Sweden, while Courland remained with Poland; the 18th century saw the Russian occupation of both regions and the beginning of the 20th century brought independence [65,67–70].

During this time of continual leadership changes, the influence and importance of Germans and Baltic Germans remained the same and was even intensified by several waves of German immigration. Among the immigrants, there were often intellectuals and academics who were immediately accepted as part of the higher social classes. Another effect of this cultural influence and the immigration of academics was that the academic language in the region was German for a long time. Also, peasants, being mostly locals and making up the majority of inhabitants, remained illiterate for a long time. Therefore, the field of science in Livonia was established by Germans and Baltic Germans, and as a result, the German language was utilized up until the end of the 19th century [71–73].

5. Conclusions

Our results demonstrate high biocultural diversity in terms of taxa and their medicinal use within a limited temporal and spatial context, especially regarding the use of local, wildly growing plants. The high overlap among the three authors and with scholarly sources as well as the use of cultivated and purchased taxa do not diminish the value of the biocultural diversity of the medicinal use of locally growing plants. The authors encourage researchers to study and re-evaluate historical ethnobotanical works from recent centuries in Europe in order to better evaluate the evolution of medicinal ethnobotany.

Author Contributions: Conceptualization, R.S. and A.P.; Methodology, R.S. and J.P.; Validation, R.S., J.P. and R.K.; Formal analysis, M.A. and R.S.; Resources, R.S.; Data curation, M.A. and R.S.; Writing—original draft preparation, M.A., J.P., R.K., B.P., A.S. and R.S.; Writing—review and editing, R.S., A.P. and R.K.; Visualization, R.S. and M.A.; Supervision, R.S. and A.P. All authors have read and agreed to the published version of the manuscript.

Funding: The research was supported by the European Research Council (ERC) under the European Union's Horizon 2020 research and innovation program (Starting Grant project "Ethnobotany of divided generations in the context of centralization", Grant agreement No. 714874).

Institutional Review Board Statement: Not applicable.

Informed Consent Statement: Not applicable.

Data Availability Statement: https://doi.org/10.5281/zenodo.6106746 (accessed on 2 February 2022).

Conflicts of Interest: The authors declare no conflict of interest. The funders had no role in the design of the study; in the collection, analyses, or interpretation of data; in the writing of the manuscript, or in the decision to publish the results.

References

1. Casas, A.; Blancas, J.; Lira, R. Mexican Ethnobotany: Interactions of People and Plants in Mesoamerica. In *Ethnobotany of Mexico*; Lira, R., Casas, A., Blancas, J., Eds.; Springer: New York, NY, USA, 2016; pp. 1–19. [CrossRef]
2. Kujawska, M.; Łuczaj, Ł.; Typek, J. Fischer's Lexicon of Slavic beliefs and customs: A previously unknown contribution to the ethnobotany of Ukraine and Poland. *J. Ethnobiol. Ethnomed.* **2015**, *11*, 85. [CrossRef] [PubMed]
3. Kujawska, M.; Klepacki, P.; Łuczaj, Ł. Fischer's Plants in folk beliefs and customs: A previously unknown contribution to the ethnobotany of the Polish-Lithuanian-Belarusian borderland. *J. Ethnobiol. Ethnomed.* **2017**, *13*, 20. [CrossRef] [PubMed]
4. Spałek, K.; Spielvogel, I.; Próćków, M.; Próćków, J. Historical ethnopharmacology of the herbalists from Krummhübel in the Sudety Mountains (seventeenth to nineteenth century), Silesia. *J. Ethnobiol. Ethnomed.* **2019**, *15*, 24. [CrossRef] [PubMed]
5. Teixidor-Toneu, I.; Kjesrud, K.; Bjerke, E.; Parekh, K.; Kool, A. From the "Norwegian Flora" (Eighteenth Century) to "Plants and Tradition" (Twentieth Century): 200 Years of Norwegian Knowledge about Wild Plants. *Econ. Bot.* **2020**, *74*, 398–410. [CrossRef]
6. Pardo-de-Santayana, M.; Quave, C.; Sõukand, R.; Pieroni, A. Medical Ethnobotany and Ethnopharmacology of Europe. In *Ethnopharmacology*; Heinrich, M., Jäger, A., Eds.; John Wiley & Sons, Ltd.: Chichester, UK, 2015; pp. 343–356. [CrossRef]
7. Svanberg, I.; Łuczaj, Ł.; Pardo-De-Santayana, M.; Pieroni, A. History and current trends of ethnobiological research in Europe. In *Ethnobiology*; Anderson, E.N., Pearsall, D., Hunn, E., Turner, N., Eds.; Wiley-Blackwell: Hoboken, USA, 2011; pp. 189–212. [CrossRef]
8. Dioscorides, P. *De Materia Medica*, 3rd ed.; Introduction by John Scarborough; Lily, T., Beck, Y., Eds.; Olms-Weidmann: Hildesheim, Germany, 2017; pp. 1–540.
9. Buenz, E.; Schnepple, D.; Bauer, B.; Elkin, P.; Riddle, J.; Motley, T. Techniques: Bioprospecting historical herbal texts by hunting for new leads in old tomes. *Trends Pharmacol. Sci.* **2004**, *25*, 494–498. [CrossRef] [PubMed]
10. Halberstein, R. Medicinal plants: Historical and cross-cultural usage patterns. *Ann. Epidemiol.* **2005**, *15*, 686–699. [CrossRef]
11. Jamshidi-Kia, F.; Lorigooini, Z.; Amini-Khoei, H. Medicinal plants: Past history and future perspective. *J. Herbmed Pharmacol.* **2018**, *7*, 1–7. [CrossRef]
12. Leonti, M. The future is written: Impact of scripts on the cognition, selection, knowledge and transmission of medicinal plant use and its implications for ethnobotany and ethnopharmacology. *J. Ethnopharmacol.* **2011**, *134*, 542–555. [CrossRef]
13. Leonti, M.; Casu, L.; Sanna, F.; Bonsignore, L. A comparison of medicinal plant use in Sardinia and Sicily–De Materia Medica revisited? *J. Ethnopharmacol.* **2009**, *121*, 255–267. [CrossRef]
14. Leonti, M.; Cabras, S.; Weckerle, C.; Solinas, M.; Casu, L. The causal dependence of present plant knowledge on herbals: Contemporary medicinal plant use in Campania (Italy) compared to Matthioli (1568). *J. Ethnopharmacol.* **2010**, *130*, 379–391. [CrossRef]
15. Sõukand, R.; Mattalia, G.; Kolosova, V.; Stryamets, N.; Prakofjewa, J.; Belichenko, O.; Kuznetsova, N.; Minuzzi, S.; Keedus, L.; Prūse, B.; et al. Inventing a herbal tradition: The complex roots of the current popularity of Epilobium angustifolium in Eastern Europe. *J. Ethnopharmacol.* **2020**, *247*, 112254. [CrossRef] [PubMed]
16. Anegg, M.; Prakofjewa, J.; Kalle, R.; Sõukand, R. Local Ecological Knowledge and folk medicine in historical Esthonia, Livonia, Courland and Galicia, 1805–1905 (1.0) [Data set]. *Zenodo* **2021**. [CrossRef]
17. Anegg, M. Medicinal Usage of Wild Plants in the Historic Regions of Livonia and Courland in the 19th Century. Master's Thesis, Ca' Foscari University, Venice, Italy, 5 March 2021.

18. Von Luce, J.W.L. *Topographische Nachrichten von der Insel Oesel, In Medicinischer Und Ökonomischer Hinsicht*; W.F.Häcker: Riga, Latvia, 1823; pp. 1–383.
19. Aronson, E. Ueber die Volksheilmittel der Letten. *Mag. Der Lettisch-Literärischen Ges.* **1891**, *19*, 185–203.
20. Alksnis, J. Materialien zur lettischen Volksmedizin. *Histor. Stud. Aus Dem Pharmakol. Inst. Der Kaiserl. Universität. Dorpat.* **1894**, *4*, 166–283.
21. Von Luce, J.W.L. *Heilmittel der Ehsten auf der Insel Oesel*; Gotthard Marquardt: Pernau, Livonia, 1829; pp. 1–126.
22. Gottzmann, C.; Hörner, P. *Lexikon der Deutschsprachigen Literatur des Baltikums und St. Petersburgs: Vom Mittelalter Bis Zur Gegenwart*; Walter de Gruyter: Berlin, Germany, 2011; pp. 1–1476.
23. Demitsch, W. Literärische Studien über die wichtigsten russischen Volksheilmittel aus dem Pflanzenreiche. Ph.D. Thesis, Dorpat University, Dorpat, Estonia, 23 March 1888.
24. Alksnis, J. Dazchas peezihmes tautas ahrztneecibas materialu krahjejeem. *"Dienas Lapa" Peelikums* **1892**, *9*, 130–133.
25. Alksnis, J. Tautas medicīniskie materiāli. *Rakstu krājums, izdots no Rīgas latviešu biedrības Zinību komisijas* **1898**, *12*, 1–46.
26. Ilsters, J. *Botanika tautas skolām un pašmācībai*; Puhzischu Ģederta: Rīga, Latvia, 1883; pp. 1–144.
27. Ilsters, J. Stahdu draugeem uz jaunu zeedu-laikmetu. *Balss* **1885**, *20*, 3–4.
28. Ilsters, J. Tautas medizina. *"Dienas Lapa" Peelikums* **1891**, *2*, 27–30.
29. Birzmans, E. Luhgums. *Latweeschu Awises* **1897**, *10*, 1.
30. Birzmanis, E. *Latvijas ārstniecības stādi*; Kalniņš & Deučmanis: Rīga, Latvia, 1896; pp. 1–111.
31. Ančesvka, I. Latviešu dziedināšanas tradīcija: Teorētiskie un praktiskie aspekti: Promocijas darba kopsavilkums. Ph.D. Thesis, Latvijas Universitāte, Rīga, Latvia, 24 April 2018.
32. Rowell, M. Russian medical botany before the time of Peter the Great. *Sudhoffs Arch.* **1978**, *62*, 339–358.
33. Glück, L. Skizzen aus der Volksmedicin und dem medicinischen Aberglauben in Bosnien und der Hercegovina. *Wiss. Mitt. Bosnien. Herceg.* **1894**, *2*, 392–454.
34. Sõukand, R.; Kalle, R. (Eds.) *HERBA: Historistlik Eesti Rahvameditsiini Botaaniline Andmebaas*; EKM Teaduskirjastus: Tartu, Estonia, 2008; Available online: http://herba.folklore.ee (accessed on 21 December 2021).
35. Vilbaste, G. *Eesti taimenimetused*; Emakeele Selts: Tallinn, Estonia, 1993; pp. 1–706.
36. Fischer, J. *Versuch einer Naturgeschichte von Livland*, 2nd ed.; Nicolovius: Königsberg, Germany, 1791; pp. 1–826.
37. Grindel, D.H. *Botanisches Taschenbuch für Liv-, Cur- Und Ehstland*; Hartmann: Riga, Latvia, 1803; pp. 1–575.
38. Fleischer, J.G. *Flora von Esth-, Liv- und Kurland*, 2nd ed.; Bunge: Mitau, Latvia, 1853; pp. 1–291.
39. Schmidt, F. *Flora des Silurischen Bodens von Ehstland, Nord-Livland und Oesel*; Laakmann: Dorpat, Estonia, 1855; pp. 1–114.
40. Friebe, W.C. *Oekonomisch-Technische Flora für Liefland, Ehstland und Kurland*; Hartmannische Buchhandlung: Riga, Latvia, 1805; pp. 1–392.
41. Lehmann, E. *Flora von Polnisch-Livland*; Mattiesen: Dorpat, Estonia, 1895; pp. 1–556.
42. Birsmanis, E. *Materiāli Latviešu Tautas Medicīnai*; Kalniņš un Deučmans: Rīga, Latvia, 1895; pp. 1–20.
43. Kalle, R.; Sõukand, R. The name to remember: Flexibility and contextuality of preliterate folk plant categorization from the 1830s, in Pernau, Livonia, historical region on the eastern coast of the Baltic Sea. *J. Ethnopharmacol.* **2021**, *264*, 113254. [CrossRef] [PubMed]
44. Kalle, R.; Pieroni, A.; Svanberg, I.; Sõukand, R. Early Citizen Science Action in Ethnobotany: The Case of the Folk Medicine Collection of Dr. Mihkel Ostrov in the Territory of Present-Day Estonia, 1891–1893. *Plants* **2022**, *11*, 274. [CrossRef] [PubMed]
45. Sile, I.; Romane, E.; Reinsone, S.; Maurina, B.; Tirzite, D.; Dambrova, M. Medicinal plants and their uses recorded in the Archives of Latvian Folklore from the 19th century. *J. Ethnopharmacol.* **2020**, *249*, 112378. [CrossRef]
46. Kalle, R.; Belichenko, O.; Kuznetsova, N.; Kolosova, V.; Prakofjewa, J.; Stryamets, N.; Mattalia, G.; Šarka, P.; Simanova, A.; Prūse, B.; et al. Gaining momentum: Popularization of Epilobium angustifolium as food and recreational tea on the Eastern edge of Europe. *Appetite* **2020**, *150*, 104638. [CrossRef]
47. Sõukand, R.; Kalle, R.; Pieroni, A. Homogenisation of Biocultural Diversity: Plant Ethnomedicine and Its Diachronic Change in Setomaa and Võromaa, Estonia, in the Last Century. *Biology* **2022**, *11*, 192. [CrossRef] [PubMed]
48. Talts, M. The role of popular science literature in shaping Estonians' world outlook. *Историко-биологические Исследования* **2013**, *5*, 59–83. Available online: https://www.digar.ee/arhiiv/nlib-digar:125277 (accessed on 21 December 2021).
49. Foster, G. *Hippocrates' Latin American Legacy. Humoral Medicine in the New World*; Gordon & Breach Science Pub: Berlin, Germani, 1994; pp. 1–235.
50. Sõukand, R.; Kalle, R. *Changes in the Use of Wild Food Plants in Estonia: 18th–21st Century*; Springer International Publishing: New York, NY, USA, 2016; pp. 1–175.
51. Sõukand, R.; Raal, A. How the name Arnica was borrowed into Estonian. *Trames* **2008**, *12*, 29–39. [CrossRef]
52. Sõukand, R. Herbal Landscape. Ph.D. Thesis, Tartu Ülikool, Tartu, Estonia, 9 November 2010.
53. Riecke, J. *Die Frühgeschichte der Mittelalterlichen Medizinischen Fachsprache im Deutschen*; Walter de Gruyter: Berlin, Germany, 2004; pp. 1–650.
54. Verein Für Computergenealogie E.V. 2004 Onwards, GenWiki, Lexika. Available online: http://wiki-de.genealogy.net/Kategorie:Krankheitsbezeichnung,_Medizinischer_Begriff (accessed on 7 December 2020).
55. International Classification of Diseases 11th Revision. WHO, 2018. Available online: https://icd.who.int/browse11/l-m/en#/http://id.who.int/icd/entity/2137507044 (accessed on 7 December 2020).

56. González-Tejero, M.; Casares-Porcel, M.; Sánchez-Rojas, C.P.; Ramiro-Gutiérrez, J.M.; Molero-Mesa, J.; Pieroni, A.; Giusti, M.E.; Censorii, C.; de Pasquale, C.; Della, A.; et al. Medicinal plants in the Mediterranean area: Synthesis of the results of the project Rubia. *J. Ethnopharmacol.* **2008**, *116*, 341–357. [CrossRef]
57. Tutin, T.G.; Burges, N.A.; Chater, A.O.; Edmondson, J.R.; Heywood, V.H.; Moore, D.M.; Valentine, D.H.; Walters, S.M.; Webb, D.A. (Eds.) *Flora Europaea*; University Press: Cambridge, UK, 1980; Volume 1–5, Available online: http://ww2.bgbm.org/EuroPlusMed/query.asp (accessed on 21 December 2021).
58. Beiche, W.E. *Vollständiger Blütenkalender der Deutschen Phanerogamen-Flora. Unter Zugrundelegung von Dr. Kittel's Taschenbuch der Flora Deutschlands*; Hahn: Hannover, Germany, 1872; pp. 1–656.
59. Genaust, H. *Etymologisches Wörterbuch der Botanischen Pflanzennamen*, 3rd ed.; Springer-Verlag: Basel, Switzerland, 2013; pp. 1–701.
60. Hiller, K.; Melzig, M.F. *Lexikon der Arzneipflanzen und Drogen*, 2nd ed.; Spektrum Akademischer Verlag: Heidelberg, Germany, 2006; pp. 1–644.
61. Dragendorff, J.G.N. *Die Heilpflanzen der Verschiedenen Völker und Zeiten*; Enke: Stuttgart, Germany, 1898; pp. 1–892.
62. Krebel, R. *Volksmedicin und Volksmittel Verschiedener Völkerstämme Russlands*; Winter: Leipzig, Germany, 1858; pp. 1–194.
63. Rosenthal, D.A. *Synopsis Plantarum Diaphoricarum*, 2nd ed.; Enke: Erlangen, Germany, 1861; pp. 1–1359.
64. Plants of the World Online. Facilitated by the Royal Botanic Gardens, Kew, UK. Available online: https://powo.science.kew.org/ (accessed on 8 December 2021).
65. Wittram, R. *Baltische Geschichte: Die Ostseelande, Livland, Estland, Kurland, 1180–1918*; Wissenschaftliche Buchgesellschaft: Darmstadt, Germany, 1973; pp. 1–323.
66. Knappe, E.; Waack, C. Die baltischen Staaten—Geographischer Überblick und Naturräumliche Gliederung. *Der Bürger im Staat* **2004**, *2/3*, 85–91.
67. von Rauch, G. *Geschichte der Baltischen Staaten*; Kohlhammer: Stuttgart, Germany, 1970; pp. 1–224.
68. Tuchtenhagen, R. *Geschichte der Baltischen Länder*; Beck: München, Germany, 2005; pp. 1–128.
69. Kasekamp, A. *A History of the Baltic States*, 2nd ed.; Palgrave Macmillan: London, UK, 2018; pp. 1–223.
70. Selart, A. Non-German Literacy in Medieval Livonia. In *Uses of the Written Word in Medieval Towns: Medieval Urban Literacy II*; Mostert, M., Adamska, A., Eds.; Brepols Publishers: Turnhout, Belgium, 2014; Volume 28, pp. 37–63. [CrossRef]
71. Drost, A. Historical Borderlands in the Baltic Sea Area: Layers of Cultural Diffusion and New Borderland Theories—The Case of Livonia. *J. Hist. for the Public* **2010**, *7*, 10–23. [CrossRef]
72. Plath, U. Heimat: Rethinking Baltic German Spaces of Belonging. *Kunstiteaduslikke Uurim.* **2014**, *23*, 55–78.
73. Tamm, M.; Mänd, A. Introduction: Actors and networks in the medieval and early modern Baltic Sea region. In *Making Livonia—Actors and Networks in the Medieval and Early Modern Baltic Sea Region*; Mänd, A., Tamm, T., Eds.; Routledge: Abingdon, UK, 2020; pp. 1–13.

Article

Plants and Other Materials Used for Dyeing in the Present Territory of Poland, Belarus and Ukraine according to Rostafiński's Questionnaire from 1883

Piotr Köhler [1], Aleksandra Bystry [2] and Łukasz Łuczaj [3,*]

1   Faculty of Biology, Institute of Botany, Jagiellonian University, ul. Gronostajowa 3, 30-387 Kraków, Poland; piotr.kohler@uj.edu.pl
2   Dzikie Barwy, ul. Pomorska 98 lokal 108, 91-402 Łódź, Poland; kontakt@dzikiebarwy.com
3   Faculty of Biology, Institute of Biology and Biotechnology, University of Rzeszów, ul. Pigonia 1, 36-100 Rzeszów, Poland
*   Correspondence: lukasz.luczaj@interia.pl

Abstract: Background: Traditional dyeing methods are practically forgotten in Poland. Józef Rostafiński included questions on the use of dyes in his ethnobotanical survey from 1883. Methods: 126 questionnaires contained information on dye plants. They were identified by the respondents using folk names or sometimes even Latin names. Folk names were analyzed by comparison with other literature. Several voucher specimens were also present. Results: 74 plant taxa were identified to genus or species level. The most commonly used were: onion (*Allium cepa*), brazilwood (*Caesalpinia brasiliensis* or *Paubrasilia echinata*), winter corn (mainly rye *Secale cereale*), black alder (*Alnus glutinosa*), safflower (*Carthamus tinctorius*), apple (*Malus domestica*), birch (*Betula pendula*), oak (*Quercus robur*), and violet flowering spring flowers (mainly *Hepatica nobilis* and *Pulsatilla* spp.). Conclusions: Most species are well known in the literature about plant dyeing, but the paper provides extra details on the picture of dyeing traditions in Eastern Europe.

Keywords: ethnobotany; natural dyes; traditional ecological knowledge; textiles; wool; flax; Easter eggs

Citation: Köhler, P.; Bystry, A.; Łuczaj, Ł. Plants and Other Materials Used for Dyeing in the Present Territory of Poland, Belarus and Ukraine according to Rostafiński's Questionnaire from 1883. *Plants* **2023**, *12*, 1482. https://doi.org/10.3390/plants12071482

Academic Editor: Laura Cornara

Received: 25 February 2023
Revised: 25 March 2023
Accepted: 26 March 2023
Published: 28 March 2023

1. Introduction

Plants have been a source of dyes since the dawn of humanity, used to decorate human bodies, textiles, containers, and for artistic and religious painting, etc. [1–4]. Species used for traditional dyeing and the techniques employed have been recorded in some areas of the world [5–14].

Before the popularization of synthetic dyes in the 19th and 20th centuries, the art of dyeing was an important craft [15–23]. Apart from dyes of plant origin, a local species of insect, *Porphyrophora polonica* (Linnaeus, 1758), sometimes called the Polish cochineal, was used to make red paint. The larvae of this scale insect live on the roots of various herbs, especially those of the perennial knawel *Scleranthus perennis* L., which is common on the sandy soils of Central Europe. Before aniline, alizarin, and other synthetic dyes were invented, the insect had been of great economic importance. It was exported to other parts of Europe, but its gathering and red dye production collapsed after the discovery of America and the introduction of cochineal red from another insect species [1,4,15,23]. A few interesting publications on dyeing plants and techniques were published in 18th-century Poland and later in the former Polish territories occupied by Russia during the partitions (1772–1918), which are now part of Poland, Lithuania, Belarus, and Ukraine [24–31]. For example, Krzysztof Kluk (1739–1796), an eminent Polish naturalist, encouraged the cultivation and use of dye plants in his textbook [29]. In addition to basic and generally known plants, such as reseda (*Reseda lutea*), woad (*Isatis tinctoria*), and madder (*Rubia tinctoria*), he paid attention to a number of field and forest plants with dyeing properties. He enumerated the local raw materials used by the villagers, such as barberry bark and twigs, apple and alder bark, bowls of acorns, and many others [29]. In the 19th century, information on traditional dyeing methods and materials could be found in Józef Gerald-Wyżycki's (1792–1868) herbal [30] and in Anna Ciundziewicka's (1803–1850) *Gospodyni Litewska* [31]. In the 20th century, several more plant dye textbooks and monographs on plant dyeing materials were published in Poland [32–35]. The contributions of the Polish ethnologist Kazimierz Moszyński (1887–1959) in his *Kultura ludowa Słowian* (*Folk Culture of Slavs*) [22] and the Polish historian Elżbieta Kowecka (1929–2001) are especially important [15]. An interesting monograph on dyeing was recently published by the botanist Adam Kapler [36].

However, due to the large availability of cheaper industrial dyes (usually of synthetic origin), the whole tradition of natural dyeing is disappearing, being either preserved only among some oldest craftsmen or becoming completely obsolete. In the case of Poland, the latter is true. Apart from the use of onions to color Easter eggs, natural dyeing is completely forgotten and does not occur even in 20th-century ethnographic publications, apart from data from the Polish Ethnographic Atlas, mainly from 1983–1990, where, analogously to Rostafiński's questionnaire, two questions were included [37]. One question concerned the kinds of bark used in dyeing, and the other was about Easter eggs [37]. The 19th-century ethnographic materials are also silent about this type of plant use, as reflected by the fact that Adam Fischer's ethnobotanical dictionary, which contains a synthesis of Polish data on the folk use of plants, does not mention them [38].

A valuable contribution to documenting the forgotten traditional dye plants is a questionnaire published in 1883 by the botanist Józef Rostafiński, professor of Jagiellonian University in Kraków [39]. He issued it in 60 editions of various periodicals in the territories of the former Kingdom of Poland (Poland was divided into Russia, Prussia, and Austro-Hungary at that time). Section VIII contained three questions concerning dyes:

VIII. Dyes

(59) Do simple people dye flax or cannabis textiles or wool or leather themselves? What plants are used and what colors do they give? I would be grateful for specimens.

(60) With what do they dye Easter eggs?

(61) Do people still gather "czerwiec polski" (Polish cochineal) and from under what plants?

Some information on dyeing was also found scattered in answers to other questions, especially question no. 46 about *Carthamus tinctorius* L. ("Do people know the name krokosz and is this herb used for?").

Rostafiński began his research career as a taxonomist. However, at the beginning of the 1880s, he became interested in the history and names of cultivated plants. In 1883, Rostafiński started his largest project connected with plant names. His concept was to collect Polish plant names and write a history of plant cultivation and use in the areas of the former Polish-Lithuanian Commonwealth.

Rostafiński's work was only partly analyzed and published. The answers to some questions, e.g., concerning wild greens or fungi, were analyzed in full detail, while others are still waiting to be properly elaborated. Basic information on the historical background of the questionnaire can be found in the works of Köhler [39–41]. The replies inspired Rostafiński to alter the scope of the questionnaire twice; however, the questions about dye plants are present in all of them. The questionnaire was published in 1883 in a few dozen Polish-language periodicals in what was then the Russian Empire, Prussia, and Austro-Hungary, as between 1795 and 1918, Poland did not exist as an independent country and was partitioned between these three empires (see Figures 1 and 2 for the geographical scope of answers).

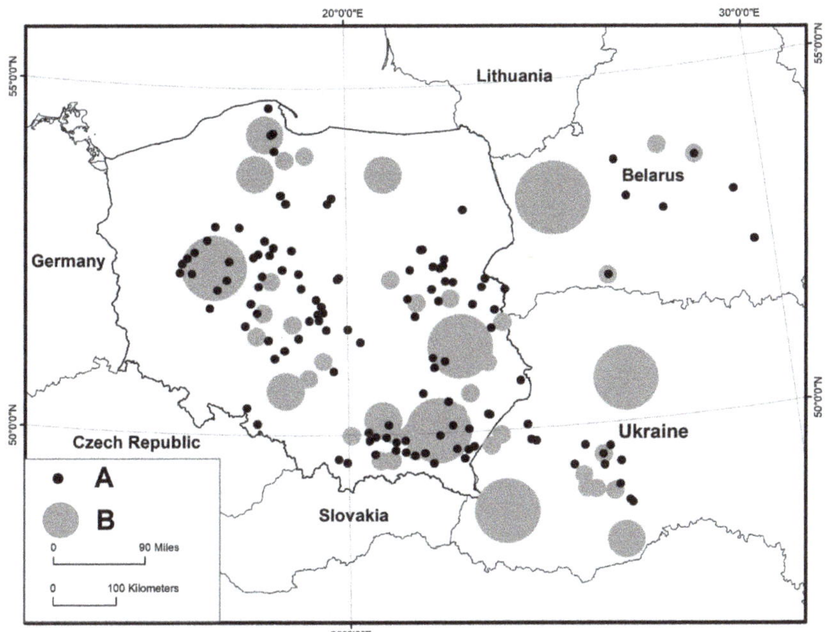

Figure 1. The distribution of records for dyes of organic origin in our study; A—places, B—regions.

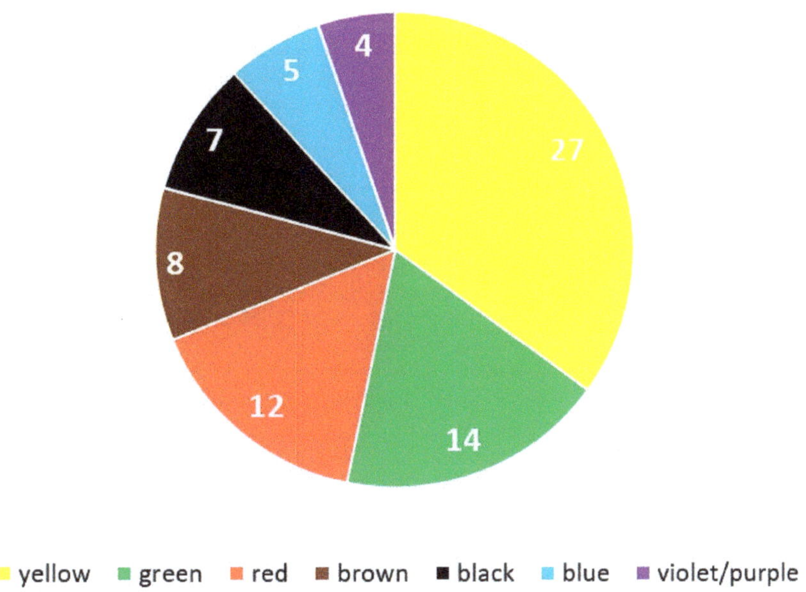

Figure 2. Number of species used for obtaining colors from plant dyes.

The aim of the present study was to analyze Rostafiński's questionnaire. We hypothesized that most of the traditional dyes recorded in the study are already known from other specialist literature. However, we hope to find at least some novel species that could be utilized.

Researching plant dyes is important not only from the point of view of recording traditional knowledge. The information is also of value to modern enthusiasts of plant dyes.

In the last two decades, there has been an immensely increasing trend in using natural products such as wild vegetables [42,43], medicinal herbs and teas [44], and plant dyes. Several handbooks on traditional dyeing have been published recently [45–48].

2. Results

The largest number of answers was given to question no. 60 (about Easter egg dyes), with 125 answers altogether and 425 use reports. A much smaller number of answers (62) was given to question no. 59 (dyes for textiles or leather), i.e., 176 use reports. Question no. 61 turned out to be a failure. Only six answers containing 14 use reports were given, mainly concerning other dyes, and not a single description of the contemporary use of *czerwiec* (Polish cochineal) was sent to Rostafiński. Thirteen use reports come from answers to other questions.

As many as 74 taxa of identified plants were recorded to species or genus level, and 13 taxa remained unidentified (Table 1), not counting a few materials of animal or human origin (e.g., dog feces and human urine). A few group categories were also distinguished (lichens, grasses, cereals, hay).

The most commonly used plants were onion (*Allium cepa*), brazilwood (*Caesalpinia brasiliensis* or *Paubrasilia echinata*), winter corn (mainly rye *Secale cereale)*, black alder (*Alnus glutinosa*), safflower (*Carthamus tinctorius*), apple (*Malus domestica*), birch (*Betula pendula*), oak (*Quercus robur*), and violet flowering spring flowers (mainly *Hepatica nobilis* and *Pulsatilla* spp.).

Hues of yellow, green, and red were the most common colors obtained from plants (Figure 2). Onion (*Allium cepa*) was the most widely used dye to obtain yellow and brownish colors, mainly on Easter eggs.

The second most commonly mentioned dye was Brazil wood, used mainly for dyeing textiles shades of red. We could not identify the exact species as the same names are used to refer to two related taxa, *Caesalpinia brasiliensis* and *Paubrasilia echinata*, in Europe (for more discussion, see footnote 4 in Table 1).

Green blades of cereals were the third in the frequency of use as dyeing materials. Mainly winter corn was used, especially rye (*Secale cereale*), which is usually sown in autumn and easily obtainable during Easter for dyeing eggs green. Oats (*Avena sativa*) and wheat (*Triticum* spp.) were also used for this purpose.

Another commonly mentioned plant was black alder (*Alnus glutinosa*), a native common tree in the area. This is a very interesting plant dye as different parts of the plant (bark, fruits, leaves, roots) used with different mordants can give various shades of brown, black, and yellow. Widely used for textiles, wool, and yarn, it was also applied to Easter eggs. However, another common native tree used for dyeing was birch (mainly the most abundant *Betula pendula*), whose leaves were applied to give yellow color to textiles, wool, and probably Easter eggs as well. The third common native tree used in dyeing was oak (mainly the most abundant *Quercus robur*), whose bark was applied to give dark (black, brown) hues to wool and flax.

Apple (*Malus domestica*) is the commonest fruit tree in central-eastern Europe, and, as such, its leaves and bark were easily available materials, mainly for dyeing wool and probably also Easter eggs yellow.

Another commonly used material was safflower (*Carthamus tinctorius*), an annual plant that was specially cultivated for dyeing purposes (for wool and Easter eggs) due to the attractive shades of yellow and pink achieved.

Table 1. Natural dyes recorded by Rostafiński in 1883.

Scientific Name	Local Name	Part	UR	Use	Main Color	Geography	Use in Other Sources
Alcea rosea L. (identification uncertain)	malwa	fl, l	4	Easter eggs?	green (sapphire)		Used by amateur dyers for pink, blue, and gray colors [46]; roots used for eggs in Kostrzyń near Białobrzegi.
Allium cepa L.	cebula, cybula, dymka	outer layers of onion	150	Easter eggs, sheepskin coats	yellow (red-brown, brick-yellow)	throughout PL and BE	Mainly used to dye wool and eggs [37]. The longer the dyeing time, the redder the wool or eggshells. With a small amount of raw material and a short immersion time, the eggs or yarn turn yellow (the color of turmeric). With alum, the stained wool achieves more luminous colors—from golden yellow to brown saturated red [32].
Allium cf *schoenoprasum* L.	szczypiór	sh	1	Easter eggs	n. sp.		
Alnus glutinosa (L.) Gaertn.	olcha, olszyna (PL), aleszyna (BE)	bk, (r, c)	37	wool, flax yarn, Easter eggs; mordants: alum and sulfate, sometimes just iron filings, beetroot kvass	black, yellow, brown	throughout PL, BE, UA	A typical folk dye, alder was not used by manufactories or industry. Used to dye fibers (thread, wool), leather, and eggs [15,37]. Brown and beige on alum mordant (warm shades), and grey to brown and deep black on iron mordant. Such colors are given by the cones and bark, while leaves and catkins (inflorescences) give shades of yellow with alum and green-brown with iron ([32] and A.B.'s own experiments).
Anethum graveolens L.	koper	n. sp.	1	Easter eggs	yellow	Olszany (Przemyśl, PL)	
Arctostaphylos uva-ursi (L.) Spreng.	muczennik, muczennicznik, muczennik, mączennik	wh, r	5	textiles; dried, pounded to powder in a wooden mortar[1].	brown, black	Słonim area, Bobrujsk, Weleśnica, Jeziora (Grodno), Ihumeń (BE)	Yellow, gray and black [32,34].
autumn-sown cereals (winter corn)	ozimina, zboże	sh	17	Easter eggs	green		Widely used for eggs [37].
Avena sativa L.	owies	sh	2	Easter eggs	green (fresh shoots), yellow (straw)	PL: Staszkówka (Ciężkowice), Tarnów	
Berberis vulgaris L.	berberys	r, bk, br	9	Easter eggs? wool, flax	yellow	throughout (PL); Lipów (Rzeczyca county), Nieśwież, Łuck, Mińsk (BE)	Widely used around the world [46]. According to Kluk [29], peasants use barberry bark to dye yellow—he did not specify what. Used for brown dye in Kłudzie near Lipsko [37].
Beta vulgaris L.	burak		4				Probably only used for dyeing eggs. There are no sources about the use of beetroot for dyeing fibers. It is a pH-sensitive dye and gives unstable dyes (A. B.'s own experience).
Betula sp. (mainly *B. pendula* Roth)	brzoza (PL), bierioza (BE)	L (bd, bk)	12	Easter eggs? wool, flax, cotton; alum mordant mentioned in 1 answer	yellow, (green, yellow-green)	throughout (PL); Naliboki (Mińsk, BE); NW Belarus; Mikulińce, Tarnopol, Zbaraż, Skałat, Nowesioło (UA)	Widely used. Good for dyeing eggs, fabrics, and yarns. The leaves give yellow colors (leaves) with alum and greenish with iron. According to sources [32,24,37,45,46], birch bark dyes brown; A. B.'s experience shows that it also gives shades of pink and dirty pink, and pinkish red can be reached in an alkaline environment.
Brassica oleracea L. (probably var. *capitata* f. *rubra*)	kapusta	juice	2	flax textiles, Easter eggs?	n. sp.	Central Poland —lax textiles; Grąziowa nad Wiarem (SE PL)—probably Easter eggs	Doubtful for linen, as anthocyanin dyes do not bind permanently to cellulose fibers. Good for eggs, as its pH is sensitive and gives many shades, from pink to green (own experience, A.B.).

Table 1. *Cont.*

Scientific Name	Local Name	Part	UR	Use	Main Color	Geography	Use in Other Sources
Caesalpinia brasiliensis L. or *Paubrasilia echinata* (Lam.) Gagnon, H.C.Lima and G.P.Lewis [4]	brezylia, brazylia, bryzylia, farnebuk, farembak		73	Easter eggs	red (black, brown, dark blue, violet)	throughout	Imported. See footnote no. 4.
Caltha palustris L.	łotoć		1	Easter eggs	yellow		
Cannabis sativa L.	konopia	n. sp.	1	Easter eggs?	green	Pińsk (BE)	A.B. once used the hemp herb as a dye, and a greenish tint appeared on the wool fibers. Wool often reacts similarly to eggshells (due to protein content), so shades of green are possible on eggs.
Carthamus tinctorius L.	krokosz (mainly), also: krokost (PL), krokos (PL and BE), ćwitłuszka, ćwitłycia (UA)	n. sp.	30	Easter eggs? wool	yellow (red or pink in acid pH)	throughout (PL); SW UA	Safflower is still used in many countries [46] but is a difficult raw material, so in folk culture, it probably served only as a yellow dye for wool or eggs. According to Kluk [29], safflower was used to create varnishes and thick paints. Abroad, it was most often used to dye silk pink. By working with it for a long time and in stages, a bright pink or red color can be achieved from the flower stamens [46].
Centaurea cyanus L.	haber, modraki (PL); wołoszka, wałoszka (BE)	fl—for blue, l, r—for yel-low or green	4	Easter eggs	blue, yellow, dirty green	Pleszew (Wielkopolska), Brus (Włodawa) (PL); Nieśwież, Słuck, Mińsk (BE), NW Belarus	Not mentioned in the literature, although there are modern recipes for dyeing with cornflower on the Internet.
Cichorium intybus L.	cykorya	n. sp.	2	Easter eggs?	n. sp.	NW Poland: Wielkopolska and Bielno (Świecie)	Leaves are used for dyeing in Iran [49].
Coffea sp.	kawa	n. sp.	2	Easter eggs?	n. sp.	Pomerania: Wysin (Kościerzyna), Kociewie Kaszuby (Pomerania)	Gives beige and brown. Probably coffee dregs were used.
Corylus avellana L.	leszczyna	bk	1	textiles?	probably brown		Hazelnut bark with alum gives a saturated shade of yellow ([32] and A.B.'s experience); green-brown was also reported [37].
Crocus sativus L. (rarely) (or frequently *Carthamus tinctorius* L.)	szafran	n. sp.	8	Easter eggs	yellow	throughout (PL), Mościska, Jaworów, Lwów area (UA)	Saffron is a widely known, expensive source of yellow.
Dianthus sp.	goździki leśne	n. sp.	1	textiles?	yellow	Pińsk area (BE)	
Canis familiaris L.	pies (psie łajno)	feces	1	collected by boys for dyeing Morocco leathers (safiany)	red	SW Ukraine	
Fagopyrum esculentum Moench	hreczka	chaff	1	textiles?	yellow	Mikulińce, Tarnopol, Zbaraż, Skałat, Nowesioło (UA)	Works very well on cellulose fibers, turning them brown, beige, or yellow (A.B.).
Fagus sylvatica L.	buk	bk	1	textiles?	probably brown	Kaszuby (Pomerania)	Dyes brown, beige, and brown with a shade of pink very well (A.B.). Used to dye flax [37].
Frangula alnus Mill.	kruszyna	n. sp.	1	textiles?	blue	Lipów (Rzeczyca, BE)	Used to dye fabrics, wool, yarn, eggs (yellow, red, and green, depending on the mordant used) [37,46], and fruits, especially unripe ones. They give shades of green on copper mordant [32].

Table 1. *Cont.*

Scientific Name	Local Name	Part	UR	Use	Main Color	Geography	Use in Other Sources
Gallus gallus domesticus (Linnaeus, 1758)	żółtko [chicken yolk]	yolk	1	sheepskin coats	red (one of the ingredients)		
Genista tinctoria L.	janowiec (PL), zinowat, zinówka, żołtucha (BE)	Fl	2	n. sp., probably textiles; alum mordant (1 answer)	yellow	Międzyleś, near Radzymin (PL); Weleśnica (BE)	Flowering shoots dye yellow. Widely used in Europe [32,46].
Haematoxylum campechianum L.	kempesz, kampesz	wd	4	Easter eggs?		PL: Biecz, Janów Lubelski, Mazury, Łyczkowice (Skierniewice)	Imported. Mainly used for textiles and yarn, cited by most dyeing handbooks [32,46]. Treated twice with alum Turns violet with alum, blue or grey with iron, and blue with sodium bicarbonate.
hay	siano	hay	3	Easter eggs	dark yellow	Zabłocie (Łask), Kalisz and Wieluń county, Wojnicz Siedlce	
Helianthus annuus L.	słonecznik	n. sp.	1	textiles?	yellow		
Hepatica nobilis Schreb. & *Pulsatilla* sp. (probably both used but mainly the former)	kocanki, sasanki, podlaszczka (PL); son, przelaszczka, sonczyki, praleski, wiośnianki, pierwiosnki (BE)	fl	11	Easter eggs (mainly), textiles; the dye was made by the extraction of flowers with vodka	blue, violet	central-western Poland: Silnica (Radomsko), Lisice nad Nerem (Koło), Częstochowa county, Wieluń county, Bielice (Kutno) Ostrzeszów, Zabłocie (Łask); Nieśwież, Słuck, Mińsk, Bobrujsk (BE)	*Pulsatilla* has been used for amateur dyeing [50].
Homo sapiens Linnaeus, 1758	mocz [urine]	urine	1	as a preservative for blue dye in textiles [3]	blue		Urine was commonly used as a mordant in folk dyeing, especially for building indigo vats of indigo or woad (fermentation vats with settling clarified urine) [15].
Hordeum vulgare L.	jęczmień	n. sp.	1	Easter eggs?	n. sp.	Grąziowa nad Wiarem (SE Poland)	
Hypericum perforatum L.	trojca świataja	fl	1	vodka	n. sp.	Przemyskie (PL); Czartkowskie i Tarnopolskie (UA)	Mostly used with oil and alcohol but also for dyeing textiles yellow [46].
Juglans regia L.	orzech włoski	fruit rind	2	Easter eggs	n. sp.	Jarosław, Cieszanów (PL)	A classic dye for wool that is also used for textiles [32,34,37,46]. Its surprising absence in the materials can be attributed to the small prevalence of walnuts in the 19th c. in the area. The pseudofruits dye brown [32].
Juniperus communis L.	jałowiec	fr	1	wool	n. sp.		
Lichens	mech z kamieni, wilcz	wh	6	wool	brown (olive, red—depending on the kind of lichen)	northern Poland (Kaszuby; Polskie Brzozie near Brodnica; Kociewie; Suszyn, Stępów, Wola Stępowska, all near Gostyń)	Lichens were widely used in NE Europe [51]. Reported as "mech" from Regnów near Rawa Maz. and Kocierzew near Łowicz [37].
Linum usitatissimum L.	len	n. sp.	1	Easter eggs?	n. sp.	Jeżewo near Borek	
Lithospermum cf *officinale* L.	wróble proson. sp.	roots	3	Easter eggs, wool, yarn, flax	red	Ozorków, Zgierz, Konstantynów, Pabianice, Tomaszów (PL)	Another species of the genus, *L. arvense*, was used as a dye in Poland [36].

Table 1. *Cont.*

Scientific Name	Local Name	Part	UR	Use	Main Color	Geography	Use in Other Sources
Lupinus sp.? (uncertain identification by Rostafiński)	nogurje, zonagurje, nagwięć, zonagwięć	n. sp.	1	textiles	n. sp.	Niedźwiadka near Łuków	
Lycopodium complanatum L., and *L. clavatum* L. (and possibly other species from this genus)	zielenica, zeglen, swarzybaba	n. sp.	4	textile (flax, cotton), Easter eggs	green, yellow, red	Eastern Poland: Łuków, Sokołów Podlaski, Węgrów (esp. in Rażny), Włodawa (e.g., in Brus), Janów Lubelski (PL); Weleśnica (BE)	Used to dye wool [32]. In Iron Age Finland, it was boiled for several days to produce an alum mordant [52]. *Zielenica* was reported as used for eggs in Gnieczyna near Przeworsk and Zalesie near Siemiatycze [37].
Malus domestica (Suckow) Borkh.	jabłoń [mainly], also: jabłonnoka, kwaśnica	l, bk	19	wool, Easter eggs?	yellow, (green, red)	throughout (PL); Bobujsk (BE); Mościska, Jaworów, Mikulińce, Tarnopol, Zbaraż, Skałat, Nowesioło (UA)	Obtaining red from this species is difficult, but there are recipes from the early Middle Ages. In general, most sources on the practical side of dyeing [32,34,46] mention apple leaves as a raw material for dyeing yellow with alum and green with iron, and the bark as a raw material giving orange shades of yellow with glow or slightly pink colors for wool, as confirmed by A.B.'s experience. Widely used in north Slavic territories for red dye [22,29,31]. For example, Kluk writes, "village women boil it with alum to obtain red dye".
Origanum vulgare L.	lebiodka, lebioda pusząca, lebioda (PL); macier-duszka (BE)	flowering tops	8	textiles, yarn, wool	(dark) red	Eastern Poland: Międzyleś (Radzymin), Niedźwiadka and the whole Łuków area, Rażny near Sadowne (Węgrów); NW Belarus and Kuchcice (Ihumeń) (BE)	
Papaver rhoeas L.	polny mak, zajęczy mak, glapi mak	n. sp.	3	textiles	n. sp.	Pleszew area (western Poland)	Flowers formerly used to dye dark blue [29].
Phaseolus vulgaris L.? (with dark seeds)	fasola czarna	n. sp.	1	Easter eggs?	violet	Mikulińce, Tarnopol, Zbaraż, Skałat, Nowesioło (UA)	
Poaceae	trawa	sh	10	Easter eggs	green	throughout (PL)	
Populus nigra L. and *P.* x *canadensis* Moench	topola, jabrzędź	catkins	1	Easter eggs	green, (blue)	Maciejowice (Garwolin), Romanów (Włodawa)	Poplar catkins used to dye eggs brown in Starosiedlce near Iłża, Kostrzyń near Białobrzegi [37].
Porphyrophora polonica (Linnaeus, 1758)	czerwiec, czerwiec polski, maściki	insect body	4	textiles; e.g., flax and cannabis yarn (BE)	red	Podpniewki (Pniewy), Ostroróg (Szamotuły), Buk county, Kościelna Wieś (Kujawy) (PL); Ihumeń county (BE); Quote from Szamotuły (W. Zentkeler): they gathered worms called "maściki" and sold to apothecaries not only in Szamotuły county but also in Buk county	According to Jakubowski [23], use ceased before the 19th century. At that time, it was already an uneconomic raw material, as cochineal was imported instead.
Potentilla erecta L.?	termentyla	n. sp.	1	walking sticks, red stripes (together with alder phloem bitten in the mouth)	red	Jarosław, Cieszanów (SE PL)	

Table 1. *Cont.*

Scientific Name	Local Name	Part	UR	Use	Main Color	Geography	Use in Other Sources
Prunus domestica L.	śliwa	bk	1	Easter eggs?	n. sp.	unknown location	Pink color [46].
Prunus spinosa L.	tarnina, tarń	bk, r	2	Easter eggs	black (roots), green (bark)	Grąziowa nad Wiarem—root, Rymanów—bark (SE Poland)	Brown color [32].
Pterocarpus sp.?	kraska, drzewo sandałowe	wd	1	Easter eggs	red	n. sp.	Imported. Formerly used for textiles and yarn [32,45,46], not to be confused with *Santalum*.
Pyrus communis s.l.	gruszka, przycierpka, lasówka	dried fr	1	leather	n. sp.	Rojówka near Tęgoborze (S Poland)	
Quercus spp. (mainly *Q. robur* L.)	dąb	bark	12	Easter eggs? wool, flax yarn, aprons, skirts, and corsets; with iron salts as	black, (brown, reddish)	throughout (PL); Ihumeń county, Lipów, Rzeczyca county (BE); Czortkowskie i Tarnopolskie (UA)	Was used for dyeing fabrics, wool yarn, and mordanting, as well as wood, hair, and fishing nets [37]. Produces shades of brown and beige with alum mordant and shades of gray and black with the addition of iron/iron sulfate in a different form. Black is the easiest to achieve on wool [32].
Rhamnus cathartica L.	sakłak, szakłak, szkłak	fr	4	flax, cotton, wool	brown, yellow, dark green, (black 2).	PL: Młyny (Strzelno), Sokołów, Węgrów, Włodawa, Janów Lubelski; UA: between Lwów and Żółkiew	Unripe buckthorn berries were used to dye fibers green instead of the in-house method (first dyeing blue, then yellow) [15]. *R. cathartica* and *R. frangula* were probably confused and used interchangeably.
Rubia cf *tinctorum* L., though some *Galium* sp. cannot be excluded	marzanna, mazonna	r	3	textiles, Easter eggs, cultivated	n. sp. [probably red]	Łódź, Pabianice, Ozorków, Janów Lubelski, also: Lithuania and Samogitia	People used madder very rarely. It was a typical plant of the dyeing industry [15], usually used to give red, pink, orange, brick, and brown colors. It is the most durable of the natural reds of plant origin, with no equal (e.g., brazilwood is photosensitive; all practical sources say that it is used for dyeing evening clothes rather than day clothes) [32,34,45,46].
Rumex sp.	szczaw dziki	n. sp.	1	textiles	yellow	Mikulińce, Tarnopol, Zbaraż, Skałat, Nowesioło (UA)	Used for wool [32]. A.B.'s experience suggests a bright yellow color with alum mordant.
Salix sp.	wierzba	bk from young twigs	1	Easter eggs		Tarnów county (Dębica, Dąbrowa, Żabno, Ropczyce)	Gives a light green color [37].
Salvia officinalis L.	szałwija	n. sp.	1	Easter eggs	n. sp.	Malbork (PL)	Used for dyeing rugs in Turkey [53].
Sambucus nigra L.	bez (PL), baznyk (UA)	fr, (bk)	4	Easter eggs	violet, dark	Jarosław, Cieszanów; Brus (Włodawa) (PL); Winniki (Sambor, UA)	Widely known, e.g., [32,34,46].
Saponaria officinalis L.	kukułyca	n. sp.	1	washing clothes before dyeing		Weleśnica (BE)	
Secale cereale L.	żyto	sh	39	Easter eggs	green	throughout	Easter eggs [37,54].
Serratula tinctoria L.	żółkwiło	n. sp.	1	wool	yellow	SE part of Biłgoraj area (E Poland)	Good yellow dye for fibers [32,34]
Solanum tuberosum L.	ziemniak	peel from tubers	1	Easter eggs?	n. sp.	Kartuzy, Kaszuby (PL)	

Table 1. *Cont.*

Scientific Name	Local Name	Part	UR	Use	Main Color	Geography	Use in Other Sources
Sorbus aucuparia L.	jarzębina	bk	1	Easter eggs?	dark yellow	Wołyń (PL)	Bark is a good olive-colored dye for fibers [34].
Spinacia oleracea L.	szpinak	n. sp.	1	yarn	n. sp.	Tykocin, Zambrów (NE PL)	
Tagetes sp.	kupczaki pełne	n. sp.	1	wool	red	Iwanków near Borszczów (UA)	Depending on the mordant used, it dyes olive, yellow, brown, or orange. A good dye for both plant and animal fibers [34].
Thymus pulegioides L., *T. serpyllum* L. or *Origanum vulgare* L.	macierzanka (PL and BE), materynka (PL), matyrynka (UA)	n. sp.	3	wool	dark brown	Brus (Włodawa) (PL); Lipów (Rzeczyca) (BE); Czortków area (UA)	The name can be applied to both genera, though *Origanum vulgare* is most likely.
Thymus pulegioides L.	czabor, cząber	n. sp.	1	Easter eggs	green	NW Belarus	Identification confirmed by voucher specimen. *T. serpyllum* L. was probably used in a similar fashion for eggs [37].
Triticum sp.	pszenica	sh	6	Easter eggs	green	throughout (PL)	
Urtica dioica L. and *U. urens* L.	pokrzywa (PL and BE), krapiwa piekuszcza (BE, for U. dioica), rzeszka (BE for U. urens)	rt	3	Easter eggs	bright yellow	NW Belarus, esp. Jeziora near Grodno; SE part of the Biłgoraj area (PL)	Dyes bright green or yellow. Best used with mordants [46].
Vaccinium myrtillus L.	czarne jagody (PL), czernice (BE)	fr (dried or juice)	4	flax and cannabis textiles, e.g., kerchiefs	black	Kalisz area, Kcynia (Szubin) (PL); Naliboki (Pińsk) and Weleśnica (Mińsk) (BE)	The color is sensitive to light and pH change; most suitable for wool (A.B.'s own observations). Mentioned by [54] as a source of blue dye.
Vaccinium sp.	[only Latin name given]	fr	1	dried fruit in vodka for wool and flax; alum mordant	black		
Vaccinium uliginosum L.	maczało	n. sp.	1	threads	n. sp.		
Viburnum opulus L.	kalina	fr	1	decocted	red	Winniki (Sambor) (UA)	Fruits and bark used as a dye in Turkey [55].
Vinca minor L.	barwinek	sh	1	Easter eggs	green	SW Ukraine	
Viola sp. (probably *V. odorata* L.)	fiołki	f	2	Easter eggs	violet, blue	Rymanów (Krosno), Grąziowa (Ustrzyki Dolne) NW Ukraine	Flowers are used as an amateur dye [56].
Silene latifolia ssp. *alba* (Mill.) Greuter and Burdet	sabaczeje mydło	n. sp.	1	washing clothes		NW Ukraine	
unidentified, either *Isatis tinctoria* L. or *Indigofera tinctoria* L.	indygo, indyk	n.sp.	3	textile, maybe also Easter eggs	n. sp.	Janów Lub., Maciejowice near Garwolin, Rabka	Imported; both widely used dye species.
unidentified	jabłonnik [not apple]	n. sp.		Easter eggs?	n. sp.		
unidentified	kacanki (pierwiosnek)	n. sp.		textiles?	n. sp.		

Table 1. *Cont.*

Scientific Name	Local Name	Part	UR Use	Main Color	Geography	Use in Other Sources
unidentified (maybe *Berberis vulgaris* L. though *Malus domestica* may not be excluded)	kwaśnica	bk	1 textiles	yellow	Mikulińce, Tarnopol, Zbaraż, Skałat, Nowesioło (UA)	
unidentified	leśnica	bark	textiles?	yellow		Probably hazel, which dyes yellow with alum (A.B.'s experiments).
unidentified	mieszalnik	n. sp.	Easter eggs?	n. sp.	Nieśwież, Słuck, Mińsk (BE)	Described as "forest grass."
unidentified	mydło	n. sp.	textiles?	n. sp.	Tyłowo near Puck	Maybe a saponin-rich plant used as an element in preparing fabrics for dyeing. On the other hand, "gapie mydło" suggests *Lithospermum arvense*.
unidentified	pietruszka wodna	n. sp.	Easter eggs?	n. sp.	Łuków area	
unidentified	popawka	n. sp.	Easter eggs?	n. sp.	Lipów near Rzeczyca (BE)	
unidentified	wiluk	n. sp.	Easter eggs?	n. sp.	Lipów near Rzeczyca (BE)	
unidentified	zanogięć, zanagięć	n. sp.	Easter eggs?	yellow	Niedźwiadka near Łuków	
unidentified	zanowica	n. sp.	Easter eggs?	yellow, brown	Brus near Włodawa	
unidentified (could be *Lycopodium* sp.)	zielonka	n. sp.	Easter eggs?	n. sp.	Lipów near Rzeczyca (BE)	Described as "grass in conifer woods with branches similar to cypress, bright-green." Also mentioned in *Gospodyni Litewska* [31].

bd—buds, bk—bark; br—branches; c—cones; fl—flowers, fr –fruits; l—leaves, r—roots; sh—young shoots; n. sp.—not specified; wd—wood; wh—whole plant with roots. More details: [1]—After soaking textiles overnight, they become brown. They turn black if left in an iron-rich meadow soil. [2]—One of the respondents said: "The old woman says that her parents dyed their wool and yarn black with buckthorn, but this is probably a mistake, for unripe buckthorn berries give a permanent yellow, ripe dark green". [3]—"in blue—they painted with commercial blue paint, but because this paint faded from the sun and was washed out by the rain, the housewives came up with an experiment to fix the color by soaking the yarn after dyeing in urine. For this purpose, a large bowl with two handles was ordered by the stove fitter, carefully stored from year to year in every peasant house for known use. All the elders participated in this activity, because a lot of fresh liquid was needed, and children were excluded, because if one of them said 'it stinks' the color would wear off. In general, dyeing activities were kept secret and not revealed to profane eyes". [4]—Podbielkowski [57] lists many synonyms for the species, e.g.,: "drzewo brazylijskie", "drzewo fernambukowe", "drzewo pernambukowe" (for *C. bras* and *C. echinata*). A related species, called sappanwood or eastern brazilwood (*Biancaea sappan* (L.) Tod.), is also used as a textile colorant to this day. Wood, bark, and roots are used, usually heartwood. The dye gives shades of red (orange, pink), depending on the mordants used and the pH of the dye bath (according to A.B.'s experience and [58]). In the 19th century, the names "drzewo fernambukowe" and "brazylka" were used [54]. We also suspect that in some sources from the 19th century, "brazylia" was used for *Haematoxylum campechianum* L., which gives blue and purple colors, e.g., according to [54] "Dyeing Easter eggs (. . .) red: decoction of 'fernambuk' with alum. For blue: decoction of 'brazylia'".

3. Discussion

3.1. Comparison with Other Studies

As shown in Table 1, most of the listed taxa are widely known in the dyeing industry or have been recorded by other studies. Nevertheless, this study is a valuable and large-scale documentation of dyeing practices in the Eastern European countryside. The prominence of a few well-known materials is visible. The studies of the Polish Ethnographic Atlas list only 19 species of dye plants, compared to 74 (plus unidentified taxa) in Rostafiński's questionnaire from over a century before. This well illustrates the decrease in dyeing traditions between the 19th and 20th centuries.

There are very few records of the cultivation of *Rubia* in Poland, which is consistent with the observations of other authors that this plant was used in the dyeing industry but was not part of the Polish dyeing tradition [15,22].

The low presence of blue dyes (Figure 2) is no surprise. This was a sought-after color, but very few plants can provide it, in contrast to yellow, brown, or even red, often present in nature [21].

Rostafiński's study confirmed that the tradition of using Polish cochineal red dye was dead by the end of the 19th c. No further ethnographic studies in Eastern Europe ever reported the use of this species as a dye.

No mushrooms were used as dyes in the 19th century, though the use of lichens was reported in some parts of Poland (Table 1). Unfortunately, there are no details or specimens that would enable their identification. Lichens have been used as a dye in some parts of the world, e.g., in the UK [51], and the information about their use in Poland is the only such record from Polish ethnographic literature. This can be counted as an achievement of Rostafiński's study.

3.2. Identification Problems

Historical data usually do not have voucher specimens attached [59]. That is why some species were not identified either at the genus or species level. We faced this problem for most questionnaires, apart from Federowski's [60], which included a detailed herbarium that also enabled the identification of taxa in other questionnaires.

One of the problems was distinguishing *Carthamus tinctorius* and *Crocus sativus*. The former served as a cheap alternative to the latter. Another problem was distinguishing *Origanum vulgare* and *Thymus* spp. Both species can be called by folk names starting from *macier-*, *mater-*, meaning mother. Usually, when the red dye is concerned, we are dealing with *Origanum vulgare* [22], but possible identification problems may occur. We also had a problem distinguishing *Origanum vulgare* and *Chenopodium*, which tend to have similar names (lebiodka, lebioda). We were also unsure which species of Caesalpinieae was used for red paint (see Table 1). Another issue is distinguishing *Hepatica nobilis* and *Pulsatilla* spp. Both of these taxa were probably used.

3.3. Weeds and Woody Species as Dyeing Plants

It must be noted that many dye plants were "weeds" in pastures, e.g., *Origanum vulgare*, *Arctostaphylos uva-ursi*, or woodland plants of little fodder value, e.g., *Lycopodium* spp., *Hepatica nobilis*, *Rhamnus cathartica*, *Berberis vulgaris*, *Prunus spinosa*. Others were cereal weeds (*Centaurea cyanus*, *Papaver rhoeas*). Thus, the harvesting of the abovementioned species came with an extra benefit for farming by removing weedy or inedible species.

Tree leaves, bark, and roots are another important category of dyeing plants. Out of the 10 most commonly used dyeing ingredients, three taxa were native common tree species: alder, birch, and oak; apple was the most common fruit tree. Given their abundance, they could easily be utilized as a dye. Trees in gardens and small woods often have their lower branches chopped off to increase the growth of the main stem, so the parts used for making dyes could have been just farming by-products.

3.4. Easter Eggs

This study is an important contribution to the issue of dyeing Easter eggs. Dyeing Easter eggs has been a custom widespread in Eurasia since the early ages of Christianity. Depending on the local tradition, the eggs could be later eaten, or only empty shells were used for dyeing. Examples of ritual egg dyeing are also known from pre-Christian times [61,62]. The decorations of Easter eggs were a subject of ethnographic research in Poland, but mainly in the context of their patterns and customs associated with them, not the species of plants used for dyeing [62,63], apart from the study of the Polish Ethnographic Atlas [37]. Recording the use of 53 taxa for dyeing Easter eggs may help in preserving this tradition. This number is quite impressive, considering that, for example, Guarrera recorded only three species used from a few regions of Italy [64], and only 13 taxa were recorded by the studies of the Polish Ethnographic Atlas for dyeing eggs in Poland [37].

4. Materials and Methods

We extracted the information concerning the researched questions from a database of letters written to Rostafiński in response to his questionnaire, which was created by the first author (P.K.). Additionally, we included two published works of imminent Polish ethnographers, Michał Federowski (1853–1923) [60] and Zygmunt Gloger (1845–1910) [65], which were structured using Rostafiński's questionnaire and can be treated as the conceptual part of this project, i.e., responses to the questionnaire which were never sent to Rostafiński. Altogether, our database included 640 records from 126 respondents who provided meaningful information on plant dyes.

Plants were identified using standard methods applied in historical ethnobotany (compare the studies listed by da Silva et al. [66] and summarized by Lardos [67]), i.e., comparing available voucher specimens, folk names recorded in other sources, uses reported in previous publications and geographical distribution and abundance of the taxa used. The credibility of such historical identifications was also extensively discussed by Łuczaj [59]. A similar methodology has been used in other historical ethnobotany publications in the same special issue, *Historical Ethnobotany: Interpreting the Old Records* (e.g., [68–70]. Identification was facilitated by the voucher specimen collection supplied by Federowski. Twenty-six informants supplied 53 scientific names of plants, which seemed trustworthy in most cases (see the discussion on *Hepatica nobilis* and *Pulsatilla* in the Discussion). Common plants were also identified by comparing the folk names supplied with other sources on folk botany, e.g., Fischer's dictionary [38].

When the provided local names of plants have been exclusively and commonly used for a certain genus or species throughout the study area, the scientific name of the genus or species was assigned to the local name. For example, "cebula" is the main name for *Allium cepa* L. used exclusively for this species. "Olcha" is used for the genus *Alnus*, "dąb" for oak (*Quercus* sp.), and no other plants have ever been called these names. Additionally, distribution maps of the taxa were checked to ensure they occurred in the studied localities, though most of the taxa used are very common species with large ranges. In the case of 13 plant names, trustworthy identification was impossible (Table 1).

5. Conclusions

Remnants of traditional plant dyeing knowledge were saved by Rostafiński in his 1883 study. His work revealed that the tradition of plant dyeing in Poland, Belarus, and Ukraine was already disappearing in the 19th century. The plants and other ingredients used to make dyes reported in this study are usually widely known species and ingredients used for dyeing. However, the data provided may be helpful in restoring traditional textile production and preserving the tradition of dying Easter eggs in Poland, Belarus, and Ukraine.

Supplementary Materials: The following supporting information can be downloaded at: https://www.mdpi.com/article/10.3390/plants12071482/s1. Table S1: The original data matrix.

Author Contributions: Conceptualization, Ł.Ł. and P.K.; methodology, Ł.Ł.; writing—original draft preparation, Ł.Ł.; writing—review and editing, Ł.Ł., P.K. and A.B.; historical analysis, P.K.; discussion concerning comparison with literature, A.B., references, Ł.Ł. and A.B. All authors have read and agreed to the published version of the manuscript.

Funding: This research received no external funding.

Data Availability Statement: The original data matrix is enclosed as Supplementary Material with the paper.

Acknowledgments: The authors of this paper would like to thank Michał Wegrzyn (Kraków) for making the map. We also thank Zygmunt Kłodnicki (Cieszyn) for helping in the literature search.

Conflicts of Interest: The authors declare no conflict of interest.

References

1. Brunello, F. *The Art of Dyeing in the History of Mankind*; Neri Pozza Editore: Vicenza, Italy, 1973.
2. Nieto-Galan, A. *Colouring Textiles: A History of Natural Dyestuffs in Industrial Europe*; Springer: Dodrecht, Germany, 2001.
3. Burgio, L.; Ciomartan, D.A.; Clark, R.J. Pigment identification on medieval manuscripts, paintings and other artefacts by Raman microscopy: Applications to the study of three German manuscripts. *J. Mol. Struct.* **1997**, *405*, 1–11. [CrossRef]
4. Bechtol, T.; Mussak, R. (Eds.) *Handbook of Natural Colorants*; Wiley: Chichester, UK, 2009.
5. Prigioniero, A.; Geraci, A.; Schicchi, R.; Tartaglia, M.; Zuzolo, D.; Scarano, P.; Marziano, M.; Postiglione, A.; Sciarrillo, R.; Guarino, C. Ethnobotany of dye plants in Southern Italy, Mediterranean Basin: Floristic catalog and two centuries of analysis of traditional botanical knowledge heritage. *J. Ethnobiol. Ethnomed.* **2020**, *16*, 5. [CrossRef] [PubMed]
6. Liu, Y.; Ahmed, S.; Liu, B.; Guo, Z.; Huang, W.; Wu, X.; Li, S.; Zhou, J.; Lei, Q.; Long, C. Ethnobotany of dye plants in Dong communities of China. *J. Ethnobiol. Ethnomed.* **2014**, *10*, 23. [CrossRef] [PubMed]
7. Hu, R.; Li, T.; Qin, Y.; Liu, Y.; Huang, Y. Ethnobotanical study on plants used to dye traditional costumes by the Baiku Yao nationality of China. *J. Ethnobiol. Ethnomed.* **2022**, *18*, 2. [CrossRef] [PubMed]
8. Mostacero León, J.; López Medina, S.E.; Yabar, H.; De La Cruz Castillo, J. Preserving traditional botanical knowledge: The importance of phytogeographic and ethnobotanical inventory of Peruvian dye plants. *Plants* **2017**, *6*, 63. [CrossRef]
9. Hart, K.H.; Cox, P.A. A cladistic approach to comparative ethnobotany: Dye plants of the southwestern United States. *J. Ethnobiol.* **2000**, *20*, 303–325.
10. Luu-Dam, N.A.; Ninh, B.K.; Sumimura, Y. Ethnobotany of colorant plants in ethnic communities in Northern Vietnam. *Anthropology* **2016**, *4*, 2332-0915. [CrossRef]
11. Quizon, C.; Magpayo-Bagajo, F. Botanical knowledge and indigenous textiles in the Southern Mindanao highlands: Method and synthesis using ethnography and ethnobotany. *South East Asia Res.* **2022**, *30*, 89–105. [CrossRef]
12. Sutradhar, B.; Deb, D.; Majudmar, K.; Datta, B.K. Traditional dye yielding plants of Tripura, Northeast India. *Biodiversitas J. Biol. Divers.* **2015**, *16*, 121–127. [CrossRef]
13. Malaisse, F.; Claus, W.; Drolkar, P.; Lopsang, R.; Wangdu, L.; Mathieu, F. Ü Ethnomycology and Ethnobotany (South Central Tibet). Diversity, with emphasis on two underrated targets: Plants used for dyeing and incense. *Geo-Eco-Trop* **2012**, *36*, 185–199.
14. MacFoy, C. Ethonobotany and sustainable utilization of natural dye plants in Sierra Leone. *Econ. Bot.* **2004**, *58*, S66–S76. [CrossRef]
15. Kowecka, E. *Farbiarstwo Tekstylne na Ziemiach Polskich (1750–1870)*; Wydawnictwo Ossolineum: Wrocław, Poland, 1963.
16. Kamińska, J.; Turnau, I. (Eds.) *Zarys Historii Włókiennictwa na Ziemiach Polskich do Końca XVIII Wieku*; Wydawnictwo Ossolineum: Wrocław, Poland, 1966.
17. Mączak, A. *Sukiennictwo Polskie XIV-XVII Wieku: Badania z Dziejów Rzemiosła i Handlu w Epoce Feudalizmu*; Instytut Historyczny PAN: Warszawa, Poland, 1955.
18. Wyrozumski, J. Urzet farbiarski w Polsce średniowiecznej. *Kwart. Hist. Kult. Mater.* **1966**, *14*, 437–448.
19. Wyrozumski, J. Średniowieczne kompendium wiedzy o barwnikach (ze zbiorów Biblioteki Czartoryskich w Krakowie). *Kwart. Hist. Kult. Mater.* **1974**, *22*, 663–671.
20. Faust, J. Farbiarstwo roślinne—Zagadnienia ogólne oraz zastosowanie w konserwacji zabytkowych tkanin. *Ochr. Zabyt.* **1983**, *3–4*, 215–218.
21. Pastoureau, M. *Blue: The History of a Color*; Princeton University Press: Princeton, NJ, USA, 2001.
22. Moszyński, K. *Kultura Ludowa Słowian: Kultura Materialna*; Książka i Wiedza: Warszawa, Poland, 1967; Volume 1.
23. Jakubski, A.W. *Czerwiec Polski*; Wyd. Kasy im. Mianowskiego—Instytutu Popierania Nauki: Warszawa, Poland, 1934.
24. Jabłonowska, A. *Porządek Robót Miesięcznych Ogrodnika Na Cały Rok Wypisany*; W drukarni pokojowey Siemiatyckiey: Siemiatycze, Poland, 1787.
25. Hermstadt, Z.F. *Nauka o Sztuce Farbowania Dobrze i Trwale Materyi Jedwabnych, Wełnianych, Bawełnianych i Lnianych*; A. S. Translator, Księgarnia Zawadzkiego i Węckiego: Warszawa, Poland, 1819.

26. du Monceau, D. *Dzieło o Rolnictwie*; Wilno, Lithuania; pp. 1770–1773.

27. Oczapowski, M. *Gospodarstwo Wiejskie—Uprawa Roślin Fabrycznych*; SH Merzbach: Warszawa, Poland, 1837.

28. Bradley, R. *Kalendarz Rolniczy i Gospodarski*; Drukarnia J. K. Mci i Rzplitey u XX. Scholarum Piarum: Wilno, Lithuania, 1770.

29. Kluk, K. *Dykcyonarz Roślinny*; Drukarnia Pijarów: Warszawa, Poland, 1787–1788.

30. Gerald-Wyżycki, J. *Zielnik Ekonomiczno-Techniczny*; Józef Zawadzki: Wilno, Lithuania, 1845.

31. Ciundziewicka, A. *Gospodyni Litewska Czyli Nauka Utrzymywania Porządnie Domu i Zaopatrzenia Go*; Józef Zawadzki: Wilno, Lithuania, 1873.

32. Tuszyńska, W. *Farbowanie Barwnikami Naturalnymi*; Watra: Warszawa, Poland, 1986.

33. Turski, J.S.; Więcławek, B. *Barwniki Roślinne i Zwierzęce: Chemia Stosowana*; Państwowe Wydawnictwa Techniczne: Warszawa, Poland, 1952.

34. Schmidt-Przewoźna, K. (Ed.) *Barwienie Metodami Naturalnymi. Rośliny Barwierskie i Ich Potencjał*; Wydawnictwo Uniwersytetu Przyrodniczego w Poznaniu: Poznań, Poland, 2020.

35. Dominikiewicz, M. *Podstawy Farbiarstwa: Zasady Barwienia Włókien i Tkanin*; Cz. 1: Podstawy Teoretyczne; Naukowo-Badawczy Instytut Włókiennictwa: Łódź, Poland, 1947.

36. Kapler, A. Dzikie gatunki pokrewne roślinom uprawnym o znaczeniu barwierskim i włóknodajnym—Crop wild relatives of significance for dyeing and fibre. In *Dzikie Gatunki Pokrewne Roślinom Uprawnym Występujące w Polsce. Lista, Zasoby i Zagrożenia*; Dostatny, D.E., Dajdok, Z., Eds.; Wydawnictwo Kontekst: Poznań, Poland, 2020; pp. 107–114.

37. Lebeda, A. *Wiedza i Wierzenia Ludowe. Tom 6. Komentarze do Polskiego Atlasu Etnograficznego*; Polskie Towarzystwo Ludoznawcze: Wrocław, Poland, 2002.

38. Kujawska, M.; Łuczaj, Ł.; Sosnowska, J.; Klepacki, P. *Rośliny w Wierzeniach i Zwyczajach Ludowych: Słownik Adama Fischera*; Polskie Towarzystwo Ludoznawcze: Wrocław, Poland, 2016.

39. Köhler, P. Józef Rostafiński's ethnobotanical enquiry of 1883 concerning Polish vernacular names and uses of plants. *Arch. Nat. Hist.* **2015**, *42*, 140–152. [CrossRef]

40. Köhler, P. The Romantic myth about the antiquity of folk botanical knowledge and its fall: Józef Rostafiński's case. *Acta Balt. Hist. Et Philos. Sci.* **2015**, *3*, 99–108. [CrossRef]

41. Köhler, P. An involuntary ethnobotanist? Józef Rostafiński (1850–1928) and his research in Poland. In *Pioneers in European Ethnobiology*; Svanberg, I., Łuczaj, Ł., Eds.; Uppsala University Press: Uppsala, Sweden, 2014; pp. 149–180.

42. Łuczaj, Ł.; Pieroni, A.; Tardío, J.; Pardo-de-Santayana, M.; Sõukand, R.; Svanberg, I.; Kalle, R. Wild food plant use in 21st century Europe, the disappearance of old traditions and the search for new cuisines involving wild edibles. *Acta Soc. Bot. Pol.* **2012**, *81*, 359–370. [CrossRef]

43. Łuczaj, Ł.; Wilde, M.; Townsend, L. The ethnobiology of contemporary British foragers: Foods they teach, their sources of inspiration and impact. *Sustainability* **2021**, *13*, 3478. [CrossRef]

44. Sõukand, R.; Mattalia, G.; Kolosova, V.; Stryamets, N.; Prakofjewa, J.; Belichenko, O.; Kuznetsova, N.; Minuzzi, S.; Keedus, L.; Prūse, B.; et al. Inventing a herbal tradition: The complex roots of the current popularity of Epilobium angustifolium in Eastern Europe. *J. Ethnopharmacol.* **2020**, *247*, 112254. [CrossRef]

45. Liles, J.N. *The Art and Craft of Natural Dyeing: Traditional Recipes for Modern Use*; Univ. of Tennessee Press: Knoxville, TN, USA, 1990.

46. Dean, J.; Casselman, K.D. *Wild Color, Revised and Updated Edition: The Complete Guide to Making and Using Natural Dyes*; Clarkson Potter/Ten Speed: Berkeley, CA, USA, 2010.

47. Desnos, R. *Botanical Colour at Your Fingertips*; Self-Published: UK, 2016.

48. Bystry, A. *Dzikie Barwy*; Wydawnictwo Dzikie Barwy: Łódź, Poland, 2019.

49. Tehrani, M.; Bayegan, Y.; Shahmoradi Ghaheh, F. Investigation of parameters affecting dyeing of merino wool yarn with chicory leaves plant (*Cichorium intybus* L.). *J. Text. Sci. Technol.* **2020**, *9*, 17–23.

50. Damakos, J. Flower of the Month: Pasque Flower. Available online: https://www.juliadimakos.com/flower-of-the-month-pasque-flower/ (accessed on 7 February 2023).

51. Casselman, K.D. *Lichen Dyes: The New Source Book*; Courier Corporation: North Chelmsford, MA, USA, 2001.

52. Vajanto, K. *Dyes and Dyeing Methods in Late Iron Age Finland*; Doctorate Department of Philosophy, History, Culture and Art Studies University of Helsinki: Helsinki, Finland, 2016.

53. Üzeri, B.R.B.H.D. A research on the colors obtained from sage (*Salvia officinalis* L.) and their fastness values. Yuzuncu Yıl. *Univ. J. Agric. Sci.* **2002**, *12*, 31–36.

54. Rewieński, S. *Teka Oszczędnych Wskazówek*; Maurycy Orgelbrand: Warszawa, Poland, 1887.

55. Yılmaz, F.; Koçak, Ö.F.; Özgeriş, F.B.; Şapçı Selamoğlu, H.; Vural, C.; Benli, H.; Bahtiyari, M.İ. Use of *Viburnum opulus* L. (Caprifoliaceae) in dyeing and antibacterial finishing of cotton. *J. Nat. Fibers* **2020**, *17*, 1081–1088. [CrossRef]

56. Dean, J. Ice Flowers. 2011. Available online: https://www.jennydean.co.uk/ice-flowers/ (accessed on 7 February 2023).

57. Podbielkowski, Z. *Słownik Roślin Użytkowych*; PWN: Warszawa, Poland, 1989.

58. Teresinha, Brazilwood Dyeing. Available online: http://www.wildcolours.co.uk/html/brazilwood_dyeing.html (accessed on 7 February 2023).

59. Łuczaj, Ł.J. Plant identification credibility in ethnobotany: A closer look at Polish ethnographic studies. *J. Ethnobiol. Ethnomed.* **2010**, *6*, 36. [CrossRef] [PubMed]

60. Graniszewska, M.; Leśniewska, H.; Mankiewicz-Malinowska, A.; Galera, H. Rośliny użyteczne . . . Michała Fedorowskiego—Dzieło odnalezione po 130 latach. Useful plants by Michal Fedorowski—The work found after 130 years. *Etnobiol. Pol.* **2013**, *13*, 63–118.
61. Thompson, K. *Culture & Progress: Early Sociology of Culture, Volume 8*; Routledge: Abingdon-on-Thames, UK, 2013.
62. Rucki, M.; Abdalla, M.; Benyamin, A. Jajka wielkanocne w początkach chrześcijaństwa. *Teol. Człowiek* **2015**, *29*, 295–312. [CrossRef]
63. Gawełek, F. *Palma, Jajko i Śmigus w Praktykach Wielkanocnych Ludu Polskiego*; Polskie Towarzystwo Ludoznawcze: Lwów, Ukraine, 1911.
64. Guarrera, P.M. Household dyeing plants and traditional uses in some areas of Italy. *J. Ethnobiol. Ethnomed.* **2006**, *2*, 9. [CrossRef]
65. Graniszewska, M.; Kapler, A. Odpowiedź Zygmunta Glogera (1845–1910) na ankietę etnobotaniczną Józefa Rostafińskiego ogłoszoną w 1883 r., dotycząca pogranicza Mazowsza i Podlasia. Zygmunt Gloger's (1845–1910) response to Józef Rostafiński's (1850–1928) ethnobotanical questionnaire, published in 1883, concerning the Mazovia and Podlachia borderland. *Etnobiol. Pol.* **2017**, *17*, 15–32.
66. da Silva, T.C.; Medeiros, P.M.; Balcázar, A.L.; Sousa, T.A.D.A.; Pirondo, A.; Medeiros, M.F.T. Historical ethnobotany: An overview of selected studies. *Ethnobiol. Conserv.* **2014**, *3*, 4.
67. Lardos, A. *The Herbal Materia Medica in Greek Texts—A Botanical and Pharmacognostic Perspective*. BSHP British Society for the History of Pharmacy—Talks and Conference Lectures Recorded at BSHP Meetings. Available online: https://www.youtube.com/watch?v=QMHBn5llxBQ (accessed on 19 March 2023).
68. Sõukand, R.; Kalle, R. The Appeal of Ethnobotanical Folklore Records: Medicinal Plant Use in Setomaa, Räpina and Vastseliina Parishes, Estonia (1888–1996). *Plants* **2022**, *11*, 2698. [CrossRef]
69. Dal Cero, M.; Saller, R.; Leonti, M.; Weckerle, C.S. Trends of Medicinal Plant Use over the Last 2000 Years in Central Europe. *Plants* **2023**, *12*, 135. [CrossRef]
70. Prakofjewa, J.; Anegg, M.; Kalle, R.; Simanova, A.; Prūse, B.; Pieroni, A.; Sõukand, R. Diverse in Local, Overlapping in Official Medical Botany: Critical Analysis of Medicinal Plant Records from the Historic Regions of Livonia and Courland in Northeast Europe, 1829–1895. *Plants* **2022**, *11*, 1065. [CrossRef]

Article

Early Citizen Science Action in Ethnobotany: The Case of the Folk Medicine Collection of Dr. Mihkel Ostrov in the Territory of Present-Day Estonia, 1891–1893

Raivo Kalle [1,*], Andrea Pieroni [1,2], Ingvar Svanberg [3] and Renata Sõukand [4]

1 University of Gastronomic Sciences, Piazza Vittorio Emanuele II 9, 12042 Pollenzo, Italy; a.pieroni@unisg.it
2 Medical Analysis Department, Tishk International University, 100 Meter Street & Mosul Road, Erbil 44001, Iraq
3 Institute for Russian and Eurasian Studies, Uppsala University, Box 514, SE-751 20 Uppsala, Sweden; ingvar.svanberg@ires.uu.se
4 Department of Environmental Sciences, Informatics and Statistics, Ca' Foscari University of Venice, Via Torino 155, Mestre, 30172 Venice, Italy; renata.soukand@unive.it
* Correspondence: raivo.kalle@mail.ee

Abstract: Presently, collecting data through citizen science (CS) is increasingly being used in botanical, zoological and other studies. However, until now, ethnobotanical studies have underused CS data collection methods. This study analyses the results of the appeal organized by the physician Dr. Mihkel Ostrov (1863–1940), which can be considered the first-ever internationally known systematic example of ethnopharmacological data collection involving citizens. We aim to understand what factors enhanced or diminished the success of the collaboration between Ostrov and the citizens of that time. The reliability of Ostrov's collection was enhanced by the herbarium specimens (now missing) used in the identification of vernacular names. The collection describes the use of 65 species from 27 genera. The timing of its collection coincided with not only a national awakening and recently obtained high level of literacy but also the activation of civil society, people's awareness of the need to collect folklore, the voluntary willingness of newspapers to provide publishing space and later to collect data, and the use of a survey method focusing on a narrow topic. While Ostrov's only means of communication with the public was through newspapers, today, with electronic options, social media can also be used.

Keywords: history of ethnobotany; early citizen science studies; history of ethnomedicine; archive data; ethnopharmacology; plant identification

Citation: Kalle, R.; Pieroni, A.; Svanberg, I.; Sõukand, R. Early Citizen Science Action in Ethnobotany: The Case of the Folk Medicine Collection of Dr. Mihkel Ostrov in the Territory of Present-Day Estonia, 1891–1893. *Plants* **2022**, *11*, 274. https://doi.org/10.3390/plants11030274

Academic Editor: Stephen O. Amoo

Received: 23 December 2021
Accepted: 17 January 2022
Published: 20 January 2022

Publisher's Note: MDPI stays neutral with regard to jurisdictional claims in published maps and institutional affiliations.

1. Introduction

Presently, a large number of wildlife observations in Europe are collected every year with the participation of volunteers, especially within ornithology. Amateurs have been of immense importance to many flora projects (national, provincial, and local floras) since the late nineteenth century and still are in many European countries. Based on this data collected through citizen science (CS), many distribution atlases (e.g., plants, mammals, and birds) are issued. Today, the importance of CS is growing exponentially, and it is already considered as an important pillar of science popularization and communication. For instance, on the initiative of the Swedish parliament in 1991, there is an on-going national inventory of all species of animals, vascular plants and mosses, fungi and lichens, and algae in order to map the biodiversity and this involves NGO's, volunteers, and, in fact, anyone interested in nature in Sweden [1]. The inventory project (Artdatabanken = Swedish Species Gateway) is organised by the Swedish Species Information Centre at the Swedish Agricultural University, and they are working full-time on accumulating, analysing, and disseminating gathered information concerning the species and habitats occurring in

Sweden. Citizen science is an important part of this project. Similar projects are underway in many other countries cf. [2], for example, in 2008, the Estonian Environmental Board started collecting people's wildlife observations, which thousands of individuals submit every year (see https://lva.keskkonnainfo.ee/ accessed on 21 December 2021). Citizen science is also receiving increasing attention as a mean of collecting indigenous and local knowledge to better manage and conserve ecosystems [3]. For example, an initiative was launched in Spain in 2017 where non-scientists are able to add traditional ecological knowledge to the electronic database CONECT (www.conecte.es accessed on 15 January 2022) [4,5].

However, after almost twenty years of existence as a concept, the definition of CS still depends on the context [6]. Therefore, we are inclined to view this method of involving citizens in science as something new, but history tends to repeat itself, as human nature does not change as quickly as the current technological progress would allow us to assume.

The roots of CS are sometimes traced as far back as Aristotle, while some refer to Swedish botanist Carl Linnaeus (1707–1778) or the amateur documentation of the flowering of cherry trees in Japan [7,8]. If today the CS method is primarily related to the supplementation of institutional (e.g., research, memory, or state institutions) collections, then in the 18th and 19th centuries, methods similar to what we assume in some contexts to belong to CS were used to supplement private collections. To record folk knowledge among the peasantry, the Sami, local healers, etc., was the normal process for Linnaeus and his pupils. The reason for this was mainly economic and to make botany useful. Why import expensive medicinal plants from abroad when equivalent plants could be found among the local inhabitants? Why import dye plants when the peasantry already successfully used lichens and plants to colour their fabrics? Why not increase the local authorities' knowledge of plants that could be used as a flour substitute in times of bad harvest? Linnaeus developed a research program with a questionnaire for travelling scholars [9] that they could use to gather local knowledge about these things. His own travels in the 1740s were expeditions sponsored by the government to record knowledge about wild plants and lichens that could be used as medicinal plants, dye plants, and food plants [10]. The first professorship in national economy was actually held by botanists and there were many publications dealing with the economic aspects of plants. For Linnaeus, this was an important way to show the usefulness of botanical knowledge for the university and the authorities, and for his many followers this knowledge was a way to improve the economy of Swedish communities [11]. In Sweden, this data collection included hundreds of locals, such as vicars, physicians, pharmacists, noble landowners, bailiffs, and even curious peasants, who participated in collecting plants and sending information. Linnaeus himself had around 600 correspondents, not only in Sweden (including Finland) but all over the world, who provided him with plants and information on the use of plants and animals cf. [12,13].

Some of these correspondents (for instance, Pehr Kalm in Finland; Johan Peter Falck, Johann Gottlieb Georgi, Johan Georg Gmelin, and Peter Simon Pallas in Russia) had their own networks of locals who gathered data for them [14]. One of these correspondents was the Norwegian Bishop Johan Ernst Gunnerus (1718–1773), who, very much inspired by Linnaeus, for his project on the Flora Norvegica (published in Latin in two volumes in 1766 and 1776) involved a wide network of clerks who collected information and plant specimens from all over Norway [15]. In Italy, Vincenzo de Romita (1838–1914) used a worldwide network of voluntary professionals to supplement his private collection with wildlife exhibits and data and was one of the first to use such an approach [16]. De Romita's nature collection is located today in a museum bearing his name (http://www.centrostudideromita.it/ accessed on 21 December 2021).

Yet, the essence of CS, as it is perceived today, is the thoughtful participation of citizens with no specific science-related training in a joint effort to better understand a particular scientific or social phenomenon, often from an interdisciplinary perspective [17]. A more recent and thoroughly researched example is that of the publication of questionnaires by a botanist from Jagiellonian University, Józef Rostafiński (1850–1928), which, mainly in 1883, attracted numerous correspondents who sent herbarium specimens, some of which are still preserved, along with their responses regarding the use of wild food plants [18,19]. Indeed, from the middle of the 19th century, influenced by romantic nationalism, many early (then often amateur) folklorists in Nordic and Central-European countries started campaigns to collect lyric and practical folklore through newspaper advertisements calling on people to contribute. For example, the "first appeal to initiate the collection of Swedish-speaking folklore appeared in the Vaasa-based newspaper Ilmarinen on 5 April 1848", although without success [20]. However, many diverse calls yielded a significant number of collected records, supported by intensified peasant education [21], and those campaigns had a strong component of CS. In 19th-century northern Europe, ethnobotany, the specific aspect of local practices that are of interest to us, was considered a grey area between folklore and the natural sciences (it was not until 1895 that John Harshberger coined the term ethnobotany). The majority of those numerous works collected in the 19th century were buried under the tons of never thoroughly analysed folklore, and for a valid reason: in folkloristic texts the local name itself, due to its ambiguity and fluidity, is rarely sufficient to make a direct connection between the botanical plant taxa and its use. In Europe, the 20th century was more productive in terms of plant-lore collection (including the involvement of students in the collecting work, as is seen in references to collections from the 1930s in Estonia [22] or Ireland [23]). Such early ethnobotanical collections are still to be explored and analysed, not only for the reported plants and uses, but also for the methodology with respect to the successful involvement of citizens in the preparation of a scientific study.

1.1. Mihkel Ostrov and Ethnopharmacology in Estonia the 19th Century

An excellent study ground for the history of CS in ethnopharmacology is provided by the collection of Estonian medical student and later doctor Mihkel Ostrov (1863–1940), who gathered information for a few years beginning in 1891. Through newspaper appeals, following the example of folklore collectors of his time, he succeeded in amassing a remarkable number of responses, including herbarium specimens. Although the original samples were not preserved, we still have the interpretations provided by Ostrov, along with his identifications of the plants. Ostrov, although studying to be a medical doctor, was very open to people's use of plants, as he had experience collecting them himself as a mentee of the pastor and folklorist Jakob Hurt (1839–1906).

Before Ostrov, Estonian ethnopharmacology had already been documented with proper botanical identification but only in very specific locations. It was first documented by the Baltic German doctor Johann Wilhelm Ludwig von Luce (1756–1842), who published the results of his research in a work titled "Heilmittel der Ehsten auf der Insel Oesel" [24], which can now be considered one of the first local pharmacological works in Europe, and it was documented a few years later, in 1831, by the amateur botanist and pastor Johann Heinrich Rosenplänter (1782–1846), who wrote a manuscript that remained unpublished [25]. They were both Baltic Germans and Estophiles, yet the Estonian language was not their mother tongue. Unlike Rosenplänter and von Luce, Ostrov shared the same cultural code and language with the people who sent him the information. Neither Rosenplänter nor von Luce used a CS approach *sensu stricto*, as they questioned people and collected plants themselves (although being amateur scientists), without using mediators, as far as we know. Ostrov, however, received a large part of his information through a network of correspondents, which is comparable to the current methods of data collection employed in CS.

The national awakening of Estonians, which began in the middle of the 19th century, reached its peak at the end of that century. This was accompanied by the emphasizing of the value of "antiquities" and, above all, the collection of lyrical folklore. The first attempt to involve the wider population in the collection of "antiquities" was made by the pastor and folklorist Matthias Johann Eisen (1857–1934). Starting in 1883, he began to make thematic appeals in newspapers. He was particularly active in putting forward calls to the general population from 1887. His interest was initially in fairy tales and later widened to general "antiquities" [26].

The largest action to involve the general public was the call made in 1888 by Jakob Hurt: "A couple of requests to Estonia's most alert sons and daughters". In almost 18 years, nearly 1400 correspondents sent him data. Hurt appealed to his collaborators, asking if they could, in addition to providing their knowledge, question people living in their village. Hurt's emphasis was also on the old and the archaic, and especially on lyrical folklore. Everyday practices, where herbal treatments belonged in those days, were not of great interest, and the methods selected for collection were not adequate. Although dedicating a special chapter to plant use, Hurt classified herbal treatments as "beliefs and customs". Hurt's attitude and classification also gave direction to all other subsequent folklore researchers [27,28].

Jakob Hurt regularly reported through newspapers the sources and quantity of information and what kind of folklore was sent to him. In addition to the reports, he also made repeated calls to encourage new collectors. In total, Hurt published more than 150 reports during his lifetime.

Before the appeals, Jakob Hurt had been collecting folklore privately for about ten years. He also used scholarships for this purpose. At the end of 1886, he turned to the Estonian Students' Association (EÜS) for help in collecting folklore. The society discussed it at the beginning of 1887, and among the first to agree to do fieldwork was a medical student at Dorpat (Tartu) University named Mihkel Ostrov. Hurt also sent a very comprehensive collection guide in German titled: "Bemerkungen zur Richtschnur beim Samm[e]ln alter estnischer Volkslieder, Märchen, Sprichwörter, Sagen etc". Since Hurt's home and school language was German, it was probably more convenient for him to make a guide in that language. Hurt promised to cover the fieldwork costs and salary of the scholarship holders. In 1887, Ostrov and his companion Oskar Kallas (later an Estonian diplomat and folklorist, 1868–1946) collected a very rich sample of folklore in the parishes of Laiuse, Torma, Simuna, and Põltsamaa (central Estonia) [29]. Ostrov also conducted expeditions on behalf of Jakob Hurt in Alutaguse (north-eastern Estonia) in 1888 and in Läänemaa (north-western Estonia) in 1889, where he also collected valuable material. In 1890, however, he went alone to Läänemaa to collect folklore. While collecting in Läänemaa, he also wrote down the first use of medicinal plants: "From the flowers of *liivatee* (*Thymus serpyllum*) they make a medicine against coughs and lung diseases" (EKS, c, page 63). The second impetus for Ostrov's personal collection was certainly that he was a member of the most progressive society of his time, the Society of Estonian Literati (active from 1871 to 1893), which was actively involved in collecting "antiques". From 1892 to 1893, Ostrov was also a board member of that society.

Ostrov therefore already had extensive experience in the field of folklore. Apparently because of this, he made his first public appeal while still a student (he graduated in the second half of 1891) on 6 April 1891, in the newspaper Postimees. At the same time, Jakob Hurt was communicating his appeals through the same newspaper. Ostrov states in the call, "When I collected old songs among the people, I saw that there are still many folk healers everywhere who collected a lot of medicinal plants from nature and use them to treat many diseases" [30]. His early ambition is clearly shown by the title of the appeal, which aims to gather "general information about Estonian folk medicines". The newspaper, for its part, added a request for active participation and for other newspapers to publish the call as well. This was what the newspapers Olevik [31] (see Figure 1) and Sakala [32] did on 8 and 26 April respectively. His call was scheduled for early spring so that collectors could pick the first plants in spring and summer. The call contained detailed instructions on how to pick and dry the plants and how to send them by post in a wooden box so that they would not be damaged. He also asked participants to write in detail how these plants were used and for what purpose, as well as their popular names. It is worth noting that, in the same newspaper, Ostrov also gave a positive review of the first Estonian medical textbook [33].

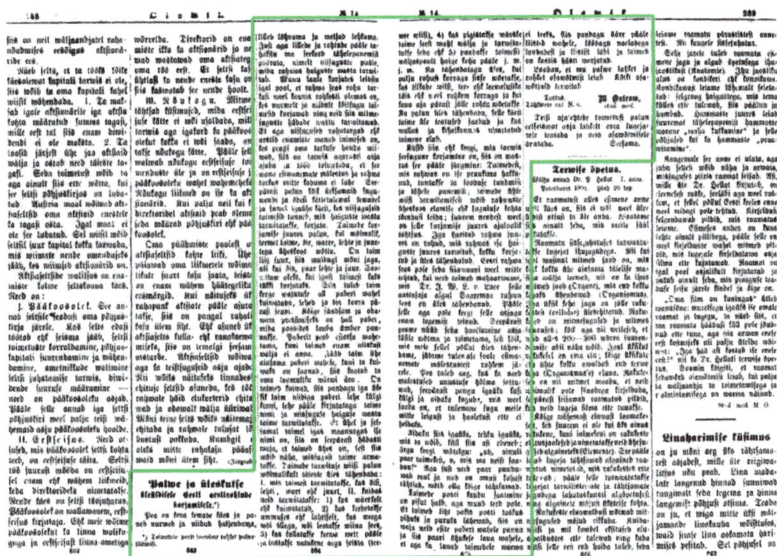

Figure 1. Ostrov's first survey plan and its placement on two pages in a newspaper (Olewik, 1891 April 8, No. 14).

For feedback, Ostrov employed the same methods as did his mentor Hurt: in July of the same year, he published the first report [34]. At the end of the report, Ostrov said that medicines made from stone and animal products could also be sent. In addition, he asked for detailed descriptions of the diseases and their symptoms, as popular names were ambiguous. The report was again re-published in other newspapers [35]. The second report was published in October [36,37].

Ostrov made his next call in April 1892, again in the spring so that people would have time to prepare for plant collection. As with the first appeal, he began his call with a positive description of nature and an inspiring tone: "Winter is over, spring is here, buds in the tree, plants are emerging. Now the picking and collecting, which went into hibernation when autumn arrived, can come back to life again, and the work in progress can now be carried forward". The summons stated that the newspaper (Postimees, Sakala, and Olewik) editors agreed to accept herbarium samples and then send them to Ostrov [38–40]. The reason may be that in the intervening time he changed his place of residence and got a job in Nõo parish as a rural municipality doctor. The third report was not published until late autumn, and, for some reason, it blamed the "sad summer" for the modest collection activity without specifying what that means. "Last sad summer also seems to have had a detrimental effect on the collection of folk medicines, as this summer's collection is only a quarter of the last" [41,42]. The third and final call came on April 26, 1893 [43]. There, too, Ostrov announced that packages with plants should be sent to the newspaper's editorial office because his residence is not permanent. However, the newspaper editors kindly allowed the samples to be sent to them. Ostrov's collection work was likely interrupted as a result of his constant changes in residence: from 1892 to 1893 he worked in Nõo; in 1893 he went to Russia to work, first to Smolensk, then in 1895 to Pskov; and in 1898 to Jelgava, present-day Latvia, where he worked as a railway doctor on the Moscow-Ventspils railway line. From 1914 to 1919, Ostrov worked as a military doctor in World War I and later in the Estonian War of Independence. At that time, he became the Commander of the Estonian Army Health Care Government and earned the rank of Sanitary Major General of the Sanitary Service. Later, he was employed as a school doctor in Põltsamaa, and in 1927 he retired [44].

1.2. Work's Aim

The main aim of this work is to understand the reasons for the effectiveness or ineffectiveness of the earliest citizen science methodology in ethnobotany. To that end, we:

1. analyse the traditional medicinal plant use of 19th-century Estonia, and
2. compare it, to the extent possible, with folklore data from the same period, based on earlier publications (Jēkabs Alksnis 1894 [45] and Johann Wilhelm Ludwig von Luce 1829 [24]), manuscripts (Johann Heinrich Rosenplänter 1830s [25]), and the information contained in the ethnomedicine and ethnobotany database HERBA (19th–20th centuries) [46].

We expect to observe some specific plant uses which will not be detected in HERBA due to the ambiguity of some plant names.

2. Results

2.1. Correspondents and Their Contributions

Sixteen people were identified as correspondents of Mihkel Ostrov (Table 1). The correspondents originated from 14 parishes (Figure 2), yet there is a mismatch between the parishes mentioned in the reports and those indicated as the origin of the knowledge in the manuscript. The parish having uses (18 use reports (UR)) but no correspondent was Põlva. There is also a discrepancy between the number of plants mentioned in the reports and the UR from a specific parish; for example, in the reports Ostrov mentions that he received 36 plants from Rõuge's only correspondent, J. Orraw, yet in the manuscript we were able to find only 14 UR and 8 plant taxa.

Table 1. The correspondents of Ostrov, based on the reports published in newspapers. Correspondents of Ostrov who also sent medicinal data to Jakob Hurt (H) or Matthias Johann Eisen (E) and the references in the respective collections.

Name and Life Dates of Correspondents	Profession	Parish	First Report (1891a)	Second Report (1891b)	Third Report (1892)	Non-Ostrov Archive Reference and Location of Correspondence (Range of Pages/Year)
Peeter Metusala (1869–1950)	Tailor	Karula	7 medicines			
J. Orraw [1] (?–?)	Farmer and potter	Rõuge	24 plants and 1 medicine	12 plants and 1 medicine		
Andres Saal (1861–1931)	Teacher, writer, and journalist	Tori	16 plants and 6 medicines			H II 21, pages: 9–64/1888
Elise Torim (1868–1929) (F)	Later wife of Mihkel Ostrov	Äksi	9 plants	14 plants		
Henrik (Heinrich) Koppel (1863–1944)	Doctor and later medical researcher, lecturer, and rector of the University of Tartu	Otepää	24 plants			
H. Karu [3] (?–?)	Farmer?	Viljandi	11 plants	13 plants and 9 medicines	6 plants and 10 medicines	
Dietrich Timotheus (1859–1929)	Vodka master, manor keeper, and industrial worker	Jõhvi		9 plants		H II 7, pages: 617–702 and 715–716/1889

Table 1. *Cont.*

Name and Life Dates of Correspondents	Profession	Parish	First Report (1891a)	Second Report (1891b)	Third Report (1892)	Non-Ostrov Archive Reference and Location of Correspondence (Range of Pages/Year)
Jaan Bergmann (1856–1916)	Pastor, translator, and poet	Tartu		2 plants		
Christjan (Kristjan) Koppel (1866–1930)	Medical student; from 1897, Khabarovsk District doctor and civil servant of the Russian Empire; later Consul of the Republic of Estonia (1920–1922); from 1922, a ward doctor in Estonia	Tartu		8 plants		
H. Pärtel (?–?)		Taurida Governorate [2]		1 plant		
Hans Kosesson (1870–1944)	Assistant to the manor's vodka master	Tarvastu		30 plants and 30 medicines		H I 3, pages: 261–268 and 297–316/1892
Jaan Miländer (1866–1940)	Medical student; later professor of medical sciences at the University of Tartu and head of the women's clinic	Saarde		10 plants		
Helene Maasen (1869–1933) (F)	Elementary school teacher and journalist	Palamuse		15 plants		H II 27, pages: 255–278; H III 8, pages: 387–414 and 439–454; H III 15, pages: 143–154; E, pages: 38788–38794, 52232, and 52271/1888–1892
Jaak Kiwisäk (1868–1903)	Farmer; died tragically, shot by poachers	Karksi			3 plants	
Jaan Ostrow (1869–1919)	Manor keeper and promoter of fish farming	Rõngu			9 plants	
Elise Aun (1863–1932) (F)	Writer and poetess	Torma			33 plants and 14 medicines	

[1]: most likely Jakob (b. 1861), as he was an active correspondent of Hurt, although he did not send him any information on plant use. His brother, Jaan Orraw (b. 1852), sent Hurt (from the Vitebsk Province of Russia, where he had settled to live in the early 1880s) 7 plant uses in 1888 (H I 2, pages 561–574), none of which match the information in the Ostrov collection. Both brothers worked as farmers and potters. [2]: Many Estonians migrated to the Crimean Peninsula at that time and ordered Estonian newspapers from there. [3]: H. Karu sent Eisen a list of farm names in his municipality in 1892 (E, 1447–1449). Hans Karu (1870–1926) comes from this municipality, but it is not certain that he is the same person. Each "plant" (sample) sent was accompanied by a use. By "medicine", Ostrov meant either a text on a medicinal plant without an herbarium sample or a non-plant medicine. (F): female correspondent.

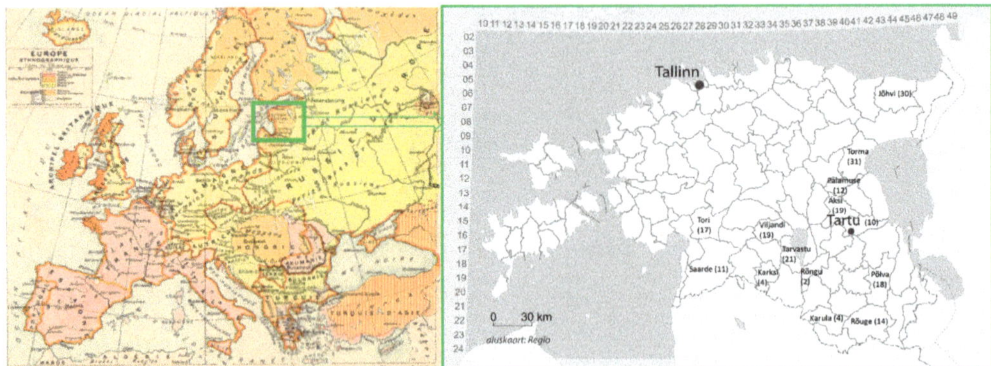

Figure 2. Historical map ("Atlas Melin Historique et Geografique" published by Andre, Paris, 1900) of the region and the parishes from which Ostrov received correspondences regarding plant uses (UR). Parish division of the territory of present-day Estonia at the end of 19th century.

With his calls, Ostrov succeeded in mobilizing the active part of society most likely also interested in medicinal plants. Among his correspondents there were several practicing doctors and students of medicine, who probably knew him from either school or the Estonian Student Society, which included all active students of that time. One of Ostrov's correspondents, Dr. Henrik Koppel, was, from 1920 to 1928, twice elected rector of Tartu University, yet at the time of correspondence he had just recently acquired his medical diploma and was preparing to defend his doctorate. Henrik Koppel was collecting folklore for Hurt while still in high school and eventually he married Hurt's niece, Sophie.

The list of correspondents also contains several farmers of whom not much is known, as well as local activists (e.g., members of charities and folk choirs, etc.) and intellectuals. Ostrov indicated that several collectors obtained data through interviewing villagers. One of them, Hans Kosesson, who worked as an assistant vodka master at Tarvastu Manor, interviewed nearly ten manor serevants, whose names were provided by Ostrov in his manuscript (EKS, c, page 34).

There were also three women included among Ostrov's correspondents. Although women were only allowed to officially study at the university in 1915, in village schools girls received an education equal to that of boys. All three of the young women that contributed to Ostrov's collection were socially active and contributed in diverse ways to the development of Estonian culture. Elise Aun and Helene Maasen knew each other through society work [47] and it is very likely that they also knew Elise Torim, the third female correspondent. Helene Maasen, who studied at a private German-language school and later also translated foreign language texts, was Hurt's most important female correspondent and one of the best correspondents overall. Elise Maria Torim also collected folk songs for Hurt but did not consider them polite enough to submit; these songs were eventually sent to Hurt by Ostrov himself (H II 25, pages 1097–1134). After at least four years of acquaintance, which also included responding to Ostrov's call in 1891, Elise and Mihkel married in the fall of 1893.

Since Hurt and Eisen also collected at the same time, half of Ostrov's correspondents were also active collaborators of those researchers. Of these correspondents (Table 1), four also reported plant uses or medicinal magic to Hurt and one did so to Eisen; however, the reports found in other collections rarely overlap with the ones reported in Ostrov's manuscript. As the source of the information is not traceable in Ostrov's report, the exact proportion of overlap cannot be determined.

2.2. The Ethnopharmacology of Ostrov's Collection

Ostrov's collection comprises 65 taxa, of which 64 were identified on the species level and one (*Betula*) on the genus level, belonging to 27 families (Table 2). The most widely used taxon was *Achillea millefolium* (with 20 UR), followed by *Plantago major* (15 UR), and *Valeriana officinalis* (10 UR). The most commonly used family was Asteraceae (12 taxa and 53 UR), followed by Lamiaceae (5 taxa and 18 UR), Valerianaceae (2 taxa and 17 UR), and Apiaceae (4 taxa and 12 UR).

Of the 219 UR, the most mentioned uses were treatments for skin diseases (56 UR), followed by general and unspecified diseases (35 UR), diseases related to the digestive tract (33 UR), and respiratory diseases (29 UR). Among the general and unspecified diseases, the most prevalent were, at that historical moment, specifically defined culture-bound diseases, such as *halltõbi* (now interpreted as malaria; it had a well-defined set of cultural rituals in folklore), *pistja* (which is some kind of sharp pain of unknown origin inside the body, often treated by poking someone with or digesting something sharp), *rabandus* (now related to stroke, but at that time considered a suddenly occurring disease brought on by the wind), and *seesthaigus* (a kind of internal pain of unknown origin). Another fairly common use (7 UR) was the symptomatic treatment of tuberculosis, which was sometimes also called *rinnahaigus* (10 UR, lung disease refers to any disease of the lungs and could also include pneumonia and severe cough). Remarkably, tuberculosis was treated with a different set of plants than was *rinnahaigus*: the only overlapping taxon was *Polygala amarella*. Cough as a treated symptom was mentioned in 11 UR, with *Achillea millefolium* being the most mentioned taxon (3 UR); this taxon was also mentioned twice in relation to the treatment of colds (*külmetus*). Still very common among musculoskeletal diseases was the treatment of a kind of rheumatic disease, called *jooksva*, which referred to the "running" of the disease, as the pain often changed places. It was historically treated by plants "running" on the ground, many of them with names referring to the disease: *jooksvarohud/joosvarohud* [48]. Among skin diseases, the most prevalent ailment was boils (*paised*, 14 UR), for which *Tussilago farfara* (4 UR) was most often used; also in this group, a culture-bound disease *maa-alused* (different forms of urticaria; literally "from underground") was frequently mentioned (6 UR).

2.3. The Disease Reflected in the Name of Plant

The variety of names reflected in Ostrov's collection is extraordinarily high and they often refer to the disease or the symptom the plant was thought to heal. For example, all the taxa used to treat *maa-alused* had similar names, such as *maa-aluse rohi, mailaseroht, maalesehein, maaleserohi*, and *maavits*; *rinnahaigus* had the name *rinnatee* (literally "tea for lung"); chills from malaria were treated with *külmaväristuse rohud* and *värisejahein* (literally "plant against chills" and "trembling plant"). Bleeding was stopped almost exclusively by *verihein* (literally "blood grass", 4 out of 5 UR, the remaining being a linen cloth). With a few exceptions, plants that have more than one UR also have more than one name. The exceptions included *Tussilago farfara* (called *paiseleht*, boil leaf, and used predominantly against boils), *Taraxacum officinale, Menyanthes trifoliata*, trees, and a few other culturally important taxa.

Table 2. Uses of medicinal plants collected by Ostrov during his appeals and a comparison with other sources (if the plant/use is present).

Family	Taxa	MO Identification	Local Name	UR	Application (Local Name of Disease)	Uses in the Rest of HERBA (19th–20th Centuries)	Uses in Rosenplän-ter 1830s	Uses in Luce 1829	Uses in Alksnis 1894
Acoraceae	*Acorus calamus* L.	*Acorus calamus*	kalmus	2	Alcohol infusion of roots was drunk against stomach-ache (*kõhuvalu*);Fresh leaves distributed in room against fleas	Widespread Widespread			Use not specified
Amaryllidaceae	*Allium cepa* L.	*Allium cepa*	sibul	2	Bulbs eaten raw with honey against Anthrax (*villitõbi*); Bulbs roasted and juice drunk against lung disease (*rinnahaigus*)	Widespread, but not exact use Widespread		Applied roasted	Cough
Apiaceae	*Angelica sylvestris* L.	*Archangelica officinalis*	heinputk	5	Decoction of roots against urination problems (*ei saa kusta*) and internal problems (*sisikonna kinnitamiseks*); Fresh root chewed against infectious diseases (*külgehakkava tõbi*); Powder made of dried roots ingested against anxiety	Jõhvi parish* in 1937 – –			
	Carum carvi L.	*Carum carvi*	köömel	2	Strong alcohol infusion of seeds drunk against stomach-ache (*kinnine kõhuvalu*); Water infusion of flowers or seeds used against flatulence (*kõhu kobisemine*)	Widespread Widespread	Mixed with beer		Other uses
	Cicuta virosa L.	*Cicuta virosa*	mürkhain, mürk	4	Whole plant (including roots and leaves) is crushed with salt and applied on erysipelas (*roos*) or tumors (*kasuva*); Baked root is applied on boils (*paised*); Applied on abscess (umbe) when black blood appeared under the skin	– – –			Similar uses
	Levisticum officinale W.D.J.Koch	*Levisticum officinale*	lääbus	1	Leaves, stems, and some of the root crushed with a little butter and applied on closed boils (*umbes, üles aand paistetus*)	Other uses		Mixed with manure	
Asphodelaceae	*Aloe arborescens* Mill.		aloe, aloe lill	2	Sap applied on cracked lips; Split leaves applied on burns	– Widespread			Use not specified

Table 2. *Cont.*

Family	Taxa	MO Identification	Local Name	UR	Application (Local Name of Disease)	Uses in the Rest of HERBA (19th–20th Centuries)	Uses in Rosenplänter 1830s	Uses in Luce 1829	Uses in Alksnis 1894
Asteraceae	*Achillea millefolium* L.	*Achillea millefolium*	raudriarohi, raudrohi, verihain,	20	Decoction of herbs is a component of epilepsy (*langetõbi*) treatment; Inflorescence decoction against cold, pneumonia and cough, constipation, diarrhoea, excessive bleeding during menstruation, *pistja*-some sharp pain inside the body of unknown origin; Crushed leaves applied to wounds to heal and stop bleeding; Applied also on boils	- Widespread - Widespread Widespread Widespread Widespread Widespread	Present, but no similar uses	Wound plaster	Cough, tuberculosis Homeostasis
	Antennaria dioica (L.) Gaertn.	*Gnaphalium dioicum*	kassikäpp	1	Aerial parts boiled along with [*Trifolium montanum*] and drunk against endometritis (*valged*)	Widespread	Present, but no similar uses		
	Arctium tomentosum Mill.	*Lappa tomentosa*	takjas, takkäs	4	Roots ground into flour and digested with water against lung disease (*rinnahaigus*); Seeds ingested whole or ground against sharp pain (*pistja*); Juice of leaves applied on wounds	Common in earlier texts Widespread -			Other uses
	Artemisia absinthium L.	*Artemisia absinthium*	koihein, koirohi	3	Alcohol infusion of fresh herbs or only leaves used to treat stomach-ache and diarrhoea	Widespread		Similar use among other uses	Similar use among other uses
	Artemisia vulgaris L.	*Artemisia vulgaris*	puju, pojokesed	2	Decoction of roots or powdered roots ingested to treat epilepsy (*langetõbi*)	Widespread	Present, but no similar uses		Same use
	Inula helenium L.	*Inula helenium*	alandi juur	1	Roots powdered and mixed with butter or grease applied on scabies (*sügelised*)	Other uses			
	Leucanthemum vulgare L.	*Chrysanthemum leucanthemum*	arnikas	1	Inflorescence tea against straining (*venitus*)-disease obtained from too much hard work	Widespread (name-based), all plants used resemble *Arnica montana*		*Arnica montana* is present	Other uses

Table 2. *Cont.*

Family	Taxa	MO Identification	Local Name	UR	Application (Local Name of Disease)	Uses in the Rest of HERBA (19th–20th Centuries)	Uses in Rosenplänter 1830s	Uses in Luce 1829	Uses in Alksnis 1894
	Matricaria chamomilla L.	*Matricaria chamomilla*	kamelid, kummel, kummelid, obinahein, ubinhain	9	Inflorescence tea for women in labour, stomach-ache, diarrhoea, lung disease (*rinnahaigus*), cold and cough; Sap of herb applied on boils (*paised*)	Widespread all (similar) uses	Other uses	Uses unspecified	Other uses
	Tanacetum vulgare L.	*Tanacetum vulgare*	reinvarred	1	Aerial parts boiled against chest pain (*rindealt valu*)	Other uses		Other use	Other use
	Taraxacum officinale F.H.Wigg. (coll.)	*Taraxacum officinale*	võilill	4	Inflorescences dried, powdered, mixed with alcohol, and ingested against severe diarrhoea (*kõhutõbi*); Teaspoon of sap used to cure constipation; Powdered roots used against jaundice (*kollatõbi*); Aerial parts boiled if a pregnant woman gets hurt and given to her to drink	Widespread Tea widespread Widespread -		Other uses	Other uses
	Tripleurospermum inodorum (L.) Sch.Bip.	*Chrysanthemum inodorum*	krambirohi	1	Decoction of aerial parts drunk and used as bath against spasms (*krambid*)	Widespread (name-based)			
	Tussilago farfara L.	*Tussilago farfara*	paiseleht	6	Fresh leaves applied on boils; Tea made from dried leaves drunk against lung disease (*rinnahaigus*)	Widespread both uses		Same use	
Betulaceae	*Betula* spp.		[(sauna) viht], [tõkat]	3	Powder made from dried birch whisk leaves and mixed with fresh milk cream applied on scabies (*sügelised*); Birch whisk leaves stuffed into the pillow against headache; Birch tar lubricated on the scabies	- Widespread, but not exact use -		Other uses	Other uses
Boraginaceae	*Anchusa officinalis* L.	*Anchusa officinalis*	villirohi	1	Crushed fresh plant applied on anthrax (*vill*) lesions	Widespread (name-based)			
	Echium vulgare L.	*Echium vulgare*	roosirohi	1	Crushed fresh plant applied on erysipelas (*roos*) lesions	Widespread (name-based)			

Table 2. *Cont.*

Family	Taxa	MO Identification	Local Name	UR	Application (Local Name of Disease)	Uses in the Rest of HERBA (19th–20th Centuries)	Uses in Rosenplänter 1830s	Uses in Luce 1829	Uses in Alksnis 1894
Brassicaceae	*Armoracia rusticana* P.Gaertn., B.Mey. & Scherb.	*Armoracia rusticana*	mädaröigas	1	Holding a small piece of root in one's mouth believed to heal tuberculosis (*tiisikus*)	Common, but not exact use	Present, but no similar uses		Other use
	Brassica oleracea var. capitata L.		hapud kapsad	1	Sauerkraut wrapped on the forehead against headache (*peavalu*)	Widespread			Same uses
	Capsella bursa-pastoris (L.) Medik.	*Capsella bursa pastoris*	silmarohi	1	Whole plant boiled in closed vessel and diseased eyes washed (*haiged silmad*) with this water	Widespread (name-based)			
	Raphanus raphanistrum L.	*Raphanus raphanistrum*	reigas	1	Fresh root grated and applied on the neck against angina (*kaelahaigus*)	Other uses			Other use
Caprifoliaceae	*Valeriana officinalis* L.	*Valeriana officinalis*	baldrian, jungver, südamevalurohi, rinnarohi	10	Powdered roots ingested with water against cold, malaria (*halltöbi*), arthritis (*luuvalu*), and headache, or mixed with *heinputk* root powder against diarrhoea; Decoction of flowers or roots given to women in childbirth as pain relief; Tea made from leaves and flowers drunk against cough, developing tuberculosis (*tiisikus*); Tea made from leaves and flowers drunk against chest pain (*südamevalu*)	– Common, but not exact use – Widespread (name-based) Widespread (name-based)		Childbirth	Cough, many other uses
Caryophyllaceae	*Silene vulgaris* (Moench) Garcke	*Silene inflate* [Ostrov's note: "*Silene inflata* probably, but it may also be *Silene nutans*"]	põierohi	1	Additive to medicine to treat epilepsy, part of the decoction of flowering herbs collected before Midsummer's Day	Other uses		Use not specified	Other use

Table 2. Cont.

Family	Taxa	MO Identification	Local Name	UR	Application (Local Name of Disease)	Uses in the Rest of HERBA (19th–20th Centuries)	Uses in Rosenplän-ter 1830s	Uses in Luce 1829	Uses in Alksnis 1894
Cupressaceae	*Juniperus communis* L.		kadakas	2	Twigs boiled and added to bath to treat swollen legs (*paistetanud jalad*), in addition a tea made from pseudo-fruits drunk; Powdered pseudo-fruits mixed with butter and gunpowder applied on scabies (*sügelised*)	Widespread Widespread, but not exact use			
Ericaceae	*Andromeda polifolia* L.	*Andromeda polifolia*	soosassaparillad	1	Decoction of herbs drunk against rheumatism	Widespread			
	Ledum palustre L.	*Ledum palustre*	sookaislad	1	Tea made from flowers drunk against tuberculosis (*tiisikus*)	Widespread		Other uses	Same and similar uses
Fabaceae	*Pisum sativum* L.		hernes	1	Boiled with wax against rectal prolapse in children (*kui lapsel "ihu väljas käib"*)	Other uses			
	Trifolium montanum L.	*Trifolium montanum*	maarjaristikhein, rohulill, valge ristikhain, valged nupud	6	Aerial parts boiled alone or with [*Antenaria dioica*] and drunk against endometritis (*valged*); Given to a woman during childbirth to maintain strength	Widespread (name-based)			
	Trifolium spadiceum L.	*Trifolium spadiceum*	põldhumalad, rinnatee	2	Decoction of flowers and stems drunk against cough and lung disease (*rinnahaigus*).	Widespread (name-based)			
Lamiaceae	*Glechoma hederacea* L.	*Glechoma hederacea*	jooksva rohi, kassiratas, maalishein, paistus hain	9	Decoction of dried aerial parts drunk against rheumatism (*jooksva*); Decoction used to wash different forms of eczema (*mailased*); Leaves and stems ground with *Urtica urens*, applied with a linen cloth on scabs and herpes (*uhatand*) lesions; Decoction drunk to treat oedema or mixed with *Solanum dulcamara* decoction to wash scabs (*kärnad*) and edema (*paistes*) swellings	All uses widespread (name-based)		Skin inflam-mation	Other uses
	Lamium album L.	*Lamium album*	naestenõges piimanõges, ema nõges, malaise hein	3	Infertile women whisked in sauna with whisks made from aerial parts to become fertile; Decoction of herbs applied externally to eczema (*mailased*) rashes and also drunk to treat the same; Decoction of flowers used to wash eyes against eye diseases	All uses widespread (name-based)			Leukorrhea

Table 2. *Cont.*

Family	Taxa	MO Identification	Local Name	UR	Application (Local Name of Disease)	Uses in the Rest of HERBA (19th–20th Centuries)	Uses in Rosenplän-ter 1830s	Uses in Luce 1829	Uses in Alksnis 1894
	Mentha arvensis L.	*Mentha arvensis*	mündid, vesimünt	2	Tea made from herbs used to treat diarrhoea and stomach-ache	Widespread			
	Mentha spicata L.	*Mentha crispa*	münt	2	Tea made from herbs used to treat cold and cough	Widespread			Cough and diarrhoea
	Thymus serpyllum L.	*Thymus serpyllum*	kolme-korralised hainad, liivatee	2	Treat an incomprehensible disease (*arusaamata tõbi*) and sudden diseases (*rabandus*)	Widespread		Other use	
Linaceae	*Linum usitatissimum* L.		[linane riie]	4	Linen cloth used to cover different medicines in treating boils and eczema; Scraps or ashes of linen cloth used to stop bleeding	Widespread -			Other uses
Menyanthaceae	*Menyanthes trifoliata* L.	*Menyanthes trifoliata*	ubaleht	7	Strong alcohol infusion of sun-dried leaves, left overnight in bread-stove, used to treat tuberculosis (*tiisikus*) and lung disease (*kuivtõbi*); Decoction of leaves, two spoonfuls ingested every two hours to treat fever associated with cold (*külmapalavik*), stomach diseases (*kõhutõbi*), edema (*vesitõbi*); Take a small amount of dried stem powder against constipation	Widespread use, but not the specific preparation Common, but not exact preparation - Similar use from Kuusalu (1964)	Cough		Tuberculosis, cough, fewer, cramps, oedema, and other uses
Orchidaceae	*Dactylorhiza maculata* (L.) Soó	*Orchis maculata*	jumalakäpp	2	Powder of dried flowers and roots given against sudden diseases (*rabandus*) and the sudden onset of other diseases, usually associated with witchcraft (*äkkiline haigus*)	Other uses			Other uses
Parmeliaceae	*Cetraria islandica* (L.) Ach.	*Lichen islandicus*	põdrasammal	1	Boiled until the water becomes jellied and ingested against cough	Widespread			
Pinaceae	*Picea abies* (L.) H.Karst.		kuusk, [vaik/tõrv]	4	White part of the bark held between the lips to heal herpes; Fresh resin applied on boils and old wounds directly or covered with butter or boiled with rye shoots, sour cream, and grease	- Similar uses widespread			Other use
	Pinus sylvestris L.		mänd	1	Needles used to make a bath against rheumatism (*jooksva*)	Similar uses widespread			Other uses

Table 2. *Cont.*

Family	Taxa	MO Identification	Local Name	UR	Application (Local Name of Disease)	Uses in the Rest of HERBA (19th–20th Centuries)	Uses in Rosenplän-ter 1830s	Uses in Luce 1829	Uses in Alksnis 1894
Plantaginaceae	*Plantago major* L.	*Plantago major*	teeleht	15	Rubbed leaves applied on boils, and old and fresh wounds; Sap of leaves given against vomiting blood and frequent menstruation; Seeds are ingested in case of diarrhea and the risk of premature birth	Widespread Similar use found in Iisaku (1929) A few similar uses Widespread -	Wounds	Ulcers	Diarrhoea, dysentery, emerging ulcer
	Veronica officinalis L.	*Veronica officinalis*	maaleserohi	2	Boiled and washed with that water (usually with *Viola tricolor* L.); Decoction of herbs drunk against different forms of urticaria (*maa-alused*)	Widespread (name-based)	Similar use and name	Eczema and other uses	
Poaceae	*Briza media* L.	*Briza media*	külmaväristuserohud, värisajahein	2	Decoction of herbs drunk against malaria (*külmatõbi*)	Widespread (name-based)	Similar use and name		Other uses
	Secale cereale L.		rukis	1	Young shoots boiled with fresh spruce resin, sour cream, and grease applied on wounds directly	Similar uses widespread			Other uses
Polygalaceae	*Polygala amarella* Crantz	*Polygala amara*	emakajuur, jooksvakaetus, kaitused, kõõmahein, naeste päästja	5	Decoction of herbs drunk against lung disease (*rinnahaigus*), tuberculosis (*tiisikus*), and rheumatism (*jooksva*) or used to wash face (after sunset) to treat mouth scurf (*suu kõõm*); Dried and boiled, given to woman to drink with sugar before childbirth to ease giving birth	All widespread (name-based) -			Other use
Rosaceae	*Fragaria vesca* L.	*Fragaria vesca*	maasikas	4	Tea made from dried leaves and flowers drunk against cough	Widespread			Cough and other uses
	Geum rivale L.	*Geum rivale*	härjapää, karukollad	2	Decoction of leaves and inflorescences drunk against rheumatism (*jooksva*); Decoction of inflorescences drunk to induce sweating in case of cold	- Common (name-based)			
	Geum urbanum L.	*Geum urbanum*	laste kõhurohi, kõhutõbe juured	2	Decoction of roots and leaves drunk against diarrhoea and stomach-ache, especially in children	Common (name-based)			Other use

Table 2. *Cont.*

Family	Taxa	MO Identification	Local Name	UR	Application (Local Name of Disease)	Uses in the Rest of HERBA (19th–20th Centuries)	Uses in Rosenplänter 1830s	Uses in Luce 1829	Uses in Alksnis 1894
	Potentilla erecta (L.) Raeusch.	*Tormentilla erecta*	tedremadar, nabahain, tedremaranas	7	Alcoholic infusion of roots used against diarrhoea and stomach-ache; Powdered dry roots ingested with water against stomach-ache	All uses widespread			Similar uses
	Rosa canina s.l.	*Rosa canina*	kibuvits	1	Tea made from petals used against cough	*Rosa* spp. not differentiated, use of *Rosa* widespread			
	Sorbus aucuparia L.		pihlakas	1	Decoction of bark used against headache	Other uses			Other uses
Scrophulariaceae	*Verbascum thapsus* L.	*Verbascum thapsus*	vägihein, üheksamehevägi, üidismed	8	Fresh leaves applied on wounds; Rubbed leaves applied on inflammations, especially between the toes; Tea made from flowers drunk against tuberculosis (*tiisikus*)	All uses widespread		Similar uses	Other uses
	Nicotiana rustica L.	*Nicotiana rustica*	tubakas	2	Dried leaves applied on snakebites	Widespread		Other uses	Other uses
Solanaceae	*Solanum dulcamara* L.	*Solanum dulcamara*	maavits, majakad, solknamarjad, solknavarred	8	Seeds eaten against internal parasites of the *Ascarida* family (*solknad*); Tea used against internal pain (*seestvalu*) and edema (*vesitõbi*); Boiled with *Glechoma hederacea* and the decoction used to wash swellings (edema) and scabs, but also drunk against internal pain (*seesthaigus, seestvalu*) and different forms of urticaria (*maa-alused*)	widespread (name-based) – Widespread – Widespread (name-based)	Similar use and name	Large roundworm infestation	Other uses
	Solanum tuberosum L.	*Solanum tuberosum*	kartul, kartohvel	3	Bulb scrapings applied on inflamed areas; Starch mixed with *Urtica urens* seeds and ingested against diarrhoea	Widespread Common, but not exact preparation			Other uses

Table 2. *Cont.*

Family	Taxa	MO Identification	Local Name	UR	Application (Local Name of Disease)	Uses in the Rest of HERBA (19th–20th Centuries)	Uses in Rosenplänter 1830s	Uses in Luce 1829	Uses in Alksnis 1894
Urticaceae	*Urtica dioica* L.	*Urtica dioica*	nõges	1	Dried leaves smoked against tuberculosis (*tiisikus*)	Common			
	Urtica urens L.	*Urtica urens*	raudnõges	7	Seeds mixed with potato flour and ingested against diarrhea; Leaves rubbed against cheek until the toothache is gone; Seeds ingested with alcohol against malaria (*halltõbi*); Decoction of herbs used in bath to treat urticaria; Whisked in sauna with fresh herb against sudden diseases (*rabandus*); Decoction of herbs used to bathe someone with chickenpox (*tuulerõuged*)	Common, but not exact prep. Common, but not exact prep. Widespread Common (name-based) - -			Other uses
Violaceae	*Viola canina* L.	*Viola canina*	seakapsas	1	Sap used to heal fresh wounds	-			
	Viola palustris L.	*Viola palustris*	südamevalurohi	1	Decoction of leaves against heart pain (*südamevalu*)	Widespread (name-based)			
	Viola tricolor L.	*Viola tricolor*	kesalill, maaaluse rohi, mailaseroht	6	Medicine for 9 diseases; Decoction of dried herbs is used to wash different forms of urticarial (*maa-alused, mailased*) rashes especially in children, but a little of it also drunk (boiled with *Veronica officinalis* L.); Decoction of herbs drunk against rheumatism (*jooksva*) and tuberculosis (*tiisikus*)	All uses widespread (name-based)		Eczema	Cough

2.4. Comparison with Earlier and Later Sources

In comparison with historical sources, we found only a limited overlap of the taxa used with Rosenplänter's collection [25] (Table 2). Only three taxa had the same use and name (*Veronica officinalis*, *Solanum dulcamara*, and *Briza media*), while two taxa had one similar use among many (*Menyanthes trifoliata* and *Plantago major*). Five more taxa are present in both sources, yet the uses do not overlap. The reason for this is that peasants used to have strict movement restrictions due to serfdom. Therefore, there was very limited interpersonal exchange or mixing of knowledge, including on the use of plants. It was not until the 1860s that peasants were allowed to move outside their parish, and in 1868 the last form of slavery was banned (the obligation of peasants to work for landowners) [49].

Compared with von Luce [24] there are twelve taxa for which some of their uses overlap (such as the application of *Tussilago farfara*, *Achillea millefolium*, and *Plantago major* on wounds, the use of *Viola tricolor* against eczema, and *Valeriana officialis* during childbirth). Some applications or preparation modes differ slightly, such as in case of *Carum carvi* and *Levisticum officinale*. Six jointly named taxa had different uses.

There are considerably more similarities with medical student Jēkabs Alksnis's (1870–1957) article [45]: only slightly more than one third (23) of the taxa mentioned by Ostrov are not listed among those medicinally used in Alksnis. However, the majority of the overlapping taxa were either used for other aliments (26 taxa) or their exact use was not specified (2 taxa). For the remaining taxa, the uses overlap either fully (in a few cases) or partially. For some uses, the difference may be on the level of the details provided by the author; for example, sour cabbage (*Brassica oleracea*) was put on the head to treat headaches in both sources, yet Alksnis mentions boiled or fresh cabbage leaves, as whole cabbages were fermented, while Ostrov's text lacks such information. Notably, while Alksnis describes the external use of the highly toxic plant *Cicuta virosa* L., he provides very few details of its preparation, while Ostrov's manuscript describes four external uses, none of which overlaps with the other limited comments in HERBA [46], but it does overlap with the one use in Alksnis (tumor).

The similarity to the rest of HERBA [46] is quite considerable, although some of the overlaps may have several potential identification options. There were only two taxa (*Viola canina* L. and *Rosa canina* s.l.) found solely in Ostrov's collection, and this may be due to the exact identification.

2.5. The Importance of Rituals

From the report it is also clear that there were some uses where the plant species did not matter, as some of them relate to the product obtained from the plant or some other features of the plant beyond its taxon. For example, it was not specified from which tree, either *Picea abies* or *Pinus sylvestris*, resin or tar, which was applied to wounds, was extracted. It was either applied by itself or as an ointment of melted resin with salt or lard. It was also irrelevant which taxon of tree, broken by thunder, was used for toothache. A piece of wood was broken from this tree and used to pierce the tooth.

Therefore, it seems that in the majority of those few cases where a ritual was involved, the ritual was more essential than the taxa used; for example, *koeranael* (literally "dog nail", translated as boil/furuncle) was treated by going to the forest and breaking a stick from three trees to press the affected area. These sticks must then be returned to the same trees.

In 1888, outside of his collection, through his personal field experience, Ostrov wrote down three detailed treatment rites that were performed in the sauna room (in a village by the Narva River in Vaivara Parish). All of them were related to paediatric diseases whose origins were thought to be magical:

A child has a worm defect when he turns and twists his head and rotates his eyes. This defect was treated as follows: two worm trees, [the ones] with which the worm was killed, were brought from the forest. One was put on the fire in the sauna oven and burned. When the oven was hot, another worm wood tree and a rope were placed on the sauna stove and covered with steam. Inside this steam, the child was whisked, which healed the mistake. (EKS, c, page 53).

Dog disease is a disease of young children when they have a big stomach, a loose body, eat and drink a lot; but it dries out more and more in the warm weather, it does not kill. The disease is treated as follows: on three Thursday nights, two widowed women have to whisk a child and a dog in the sauna, one for the child, the other for the dog. Before whisking, the dog is washed in water, then the washing water is used for whisking, and the child is given [the same water] to drink. The one who holds the dog asks the whisker, "What are you whisking?" "I'm whisking dog disease!" "Whisk so that he will be healed!". (EKS, c, page 52).

The crying of young children at night is called "Öö itk" [cry of the night]. It is treated as follows: one old rope is put in water and a child is whisked with that water; the whisked baby is pulled 3 times through the rope twist; the water was poured out, and the rope is put back on. (EKS, c, page 52).

The sauna has historically been a very important place for treatment for Estonians. "Whisking" is the rhythmic hitting of yourself or other people with a bunch of twigs (a "whisk"), usually birch leaf twigs, in a hot steam room in the sauna. Its purpose is to massage and make the body sweat. In addition, the birch whisk used in the sauna has been considered therapeutic. This is also reflected in the texts sent to Ostrov. For example: "Whisking in the sauna with a whisk which is made from *emanõgesed* ["female nettles" *Lamium album*] makes infertile women fertile" (EKS c, page 64). This text combines the sauna as a ritual place and a symbolic plant name. Whisking in the sauna was used to treat various skin diseases and itching. Abscesses were also treated in the sauna: after whisking, wet wood ash was put on the abscesses and covered with a linen cloth. Birch leaves from the sauna were also used in ointments for skin diseases. Sleeping on a pillow of birch leaves, however, helped with headaches. Ritual whisking was performed when a child had various childhood illnesses.

According to Ostrov's correspondents, birch was used for treating different forms of urticaria (called *maa-alused*): a cross was drawn on birch bark and then tied on the skin for three days, or a pentagon was drawn, pressed three times, and thrown into the oven. Birch sticks were also used to heal warts by pressing them with salt. After that, the sticks were placed at a crossroads and left there without looking back.

The texts of Ostrov's correspondents show that other trees, such as alder (*Alnus* sp.), rowan (*Sorbus aucuparia*), and juniper (*Juniperus communis*), have been used in rituals as well. Of them, rowan and juniper have been considered sacred, because it was widely believed that their berries were marked by Jesus, an idea which was spread by the Evangelical Brotherhood and radical Christian congregations in the 18th–19th centuries. Although the first missionaries of the Evangelical Brotherhood arrived in Estonia and Livonia as early as 1720 and their activities began in the 1740s, this movement remained limited to Western Estonia and Saaremaa. It was not until the "granting of mercy" to the religious movement by Emperor Alexander I in 1817 (before that the free religious movement was forbidden in the Russian Empire) that the "new awakening" of the Evangelical Brotherhood and radical Christian congregations started to spread in the region. This became the "new time of awakening" of the Christian movement there. It is believed that at that time the so-called real Christianization and the abandonment of pagan customs also took place in Estonia and Livonia governorates. By the beginning of the 20th century, however, these religious movements had already lost their importance [50]. This is the reason why key earlier pagan customs and sacraments became associated with Christianity (such as trees or flowers, which were considered sacred). Such activities helped to bring Christianity closer to the local people.

Juniper was considered a tree of health and was used in the magic of controlling diseases. At the same time, the alder tree was important as a source to which a disease was transmitted. For example, one had to get rid of malaria by walking around the alder tree three times and each time exhale into a hole made in the tree, and then block the opening with a rowan tree (Hans Kosesson, H I 3, 312 (26)). In another example, nine pieces of alder tree on which to mark crosses were brought home from the forest and then swellings were pressed with them, after which the swellings disappeared (Hans Kosesson, H I 3, 266 (7)).

The folk calendar was also a part of the ritual. Ostrov's correspondents mention Midsummer's Day, June 24, as the anniversary of the folk calendar. Herbs harvested before or during this time were the best for treatment (H. Karu, EKS c, page 56). The plant species was not very important: all the blooming flowers harvested in the forest that day were used for healing. As Hurt had a separate question regarding the folk calendar, he received more of such information from Ostrov's correspondents. Plants were also used for divination that day: single girls picked nine flowering plants on Midsummer's Eve, braided them into a wreath, and laid the wreath under their pillows. Then a girl had to dream of her future husband (Helene Maasen, H III 8, 736 (10)). In the Midsummer evening, a special flower, *Erigeron acer*, was brought home. If at night your closed flowers are opened in the morning, then the next year will be good, but if it is still closed, then death is expected.

3. Discussion

3.1. Ostrov's Report as a Cross-Section of Plant Use during the Time of Collection

The little overlap with Rosenplänter's collection is not surprising, given both the temporal and geographical differences. However, the comparatively higher similarity with von Luce [23], a contemporary of Rosenplänter, is somewhat unexpected, as von Luce reported uses from a relatively isolated island. It may be that von Luce, himself, had already influenced some uses which he recorded as local. It can also be that that there were uses influenced by his predecessors or landlords that originated from scholarly medicine of that time. Overlapping taxa included *Achillea millefolium*, *Allium cepa*, *Artemisia absinthium*, *Matricaria chamomilla*, *Nicotiana rustica*, *Taraxacum officinale*, *Valeriana officinalis*, and *Verbascum thapsus*, although some of them were used for different purposes. All eight of these plants also overlapped with the ones listed by Jēkabs Alksnis [45] in a report published a few years after Ostrov's collecting work. We can also detect the possible presence of *Arnica montana*, widely popularized in 19th-century media [51], although the name *arnikas* was associated, symptomatically, with *Leucanthemum vulgare*. Both von Luce and Alksnis were doctors, sharing a similar medical education background. While comparing with Alksnis [45], it is important to keep in mind that future Latvian professor of medicine Jēkabs Alksnis studied medicine at Dorpat (Tartu) University from 1890 until 1895 and was familiar, not only with Mihkel Ostrov but with his collecting works, which might have even inspired him. Moreover, Alksnis made a call through the newspaper, but not much about the results is known. He specifically named one doctor who was the only one who sent him properly dried plants [45].

An interesting example is that of non-native *Matricaria chamomilla*, whose uses were still unspecified in both von Luce's report and Rosenplänter's collection and were poorly represented with different applications in Alksnis, yet widespread with overlapping uses in later folklore. The reason for this was that in the middle of the 19th century a so-called "reading revolution" took place among the Estonian-speaking population, where a large number of peasants learned to read and popular books on the natural sciences were published [52]. As very few books on nature and its uses had been published in Estonian before the end of the 19th century, the newly published books were primarily translated from German. This is also shown by the subsequently popular names of chamomile such as "German flower", "Germany Anthemis", or "German dog daisy".

The overlap of both the list of plants and their uses with HERBA [46] was even higher than expected. A major part of the information comprising the database was collected from slightly before (from 1886) to up to one century after Ostrov's collection, while the collecting methods and objectives (focusing on the "old times") remained the same. This overlapping and the presence of widespread uses and similarities detectable on the basis of names, shows that Ostrov actually received a good cross-section of the medicinal plant use of that period and raises the credibility of the method he used for that time. The methodology used by Ostrov, however, allowed for the more exact documentation of the taxa behind local plant names, and, thus, will allow better identification/interpretation of information collected later.

3.2. The Problems of Citizen Science Ethnobotany in the 19th Century: Ostrov's Shortcomings

Ostrov used a method which was employed by other folklore collectors: calls in national newspapers. Thus, Ostrov's data collection was public. He made calls to people, instructing them, and motivating their assistance by providing feedback. A drawback of Ostrov was that he did not have a fixed institution and often changed his residence. Communication went through the newspaper's editorial office. So, unfortunately, he had no direct personal contact with people. Thus, misunderstandings may have arisen. For example, during the 4th Estonian General Song Festival (15–17 June 1891, in Tartu), someone left Ostrov a package containing dried plants and their uses at the Estonian Literary Society. However, Ostrov said that the package had disappeared from the association and kindly asked his anonymous assistant to resend them the following year [36,37].

A comparison of the reports published by Ostrov in newspapers with his manuscript archive reveals a discrepancy. For example, the manuscript contains 18 plant uses from Põlva parish in 1891. However, none of newspaper reports contain any information about this parish. Moreover, Hurt and Eisen, the largest collectors, do not have any correspondent from that time providing ethnobotanical information on the use of plants from Põlva parish. The identity of the person who sent the fairly large collection is still unknown. It could have been a local pharmacist, doctor, or another village activist. Perhaps this person asked to remain anonymous so that their name would not be mentioned in the newspaper.

Both directly and indirectly, Ostrov's medical data collection was influenced by the rivalry between the great collectors Hurt and Eisen. Hurt heavily lobbied for himself. He told the editor of the newspaper Olewik that he was the right person and the only one who could collect oral antiquities in Estonia. Hurt also wrote critically in the newspaper about Eisen's collection and data analysis methodology. For this reason, the newspaper Olewik banned Eisen's calls in 1893 (see [53]). Thus, Ostrov's data collection may have been left unfinished due to a quarrel between the two largest collectors. The newspaper, which a few years before encouraged its readers to collaborate with Ostrov, had now changed direction.

3.3. Pioneering Methods in Ethnobotanical Data Collection

Ostrov's collection method was innovative at the time. Although Hurt obtained a greater number of plant use records, the collection of Ostrov can be considered higher quality, as his plant identification is reliable. It was also the first successful attempt to collect folk medicinal knowledge accompanied by herbarium specimens in Estonia through an approach currently known as CS. Earlier, such an attempt was made by the German-born professor of pharmacy Johann Georg Noël Dragendorff (1836–1898), who in 1877 made a similar appeal, yet received no responses, most likely because his appeal was in German and therefore incomprehensible to potential correspondents [54,55]. As a general call to assist researchers, the active population or people interested in the subject took part. The more generalist collection of Hurt, which attracted more than a thousand correspondents, yielded results qualitatively comparable to the finely defined questionnaire of Ostrov that was answered by only a few dozen. This was due to activists working with both men. At the same time, it also shows that outside special interests (e.g., medics and pharmacists) ethnobotanical knowledge was known or noticed by only a limited number of people.

The success of the collection of Mihkel Ostrov (see Figure 3) is the result of a combination of several factors.

- The fertile ground previously prepared by the Estonian "time of awakening" and the ongoing collection of folklore by Jakob Hurt. The enthusiasm of the correspondents and the extent of their contribution was fuelled by the general understanding of the need for the preservation of "antiquities".

- He knew what he was asking for. Having had field experience, Ostrov knew exactly what to ask and how to get people interested in the subject.
- With the call, Ostrov gave his correspondents something in exchange. He taught, in great detail, how to collect and press plants in order for them to be safely preserved. This kind of instruction may have been much appreciated.
- At the end of the 19th century and the beginning of the 20th century, the image of science and the scientist was different than it is today. For ordinary people at that time, a scientist could simply be a university-educated specialist. Moreover, the so-called researcher did not have to be working at a research institution. It was acceptable for people to give their data to a freelance scientist as well. The greatest disadvantage of a freelance researcher (with whom Ostrov could be grouped) was the lack of free time to analyse data. Therefore, the collected material waited for further researchers.
- Ostrov's collection method showed that the "less is more" rule applies when employed correctly.
- Following the successful example of Jakob Hurt, Ostrov replied to all correspondents publicly, stressing the importance of their contribution and prompting them to send repeated responses, as well as assuring prospective correspondents that their work would also be acknowledged.

Figure 3. Visual representation of Mihkel Ostrov's ethnobotanical data collection and communication with correspondents. (Credits: Johanna Lohrengel).

The time of collection of Mihkel Ostrov coincided with the rise of national awareness among Estonians, combined with a recently acquired high level of literacy (reaching over 90% by the end of the 19th century [56]) and a strong background in the basics of natural science taught at village schools [57] and through popular science literature [52]. It was a time when the active part of the peasantry and young intellectuals were searching for outlets in which to channel their energy and contribute to the development of the nation, but the majority of the various societies that later attracted them were not yet formed [58]. Ostrov motivated volunteers by appealing to this desire, namely by helping to halt the loss of traditional knowledge, and this remains one of the main motivations for cooperation today. Just as the vast majority of Ostrov's volunteers were well educated, today a higher level of education increases volunteer involvement in the collection of traditions (cf. [4]).

It is important to note that at least three women took part in Ostrov's calls. Thus, women were also involved and expected to contribute to social activity. Gender equality in basic literacy already existed at that time in the territory of present-day Estonia, although it was only in 1915 that the first woman was allowed to study in Tartu University.

4. Materials and Methods

4.1. Data

Ostrov's collection is currently housed in the Estonian Folklore Archives, among the collections of the Estonian Literary Society (EKS), folder "c". Folder "c" stores mails that were sent to the society between 1907 and 1917 and contain folk beliefs and folk medicine; Ostrov's material can be found on pages 33 to 76 (EKS, c, 33–76, see Figure 4). The collection consists of three parts, each of which is sewn together with thread. Ostrov sent them from Jelgava to Tartu in three separate parcels. As Ostrov's cover letters were not preserved, neither the year nor to whom he sent his collection is known. However, the scant remarks suggest that he knew the person intimately. Thus, it can be assumed that this person was his good acquaintance Oskar Kallas, who was one of the founders of the Estonian National Museum (ENM), established in 1909, and a member of the board of the EKS.

The Estonian Literary Society was founded in 1907 and Mihkel Ostrov and his wife immediately became members. This society became the most widespread and broad-based society, which included representatives from many walks of life. The aim of the society was to promote literature, science, and art in Estonia, as well as for members to get to know their country and people comprehensively and to make the results of completed work available to the public. After the October Revolution of 1917, the activities of the society stalled until the end of the Estonian War of Independence in 1920. In 1940, after the occupation of Estonia by the USSR, the society was disbanded, but it was re-formed in 1992. The Society issued the journal Eesti Kirjandus until 1940, and Ostrov became a journal contributor while in Jelgava. The journal also began to mediate the ENM's calls to the general public to collect "antiquities" (ENM's public calls to collect traditions continue to this day). However, there were also other calls, for example, in 1912 for the Estonian Students' Society to collect folk plant names and the call of veterinarian Johannes Kool to collect folk animal treatments. It is not known which appeal Ostrov took part in and then sent his own collection to the society.

The Ostrov collection is well systematized. Latin binominal names based on herbarium specimens have been added and plants are classified by family. The first correspondence contains general folk medicine, beliefs, and prescriptions for herbal remedies, as well as information collected by him from 1888. In that letter, Ostrov used only the folk plant names by which we identified the species (see Table 2, where there is no Ostrov identification). The second letter contains a list of species of the family Asteraceae, based on Latin names and their uses. At the end of this letter there is a note stating that he will send the next list of plant families as soon as he can write them down. The third letter describes the following families and subfamilies [names unchanged]: Rosacea, Labiatae, Umbelliferae, Scrophulariaceae, Papilionaceae, Urticaceae, Solaneae, Polygaleae, Violaceae, Cruciferae, Valerianea, Plantaginea, Aroidea, Graminea, Boragineae, Ericaceae, Liliaceae, Gentianea, Orchidea, and Lichines. At the end of the third correspondence there is an indication that "The end will be with the next letter". However, there is no end. Whether the absence of the next part was due to the disappearance of mail during difficult political times or the recruitment of Ostrov into World War I remains unknown today. Ostrov used the names of archaic families and subfamilies (e.g., Plantaginea (*Plantago major*), Valerianea (*Valeriana officinalis*), Lichines (*Cetraria islandica*), Aroidea (*Acorus calamus*), Gentianea (*Menyanthes trifoliata*), etc.), which were predominantly in circulation in the 19th century. Therefore, it can be assumed that Ostrov identified herbarium specimens as early as the end of the 19th century using a German-language plant reference book, because there were no reference books in Estonian at that time. According to recent data, *Archangelica officinalis* does not grow in Estonia. Ostrov therefore probably used a key book in which this plant was included. The folk name, *heinputk*, refers instead to the species *Angelica sylvestris*. In such cases, we used the popular name of the plant as a guide.

No herbarium specimens have survived to this day. Estonia's first ethnobotanist Gustav Vilbaste (1885–1967) points out in his book that the collection of the EKS contains about a dozen dried plants. He thought that these may be plants that were sent to Ostrov [59]. The Estonian Folklore Archives indeed holds one small box in which herbarium samples of unknown origin are stored. It may be that the specimens in this box were brought to the archives by various collectors over decades, as there is no information on the time or origin of their collection. Therefore, we cannot determine if it contains plants sent to Ostrov, as we can no longer univocally connect the specific information on plant use with the sender (which also sets certain limits to the analysis).

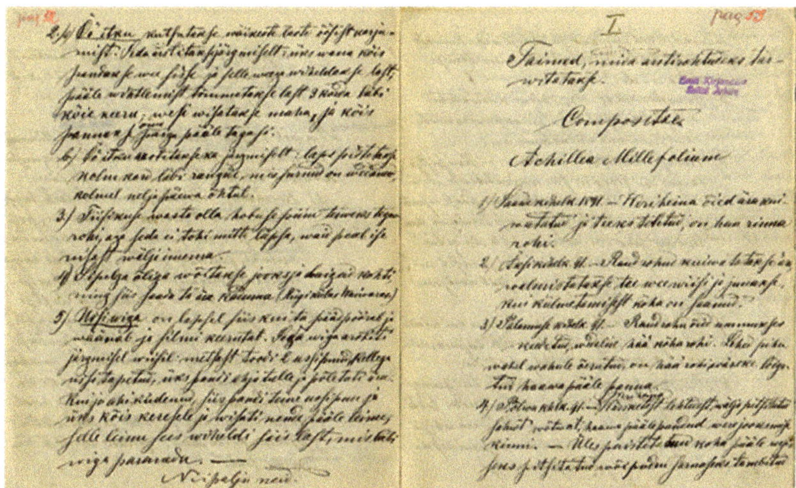

Figure 4. Sample of Ostrov's manuscript in the Estonian Folklore Archives.

The Latin plant names provided by Ostrov were adjusted to follow those listed in the Plants of the World Online [60] database and the European Flora [61] (these are presented in the "Taxa" column in Table 2); family assignments follow the Angiosperm Phylogeny Group IV [62].

4.2. Analysis and Comparisons

The names of the correspondents were provided in the three reports published in the newspaper Olewik [34,36,42]. Where possible, the first names and life dates were identified through bibliographic and biographical research. Identifications were complicated by the fact that only the initial of the first name, the surname, and the parish were known. Therefore, some people remained unidentified. The search was based on the biographical database of the analytical bibliography of the Estonian press (http://www2.kirmus.ee/biblioserver/ accessed on 21 December 2021) and biographical data from the central database of folklore collections (https://kivike.kirmus.ee/ accessed on 21 December 2021). Lastly, missing information was provided by Rein Saukas, a folklore historian who has thoroughly studied the biographies of the Estonian folklore correspondents, from his personal archive.

The data of Ostrov's manuscript was transcribed and entered into a spreadsheet. The use report (UR) of each plant was calculated by summing its different uses. The UR of each correspondent was calculated by summing all plant uses. Use reports were separated and emic disease categories correlated, as much as possible, with the current International Classification of Primary Care, 2nd edition (ICPC-2, Updated March 2003). As such correlations were not univocally interpretable, they could only be carried out on the very general scale and should be viewed with caution. In order to bring the emic diseases into line with etic diseases, we used the chapter on folk medicine from Gustav Vilbaste's book [59], which contains the Latin names of the diseases. We also used Vilbaste's book and the HERBA [46] database to describe emic diseases. However, it is impossible to identify historical diseases retrospectively with 100% certainty. Therefore, we have also shown in Table 2 the original names of the disease in Estonian (or dialect) in the manuscript.

Therefore, the results were compared qualitatively with the content of HERBA [46]. The database HERBA contains texts on the medicinal use of plants reproduced from eight major collections of the Estonian Folklore Archives collected from 1860 to 1996 (for more details see [63]). Ostrov's collection is also part of HERBA, and this was taken into account. As all archive texts in the HERBA database are coded with archival reference numbers, we separated out the texts of the Ostrov manuscript (EKS, c, 33–76) before analysis. The comparison was made from the standpoint of local plant names, taking into consideration the potential limitations of the absence of specimens for early historical texts. Qualitative comparison was made with the early historical uses recorded in Rosenplänter's (Pärnu parish) manuscript [25] and the publications of von Luce [24] covering Saaremaa and Alksnis in Latvia [45] (using for last two digitalized database [64]).

5. Conclusions

Dr. Mihkel Ostrov succeeded in accumulating, with the help of about two dozen people, the largest private medicinal plant use collection from the end of the 19th century, whose credibility is enhanced by the fact that the plants were identified on the basis of herbarium specimens sent to him by his correspondents. Ostrov's collection provides a cross-section of folk medicine and, to lesser extent, the healing rituals of herbal medicine at the end of the 19th century. It also shows the high diversity of both plants and uses known at that time in the territory of present-day Estonia, and its wealth increases the credibility of the rich medicinal plant folklore accumulated in Estonia since 1866.

The success of this CS endeavour was the result of a combination of several favourable factors, either in the environment of the time or created by the collector himself. The background factors, allowing the collection to happen at the right moment in the right place, included the recent abolishment of serfdom and access to education, the rise of national consciousness, the creation and activation of professional and student societies, and the simultaneous collection of general folklore. The factors added by Ostrov himself were the respectful approach towards his correspondents, providing specific guidelines and sharing knowledge, the public acknowledgment of their efforts, and, probably, his personal influence (as his correspondents included some of his classmates and his wife to be). The result was a fruitful and mutually beneficial collaboration where the contributions of citizens were publicly acknowledged, an approach that should serve as a good example for CS today. The collecting methods of Ostrov were pioneering for the time, yet they are still applicable from the viewpoint of modern CS. This work indicates that CS in ethnopharmacology in its wider meaning known today was born with studies of this kind, having its roots in the folklore collections of the 19th century in northern Europe.

We hope that lessons from the past can offer the modern scientist a good foundation for the development of the future involvement of citizens in studying ethnobotany.

Author Contributions: Conceptualization, R.S. and A.P.; methodology, R.S.; validation, R.S.; formal analysis, R.K.; investigation, R.K.; data curation, R.K.; writing—original draft preparation, R.S. and R.K.; writing—review and editing, A.P. and I.S.; supervision, R.S. and A.P.; project administration, R.S.; funding acquisition, R.S. All authors have read and agreed to the published version of the manuscript.

Funding: This research was supported by the European Research Council (ERC) under the European Union's Horizon 2020 research and innovation programme (Starting Grant project "Ethnobotany of divided generations in the context of centralization", Grant agreement No. 714874).

Institutional Review Board Statement: Not applicable.

Informed Consent Statement: Not applicable.

Data Availability Statement: EKS (=EKnS, Eesti Kirjanduse Seltsi rahvaluulekogu), H (Jakob Hurda rahvaluulekogu), E (Matthias Johann Eiseni rahvaluulekogu). Located at the Estonian Folklore Archives at Estonian Literary Museum, in Tartu, Estonia.

Acknowledgments: We kindly thank Rein Saukas for his help in identifying the correspondents of Mihkel Ostrov.

Conflicts of Interest: The authors declare no conflict of interest.

References

1. Gärdenfors, U. Biodiversity in Sweden. In *Global Biodiversity: Selected Countries in Europe*; Pullaiha, T., Ed.; Apple Academic Press: Oakville, ON, Canada, 2019; Volume 2, pp. 375–396.
2. Ronquist, F.; Forshage, M.; Häggqvist, S.; Karlsson, D.; Hovmöller, R.; Bergsten, J.; Holston, K.; Britton, T.; Abenius, J.; Andersson, B.; et al. Completing Linnaeus's inventory of the Swedish insect fauna: Only 5000 species left? *PLoS ONE* **2020**, *15*, e0228561. [CrossRef] [PubMed]
3. Tengö, M.; Austin, B.J.; Danielsen, F.; Fernández-Llamazares, Á. Creating synergies between citizen science and indigenous and local knowledge. *BioScience* **2021**, *71*, 503–518. [CrossRef] [PubMed]
4. Benyei, P.; Pardo-de-Santayana, M.; Aceituno-Mata, L.; Calvet-Mir, L.; Carrascosa-García, M.; Rivera-Ferre, M.; Perdomo-Molina, A.; Reyes-García, V. Participation in citizen science: Insights from the CONECT-e case study. *Sci. Technol. Hum. Values* **2021**, *46*, 755–788. [CrossRef]
5. Reyes-García, V.; Benyei, P.; Aceituno-Mata, L.; Gras, A.; Molina, M.; Tardío, J.; Pardo-de-Santayana, M. Documenting and protecting traditional knowledge in the era of open science: Insights from two Spanish initiatives. *J. Ethnopharmacol.* **2021**, *278*, 114295. [CrossRef]
6. Haklay, M.M.; Dörler, D.; Heigl, F.; Manzoni, M.; Hecker, S.; Vohland, K. What is citizen science? The challenges of definition. In *The Science of Citizen Science*; Vohland, K., Land-Zandstra, A., Ceccaroni, L., Lemmens, B., Perelló, J., Ponti, M., Samson, R., Wagenknecht, K., Eds.; Springer: Cham, Switzerland, 2021; pp. 13–33. [CrossRef]
7. Svanberg, I.; Łuczaj, Ł.; Pardo-de-Santayana, M.; Pieroni, A. History and Current Trends of Ethnobiological Research in Europe. In *Ethnobiology*; Anderson, E.N., Adams, K., Pearsall, D., Hunn, E., Turner, N.J., Eds.; Wiley-Blackwell: Hoboken, NJ, USA, 2011; pp. 191–214.
8. Prūse, B. Co-Creation of Knowledge: Supporting the Implementation of Sustainable Development Goals Through Citizen Science and Ethnobotany. Ph.D. Thesis, Latvian University, Riga, Latvia, 2020. Available online: https://dspace.lu.lv/dspace/bitstream/handle/7/53295/298-78868-Pruse_Baiba_bp15010.pdf (accessed on 21 December 2021).
9. Linnaeus, C. *Instructio Peregrinatoris, Qvam, Consent. Nobiliss. Facultat. Medica in Regia Academia Upsaliensi, sub Præsidio. Diss*; Uppsala Universitet: Uppsala, Sweden, 1759; pp. 1–13.
10. Svanberg, I. Etnobiologen Linné. In *Så varför reser Linné? Perspektiv på Iter Lapponicum 1732*; Jacobsson, R., Ed.; Skytteanska Samfundet: Umeå, Sweden, 2005; pp. 135–162.
11. Koerner, L. *Linnaeus: Nature and Nation*; Harvard University Press: Cambridge, UK, 1999; pp. 1–298.
12. Ståhlberg, S.; Svanberg, I. Frederick Hasselquist in Smyrna and Magnesia. In *Turkologica Upsaliensia: An Illustrated Collection of Essays*; Csató, É.Á., Gren-Eklund, G., Johanson, L., Karakoç, B., Eds.; Brill Academic Pub.: Leyden, The Netherlands, 2020; pp. 120–128.
13. Ståhlberg, S.; Svanberg, I. Folk knowledge in Southern Siberia in the 1770s: Johan Peter Falck's ethnobiological observations. *Stud. Orient. Electron.* **2021**, *9*, 112–131. [CrossRef]
14. Svanberg, I. *Linneaner: Carl von Linnés lärjungar i Sverige*; Wahlström & Widstrand: Stockholm, Sweden, 2006; pp. 1–372.
15. Teixidor-Toneu, I.; Kjesrud, K.; Bjerke, E.; Parekh, K.; Kool, A. From the "Norwegian Flora" (Eighteenth Century) to "Plants and Tradition" (Twentieth Century): 200 Years of Norwegian Knowledge about Wild Plants. *Econ. Bot.* **2020**, *74*, 398–410. [CrossRef]
16. Zeller, P. Naturalistic observations in Apulia during the 19th century. Vincenzo de Romita and Enrico Hillyer Giglioli. In *The Circulation of Science and Technology: Proceedings of the 4th International Conference of the European Society for the History of Science. Barcelona, Spain, 18–20 November 2010. Societat Catalana d'Història de la Ciència i de la Tècnica, Barcelona*; Roca Rosell, A.M., Ed.; Societat Catalana d'Història de la Ciència i de la Tècnica: Barcelona, Spain, 2012; pp. 201–207. Available online: http://pikaia.eu/Archivi/Pikaia/ALL/0000/380A.pdf (accessed on 21 December 2021).
17. Tauginienė, L.; Butkevičienė, E.; Vohland, K.; Heinisch, B.; Daskolia, M.; Suškevičs, M.; Portela, M.; Balázs, B.; Prūse, B. Citizen science in the social sciences and humanities: The power of interdisciplinarity. *Palg. Commun.* **2020**, *6*, 89. [CrossRef]
18. Łuczaj, Ł. Dziko rosnące rośliny jadalne w ankiecie Józefa Rostafińskiego z roku 1883. *Wiad. Bot.* **2008**, *52*, 39–50.
19. Łuczaj, Ł. Changes in the utilization of wild green vegetables in Poland since the 19th century: A comparison of four ethnobotanical surveys. *J. Ethnopharmacol.* **2010**, *128*, 395–404. [CrossRef]
20. Österlund-Pötzsch, S. Bodies in Motion: The Peregrinations of Early Finland-Swedish Folklore Collectors. *J. Folklore Res.* **2014**, *51*, 253–276. [CrossRef]

21. Coe, C. The education of the folk: Peasant schools and folklore scholarship. *J. Am. Folklor.* **2000**, *113*, 20–43. [CrossRef]
22. Kalle, R.; Sõukand, R. A schoolteacher with a mission. In *Pioneers in European Ethnobiology*; Svanberg, I., Łuczaj, Ł., Eds.; Uppsala Universitet: Uppsala, Sweden, 2014; pp. 201–217.
23. Koay, A.; Shannon, F.; Sasse, A.; Heinrich, M.; Sheridan, H. Exploring the Irish National Folklore Ethnography Database (Dúchas) for Open Data Research on Traditional Medicine Use in Post-Famine Ireland: An Early Example of Citizen Science. *Front. Pharmacol.* **2020**, *11*, 584595. [CrossRef]
24. Luce, J.W.L. *Heilmittel der Ehsten auf der Insel Oesel*; G. Marquardt: Pernau, Livonia, 1829; pp. 1–126.
25. Kalle, R.; Sõukand, R. The name to remember: Flexibility and contextuality of preliterate folk plant categorization from the 1830s, in Pernau, Livonia, historical region on the eastern coast of the Baltic Sea. *J. Ethnopharmacol.* **2021**, *264*, 113254. [CrossRef]
26. Saukas, R. Matthias Johann Eisen mõistatuste kogumise organiseerija ja publitseerijana aastatel 1869–1890. *Mäetagused* **2003**, *21*, 85–140. [CrossRef]
27. Kalle, R. Naturaliseerunud ravimtaimed etnobotaanika vaatenurgast hariliku katkujuure, hariliku siguri, aedvaagu, aedmädarõika, hariliku seebilille ja lõhnava kannikese näitel. *Mäetagused* **2007**, *36*, 105–128. [CrossRef]
28. Kalle, R. Eesti võõrliikide kajastus taimravipärimuses kuni 1944. aastani. *Akadeemia* **2010**, *22*, 83–104.
29. Pino, V. Ühest juhtnöörist ja selle taustast. *Keel ja Kirjandus* **1992**, *5*, 263–273.
30. Ostrov, M. Palve ja üleskutse üleüldisele Eesti arstirohtude korjamisele. *Postimees* **1891**, *40*, 3. Available online: https://dea.digar.ee/article/postimeesew/1891/04/06/14 (accessed on 21 December 2021).
31. Ostrov, M. Palve ja üleskutse üleüldisele Eesti arstirohtude korjamisele. *Olewik* **1891**, *14*, 288–289. Available online: https://dea.digar.ee/page/olewik/1891/04/08/4 (accessed on 21 December 2021).
32. Ostrov, M. Palve ja üleskutse üleüldisele Eesti arstirohtude korjamisele. *Sakala* **1891**, *17*, 1–2. Available online: https://dea.digar.ee/article/sakalaew/1891/04/26/3 (accessed on 21 December 2021).
33. Ostrov, M. Terwise õpetus. Wälja annud Dr. P Hellat. l. anne. Peterburis 1891. Hind 20 kop. *Olewik* **1891**, *14*, 289. Available online: https://dea.digar.ee/page/olewik/1891/04/08/5 (accessed on 21 December 2021).
34. Ostrov, M. Esimene aruanne üleüldisest rohtude korjamisest. *Olewik* **1891**, *26*, 529–530. Available online: https://dea.digar.ee/page/olewik/1891/07/01/6 (accessed on 21 December 2021).
35. Ostrov, M. Esimene aruanne Eesti arstirohtude korjamisest. *Postimees* **1891**, *75*, 3. Available online: https://dea.digar.ee/article/postimeesew/1891/07/02/14 (accessed on 21 December 2021).
36. Ostrov, M. Teine aruanne üleüldisest Eesti arstirohtude korjamisest. *Olewik* **1891**, *42*, 851. Available online: https://dea.digar.ee/page/olewik/1891/10/21/7 (accessed on 21 December 2021).
37. Ostrov, M. Teine aruanne üleüldisest Eesti arstirohtude korjamisest. *Postimees* **1891**, *160*, 3. Available online: https://dea.digar.ee/article/postimeesew/1891/10/29/18 (accessed on 21 December 2021).
38. Ostrov, M. Eesti arstirohtude korjajatele. *Postimees* **1892**, *91*, 1–2. Available online: https://dea.digar.ee/article/postimeesew/1892/04/25/4 (accessed on 21 December 2021).
39. Ostrov, M. Eesti arstirohtude korjajatele. *Olewik* **1892**, *17*, 353. Available online: https://dea.digar.ee/page/olewik/1892/04/27/5 (accessed on 21 December 2021).
40. Ostrov, M. Eesti arstirohtude korjajatele. *Sakala* **1892**, *18*, 1–2. Available online: https://dea.digar.ee/article/sakalaew/1892/04/29/3 (accessed on 21 December 2021).
41. Ostrov, M. Kolmas aruanne Eesti arstirohtude ja arstimise kombete korjamise üle. *Postimees* **1892**, *252*, 3. Available online: https://dea.digar.ee/article/postimeesew/1892/11/07/14 (accessed on 21 December 2021).
42. Ostrov, M. Kolmas aruanne Eesti arstirohtude ja arstimise kombete korjamise üle. *Olewik* **1892**, *46*, 937. Available online: https://dea.digar.ee/page/olewik/1892/11/16/5 (accessed on 21 December 2021).
43. Ostrov, M. Eesti rahwa arstirohtude ja arstimise kombete korjajatele. *Olewik* **1893**, *17*, 374. Available online: https://dea.digar.ee/page/olewik/1893/04/26/6 (accessed on 21 December 2021).
44. Eesti Ajakirjanduse Analüütiline Bibliograafia. OSTROV, MIHKEL. 2004. Available online: http://www2.kirmus.ee/biblioserver/isik/index.php?id=2662 (accessed on 21 December 2021).
45. Alksnis, J. Materialien zur lettischen Volksmedizin. *Histor. Studien aus dem Pharmakol. Institut der Kaiserl. Universität. Dorpat.* **1894**, *4*, 166–283.
46. Sõukand, R.; Kalle, R. *HERBA: Historistlik Eesti Rahvameditsiini Botaaniline Andmebaas*; EKM Teaduskirjastus: Tartu, Estonia, 2008; Available online: http://herba.folklore.ee (accessed on 21 December 2021).
47. Kikas, K. Rahvaluulekogumine epistolaarses kontekstis: Helene Maaseni kirjad Jakob Hurdale. *Keel ja Kirjandus* **2017**, *4*, 272–290.
48. Sõukand, R. "Jooksvarohud" Eesti rahvameditsiinis. *Akadeemia* **2004**, *11*, 2475–2493.
49. Kruus, H. *Eesti Ajalugu Kõige Uuemal Ajal*; Vol I. Eesti rahvusliku ärkamiseni; Loodus: Tartu, Estonia, 1927; pp. 1–176.
50. Plaat, J. Kristlike usuliikumiste mõju eestlaste ja eestirootslaste rahvakunstile ja kultuurile. *Mäetagused* **2009**, *42*, 7–32. [CrossRef]
51. Sõukand, R.; Raal, A. How the name Arnica was borrowed into Estonian. *Trames* **2008**, *12*, 29–39. [CrossRef]
52. Talts, M. The role of popular science literature in shaping Estonians' world outlook. Историко-биологические исследования **2013**, *5*, 59–83. Available online: https://www.digar.ee/arhiiv/nlib-digar:125277 (accessed on 21 December 2021).
53. Anonymous. Toimetuselt. *Olewik* **1893**, *33*, 697. Available online: https://dea.digar.ee/page/olewik/1893/08/16/9 (accessed on 21 December 2021).
54. Kalle, R.; Sõukand, R. Collectors of Estonian Folk Botanical Knowledge. *Baltic J. Eur. Stud. Tallinn Univ. Technol.* **2011**, *1*, 213–229.

55. Raal, A.; Sõukand, R. Classification of remedies and medical plants of Estonian ethnopharmacology. *Trames A J. Humanit. Soc. Stud.* **2005**, *9*, 259–267.
56. Vahtre, L. Äärmuslikkus ja äärmusetus Eesti ajaloos. In *Lehed ja Tähed: Looduse ja Teaduse Aastaraamat*; Rohumets, I., Ed.; Loodusajakiri: Tallinn, Estonia, 2004; Volume 2, pp. 99–105.
57. Paatsi, V. Eesti Talurahva Loodusteadusliku Maailmapildi Kujunemine Rahvakooli Kaudu (1803–1918). Doctoral Thesis, Tallinn Pedagogical University, Tallinn, Estonia, 2003.
58. Vilu, T. Eesti lugemisseltsid 19. sajandi lõpus ja 20. sajandi alguses. Master's Thesis, Tallinna Ülikool, Tallinn, Estonia, 2019.
59. Vilbaste, G. *Eesti Taimenimetused*; Emakeele Selts: Tallinn, Estonia, 1993; pp. 1–706.
60. Plants of the World Online. Facilitated by the Royal Botanic Gardens, Kew, UK. Available online: https://powo.science.kew.org/ (accessed on 8 December 2021).
61. Tutin, T.G.; Burges, N.A.; Chater, A.O.; Edmondson, J.R.; Heywood, V.H.; Moore, D.M.; Valentine, D.H.; Walters, S.M.; Webb, D.A. (Eds.) *Flora Europaea*; University Press: Cambridge, UK, 1980; Volume 1–5. Available online: http://ww2.bgbm.org/EuroPlusMed/query.asp (accessed on 21 December 2021).
62. Stevens, P.F. Angiosperm Phylogeny Website. 2001 onwards. Version 14, July 2017. Available online: http://www.mobot.org/MOBOT/research/APweb/ (accessed on 10 December 2021).
63. Sõukand, R.; Kalle, R. Change in medical plant use in Estonian ethnomedicine: A historical comparison between 1888 and 1994. *J. Ethnopharmacol.* **2011**, *135*, 251–260. [CrossRef]
64. Anegg, M.; Prakofjewa, J.; Kalle, R.; Sõukand, R. Local Ecological Knowledge and folk medicine in historical Esthonia, Livonia, Courland and Galicia, 1805–1905 [Data set]. *Zenodo* **2021**. [CrossRef]

Article

The Appeal of Ethnobotanical Folklore Records: Medicinal Plant Use in Setomaa, Räpina and Vastseliina Parishes, Estonia (1888–1996)

Renata Sõukand [1] and Raivo Kalle [2,3,*]

1 Department of Environmental Sciences, Informatics and Statistics, Ca' Foscari University of Venice, Via Torino 155, Mestre, 30172 Venice, Italy
2 University of Gastronomic Sciences, Piazza Vittorio Emanuele II 9, 12042 Pollenzo, Italy
3 Kuldvillane OÜ, Umbusi, Põltsamaa v, 48026 Jõgeva, Estonia
* Correspondence: raivo.kalle@mail.ee

Abstract: The historical use of medicinal plants is of special interest because the use of plants for healing is a rapidly changing, highly culture-specific and often need-specific practice, which also depends on the availability of resources and knowledge. To set an example of folkloristic data analysis in ethnobotany, we analyzed texts from the database, HERBA, identifying as many plants and diseases as possible. The research was limited to the Seto, Räpina and Vastseliina parishes in Estonia. The use of 119 taxa belonging to 48 families was identified, of which nine were identified at the genus level, four ethnotaxa were identified as two possible botanical taxa and fifteen ethnotaxa were unidentifiable. The most frequently mentioned taxa were *Pinus sylvestris*, *Matricaria discoidea* and *Valeriana officinalis*. High plant name diversity as well as high heterogeneity in the plants used were observed, especially in earlier records. The use of local wild taxa growing outside the sphere of everyday human activities, which was abandoned during Soviet occupation, signals an earlier, pre-existing rich tradition of plant use and a deep relationship with nature. Working with archival data requires knowledge of historical contexts and the acceptance of the possibility of not finding all the answers.

Keywords: historical ethnobotany; folklore collections; biocultural diversity; Estonian history; folk medicine; medicinal plants

Citation: Sõukand, R.; Kalle, R. The Appeal of Ethnobotanical Folklore Records: Medicinal Plant Use in Setomaa, Räpina and Vastseliina Parishes, Estonia (1888–1996). *Plants* **2022**, *11*, 2698. https://doi.org/10.3390/plants11202698

Academic Editor: Laura Cornara

Received: 17 September 2022
Accepted: 9 October 2022
Published: 13 October 2022

Publisher's Note: MDPI stays neutral with regard to jurisdictional claims in published maps and institutional affiliations.

1. Introduction

Historical ethnobotanical data in folklore archives are often perceived romantically as reservoirs of indigenous plant use, even in the European context. In addition, having locality-based historical data for comparison with currently existing plant uses can reveal tendencies in the evolution of plant use. The historical use of medicinal plants is of special interest because the use of plants for healing is a rapidly changing, highly culture-specific [1] and often need-specific practice, which also depends on the availability of resources and knowledge (e.g., advances in public health). Yet, studies based on historical ethnobotany are largely underrepresented in current scholarship [2]. The food-related historical ethnobotany of Northern Europe is somewhat better researched, represented by analyses of the collections of the Polish botanist, Józef Rostafiński (1850–1928) [3,4], given that they contain herbarium specimens. However, examples from the medicinal plant perspective are rather rare: there is an analysis of the first citizen science-based identification in Estonia, of which, however, only the interpretations of the contributions by the collector have survived [5]. An earlier medicinal plant study, by the pastor Johann Heinrich Rosenplänter (1782–1846) of the Pärnu parish [6], dates back to the beginning of the 19th century and can be considered as one of the first ethnobotanical studies in the

Baltic countries [6]; however, this too was based on already processed information and not folklore collections.

The difficulty in identifying plants is the primary reason why the historical ethnobotanical records, preserved among other folklore and/or local history collections in archives, are mainly neglected and excluded from current scientific discussion. The highest risk in dealing with such data lies in the absence of attached herbarium specimens, which, in a situation of ambiguity with respect to the local plant name, creates a lot of doubt regarding the reliability of the connection between the emic (local plant name) and etic (scientific name of the taxon). Such material is also often collected *en passant*, along with other folklore-related records, such as folk songs and mythology. Such records are documented only few at a time and therefore do not seem very informative.

There are quite a few databases containing ethnobotanical data [7], but only some of them have thus far focused exclusively on archival data. Databases such as HERBA in Estonia [8] and Dúchas in Ireland [9] enable the combination of data from several folklore collections. Notably, throughout the 19th and 20th centuries in Northern and Central Europe, folklore collections had specific questionnaires or campaigns which were designed keeping ethnobotanical principles in mind (although not always acknowledged during the time of collection) or, even more, the Latin names were added by botanists, so the plants have already been identified. Yet even in such cases, great attention must be paid to the details and mass analysis is rarely possible. In addition to datasets, attention needs to be given to possible misinterpretations and deliberate plagiarism, especially if the collection campaign had a competitive nature or the report was composed by schoolchildren, as is the case for several collections deriving from the 20th century, for example in Ireland [10] and Estonia [11].

The robust transformation of qualitative data in archives into quantitative data has become increasingly common, especially among researchers with a pharmaceutical background. Since today's European ethnopharmacology is largely influenced by the literature and ethnobotanical fieldwork rarely finds new uses for medicinal plants, more and more researchers have started to look at the data stored in archives. This primarily concerns unexplored archives in former socialist countries with large folklore collections; e.g., in Lithuania, the folk plant use data stored in archives have not yet been thoroughly studied [12]. However, Estonian (e.g., [13]) and Latvian [14,15] pharmacists have already published articles based on archival data without a prior thorough critical analysis of the sources. When processing archival data, there should be a strong emphasis on metadata, which pharmacologists, however, do not know how to process. It is very important to evaluate the time at which the data was collected, who collected it and with what methods, what the motivation for the collection was, etc. Such an analysis also needs to explain how plant species and diseases were identified. It is also important to describe in the Section 4, how the qualitative data, which could have been collected at different times and with different methods, were quantified and how the categorization took place. However, researchers with a pharmaceutical background do not provide such information (e.g., [13–15]). There are also no references in these above-mentioned articles revealing in which archive collections the original data are located. It can be said that there are major errors in such articles based on already published Estonian data, which gives reason to believe that there are also inaccuracies in the data for other countries.

Borderlands between cultures have been of increasing interest to ethnographers (e.g., [16,17]) and ethnobotanists (e.g., [18]) for understanding the interaction of different ethnic groups and their knowledge circulation. Therefore, we focused our case study on a limited geographic area (three historical parishes) in the Estonian and Russian borderland. In order to illustrate how to deal with the difficulties of plant identification and data interpretation, we examined seven folklore collections from the Estonian Folklore Archive, the oldest reports of which date from 1888 and the latest, from 1996. Ethnobotanical fieldwork also recently took place in this region [19] and, therefore, we used part of these data in comparison with that work.

1.1. Seto (Setomaa), Räpina and Vastseliina Parishes

Since Estonian folkloristics and ethnography are based on historical parish boundaries, we also followed this principle in our work. The reason for this is because historical parish borders remained more or less unchanged in Estonia until the end of the Estonian War of Independence. The two Estonian parishes (Räpina and Vastseliina) closest to the Russian border and the closest Russian border area (Seto) were chosen as the research area (Figure 1). After the Estonian War of Independence (1918–1920), the Setomaa area was incorporated into the Republic of Estonia, and following World War II, most of this area was incorporated into the Russian SFSR. The historical parish boundaries changed due to numerous reforms and today lie within the territories of several municipalities. However, the research area is sparsely populated; for example, in 2020, the Räpina municipality had a little over 6100 people, the Setomaa municipality (the Estonian part of the Seto area), a little over 3100 and the Võru municipality (most of Vastseliina parish belongs to it today), about 10,600 [19,20].

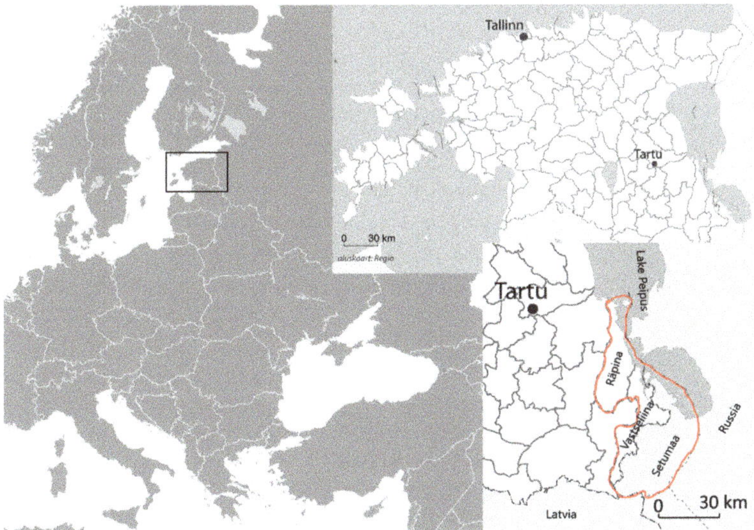

Figure 1. Map of the region.

Since these are border regions, Russians and Finno-Ugric peoples have lived there, side by side, in different villages. Räpina and Vastseliina are predominantly Lutheran areas, and Setomaa is predominantly Russian Orthodox, but paganism is also widespread. In the parishes of Räpina and Vastseliina, large households (manors) were common historically, as elsewhere in Estonia, but only small farms were common in Setomaa. Since Setomaa was located on the outskirts of Russia, it was a very poor area where farmers had little land and were mainly engaged in vegetable growing, handcrafting, trading and fishing. Nature in the region is greatly influenced by Lake Peipus, as well as the Haanja Upland (highest elevation: 318 m a.s.l.), with its hummocky landscape. Coniferous forests and, in wetter areas, swampy deciduous forests predominate. Agriculture, forestry and fishing are the main activities in the region today [19,20].

The Organization of Medicinal Support in the Region

At the end of the 19th century, there was still a large number of folk doctors or witches in Võru County. They were feared, but people went to them for help in times of illness [21]. By the beginning of the 20th century, however, in some regions of Võrumaa, folk doctors nearly disappeared. The reasons for this were the wider availability of school education, the explanatory work of the Christian Church and the greater presence of medical doctors in rural areas [22]. In addition to going to folk doctors, people in Võrumaa and Setomaa visited various healing stones, springs and trees that were found throughout the region. The most important ones, visited by people from all over the area, included Miikse Jaanikivi and Silmaallika, the Võhandu River (which is called Pühajõgi, or "holy river", near the town of Võru), and the sacred oak of Pechory. On the Republic of Estonia's side of the border, former "natural spas" have now been placed under nature or heritage protection. One of the reasons for such a popularity of natural healing objects may be the fact that in the 19th century, there were no doctors in the rural areas of the Livonian governorate (to which Võrumaa belonged), as they were located only in the towns [23], e.g., Võru and the neighboring governorate of Pechory. These towns also had rural hospitals and pharmacies, including in Võru (opened in 1827 and 1785, respectively) and Pechori (c. 1890s and 1865, respectively). Pharmacies and general stores were established in the rural areas and larger settlements of the Livonian governorate in the early 20th century, e.g., in Rõuge (opened in 1896), Räpina (opened in 1861) and Leevi (opened in 1910). The opening of pharmacies particularly increased after the Estonian War of Independence: Lepassaare (1927), Misso (1925), Osula (1924), Irboska and Värska in 1923 [24]. The Soviet Union unified the medical system and the official health care was free; however, it was not always efficient, so people also looked for alternatives. The use of plants was promoted by the state medical system as well as by the procurement of medicinal plants through pharmacies, and as a result, the use of plants was very popular throughout this time.

1.2. The Aim of the Work

This article has two main objectives:

(1) to analyze the historical material and compare the two regions; and
(2) to provide an example of how to treat archival data in ethnobotanical research, in such a way that it is fully useable in modern scientific research and comparable with currently obtained data.

To this end, we:

(a) reviewed folklore ethnobotanical texts from the HERBA database [8], and identified the various plants and diseases whenever possible; and
(b) limited our research to the whole of Setomaa and only the two parishes of Võru (Räpina and Vastseliina) in order to have a relatively comparable number of texts.

From the start, we anticipated that we would not be able to identify all the plants and diseases described in the texts. However, the remaining dataset is sufficient to contribute to the analysis of the influence of the Soviet period on the local ecological knowledge of the region.

2. Results

2.1. Cleaning the Data

All the texts initially selected from HERBA were carefully read and evaluated for their suitability for the analysis. There were specific records that were not incorporated into the analysis, even though present in HERBA, because:

- the text clearly referred to a pharmacy drug or a processed product purchased from a travelling merchant;
- no clear reference to the plant was given (e.g., make the broom from nine leafy trees);
- students supplying the records clearly misinterpreted the information and/or clearly confused the plants;
- there was no reference to the plant's application (stating that it was just medicine);
- poisonous aspects of a plant were highlighted with no medicinal application; and
- when students used both the Latin name and local name of the plant deriving from literary sources and a clear literature influence could be detected.

If the use was copied and shared by more than one student, only one text was retained for analysis. There were many such cases of students "working together". For example, in 1929, nine students from a Värska school referred to bathing with *Trifolium* to treat rheumatism, writing the sentence identically. It is noteworthy that a similar use was also described (with more details and different wording) in a text from 1937 and the same taxon was claimed to be used against typhus (which is a little suspicious, although we cannot completely rule it out).

Greatly overlapping texts also sometimes originated from two different generations and some of them could not be attributed to tradition. For example, a text from 1985, referring to the notes left by a woman who died in 1984 at age 92, is almost identical to the one written by a male student from a Pugola elementary school in 1937. The text refers to the treatment of cancer with a long list of plants that are boiled together. We were able to trace that the texts were copied from a newspaper published in September 1936 that discussed Tallinn's herb sellers and how they taught people to use medicinal plants [25].

One exception can be highlighted—*Linum ussitatisumum*—which was often used in the form of a linen cloth (not a living or dried plant), as it was stressed that the used fabric had to be made from linen, even though it was a very common fabric. Such texts were included.

2.2. The Identification of Plants

As is very common for local plant names, even in a small community, one name may refer to different species; in this situation, the description of the disease's origin and the habitat of the taxon were consulted. In the example of *kärnahain* ("scab herb"), two different *Rumex* taxa were associated with the texts. One was *Rumex crispus*, a widespread name that had already been identified in a text collected by pastor Jakob Hurt in 1903. The second text had no identification, but a description of the plant ("being dug out of the ground") was provided, which clearly refers to one growing on dry land (hence, *R. crispus*). However, the third text very likely refers to *Rumex hydrolapathum* Huds., which grows in wetlands, as the text describes a disease that derives from water.

The later the text, the more chances there were that the names had already been unified. A good example is provided by *verihein* ("blood herb"), a name that in 19th-century Setomaa could be attributed to *Argentina anserina* or *Achillea millefolium*. Here, the description of the plant is of crucial importance. For example, in a text from 1889, *verihain* is described as having leaves as those of ash (*Fraxinus excelsior*) and little yellow flowers—this refers to *Argentina anserina*. The other examples were from later times (starting from 1937), where only the use was provided (mainly cough and staunching blood). By that time, plant names were more unified and the same name could now be associated with *Achillea millefolium*, for which *verihain* was a widespread name throughout Estonia and such a use, widely known. However, one of the earliest texts (collected by Jakob Hurt in 1903) identified *Achillea millefolium* as *verihain*, wherein the use against lung disease was described.

Arnika is a local plant name that was very difficult to identify. *Arnica montana* L. (to which the name refers) does not grow in Estonia, however, according to botanist Gustav Vilbaste's book of plant names, two taxa (*Scorzoneroides autumnalis* or *Solidago virgaurea*) are most often identified with this name, although there is always the possibility for other taxa being used (see [26] for more details).

Fifteen local names (ethnotaxa) remained unidentified for various reasons. One example is *pinipussuhain* ("plant smelling like dog fart"), a name that could be related to several plants in the region and for which there was no description of the plant or any other potentially helpful details. It was used to treat lung diseases and therefore most likely refers to *Hyoscyamus niger*; however, it could also refer to *Ballota nigra* and *Ranunculus acris* (name uses unique to our region), or *Mentha arvensis* (a name widespread throughout Estonia), and so it remained unidentified. Another quite representative example is *luuvaluheinad* ("bone pain herbs"), whose specific form in the Viljandi parish referred to *Persicaria amphibia* (L.) Delarbre, while in Räpina, from where this text derived, a similar name (*luuvalurohi*) was recorded for *Polygonatum odoratum* (Mill.) Druce. The meaning of both names is the same, namely medicine for bone pain, and throughout Estonia, ten taxa had similar names. Therefore, the name is too ambiguous to be identified without further explanation or description. *Valge lill* refers to the color of the flowers (white) and without a description, the actual plant used is not identifiable and could represent various taxa even for the same use category.

The cases in which there was more than one plant name provided, helped to facilitate plant identification. For example, the local name *jumalakäpp*, refers to orchids from several different families (i.e., *Orchis* sp., *Platanthera bifolia* and *Dactylorhiza maculata*), yet in Setomaa, the name *juudakäpp*, refers to *Dactylorhiza maculata*, and therefore identification was possible.

One rather misleading plant name in the region is *takja*. In general (all across Estonia), this name refers to *Arctium tomentosum*, yet when it is used against cough, it refers to *Cetraria islandica* (L.) Ach., which is more often called *palotakja*.

2.3. The Identification of Etic Disease Categories

For establishing clear and countable use records (UR), specific ad hoc rules were followed as outlined below.

Sometimes, there were cases in which the use of one or several plants was potentially described by several symptoms referring to one disease. In the majority of cases where the use was described with two or more disease names or symptoms belonging to the same etic disease group, the text was treated as one UR. For example, foot and back pain remained symptoms of one disease belonging to the musculoskeletal etic disease group. When the disease was provided with symptoms, such as wounds or swelling and pain, it was recorded as a wound (and thus the dermatological disease category). Another example is when the plant was described as promoting sweating and against cold; since both the disease (cold) and symptom (sweating) are related to the general disease category, it was recorded as one UR.

From these rules, a few exceptions were made where the emic symptoms or diseases described together belonged to different etic disease categories, for example, in the cases of cough and lung diseases or tuberculosis. The former (cough) clearly belongs to the respiratory disease group. However, *kopsuhaigus* ("lung disease") was often a popular name for tuberculosis as well as some other infectious lung diseases, and although proper diagnostics for tuberculosis in the region were restricted, a long-lasting cough and other symptoms of tuberculosis or severe infections affecting the lungs were well differentiated from the "ordinary" cough, and we therefore attributed them to infectious diseases. We also had one example where *tiisikus* ("tuberculosis") and *tüüfus* ("typhus") were treated with the same plant—both are infectious diseases, yet very different in nature and therefore counted as two different UR. Tooth diseases and stomach diseases were also accounted for separately. Another example is that of stomach disease, liver disease and hemorrhoids, as all three belong to the gastrointestinal etic disease category, yet they were considered separately as they refer to different organs. A student from Räpina parish reported the use of a strong tea made from *arnika* to treat internal pain (*seest valu*), which was attributed to the culture-bound disease category and diarrhea (digestive category).

Diseases or symptoms with deep mythological connotations, such as *kaetus* (evil eye), *halltõbi* ("grey disease") (malaria), *seest haigus* ("internal disease") and *venitus* ("internal tension") (muscle pain, usually due to hard work), *pistja* ("stabbing pain") and *vaivaja* (nightmare), which were not univocally interpretable, were classified as belonging to the culture-bound disease category, which does not exist in the International Classification of Primary Care, 2nd edition [27]. There was a difficult decision to make in the case of *jooksva* ("runner"), which refers to pain changing its location in the body. The selection of the plants used to treat it has also historically been related to some perceived properties of the disease (such as an origin from a wet place or creeping along the ground) [28]. However, as the disease is closely related to the ailment currently known as rheumatism (and this name was also often mentioned), it was attributed to the musculoskeletal disease category (Table 1).

Table 1. The relationship between emic and etic disease categories and the number of UR. **Bold** text corresponds to etic disease categories and their summary count. Etic/emic categories: No. of UR.

Blood: 8	**Urological: 20**	**Respiratory: 134**
blood cleaning: 5	blood peeing: 3	cough: 129
blood pressure: 1	kidney and bladder disease: 10	difficulties in breathing: 1
poor blood: 2	kidney disease: 5	nose bleeding: 2
	oedema: 1	sore throat: 2
	urinary retention: 1	
Pregnancy, etc.: 3	**Female genital: 11**	**Cardiovascular: 15**
giving birth: 3	problems with menstruation: 4	heart diseases: 15
	women diseases: 7	
Culture-bound disease: 79	**Musculoskeletal: 64**	**Neurological: 49**
kaetus (evil eye): 3	backache: 4	calming: 6
halltõbi: 4	bone pain: 3	epilepsy: 1
internal disease: 38	foot diseases: 4	headache: 13
kidi (tendovaginitis or bursitis in the wrist): 11	joint disease: 1	nerve disease: 8
vaivaja, vaivajatõbi (nightmare or hernia): 3	joint dislocation: 4	paralysis: 2
pistja (stixis, pleuritis, etc.): 6	*jooksva*/rheumatic: 48	seizures: 12
riis (umbilical hernia in children): 2		sleep disorder: 7
tiir (itchy soles): 1		
tsirgutõbi (a disease in young children): 1		
ussiviga (chorea in young children): 4		
Skin: 200	**Digestive: 227**	**General: 250**
abscess: 9	appendicitis: 1	against several diseases: 15
bee stings: 1	bile disease: 1	cold: 57
bleeding: 19	constipation: 5	cholera: 3
boil: 26	diarrhea: 12	disinfection: 6
burned wound: 13	digestion problems: 10	fever: 10
skin cancer: 1	gastric disease: 13	for sweating: 5
cracked lips: 4	gastric ulcers: 2	freezing: 6
cut wound: 2	hemorrhoids: 1	good for health: 4
dandruff: 4	heartburn: 2	inflammation: 2
eruption: 4	indigestion: 3	loss of appetite: 9
erysipelas: 22	jaundice: 8	lung disease: 34
for beauty: 4	liver disease: 6	pain: 3
fresh wound: 3	mouth diseases: 1	prophylactics: 4
hair loss: 1	nausea: 2	rabies: 1
inflammation: 7	stomach disease: 53	stroke: 2
itching: 1	stomach worms: 14	throat disease: 21
local pain: 2	stomachache: 64	tiredness: 1
lump on skin: 6	tapeworm: 4	tuberculosis: 62
pimples: 6	tooth diseases: 7	typhus: 2
roos (erysipelas): 2	toothache: 18	whooping cough: 2
rotten wound: 10	vomiting: 1	
scabies: 3		
scabs: 3		
skin disinfection: 2		
skin diseases: 1		
snake bite: 5		
splinter: 1		
sunburn: 1		
warts: 8		
wound: 28		
Ear diseases: 2	**Eye diseases: 10**	

2.4. A General Overview of Plant Uses

After cleaning the data, 1072 UR were retained for the analysis (Table 2).

Table 2. The use records (UR) remaining for analysis after the cleaning of the data and identification of the plants.

Collection/Parish	Räpina	Setomaa	Vastseliina	Sum
E	5	1	2	**8**
ERA	27	45	1	**73**
ERM			3	**3**
H	22	25	28	**75**
KKI		6		**6**
RKM	37	43	11	**91**
Vilbaste	343	383	90	**816**
SUM	**434**	**503**	**135**	**1072**

The data were provided by 47 correspondents, 13 of whom provided just one or two UR. Among the correspondents were several folklorists who visited two or all three of the parishes, including the founders of the Estonian Folklore collections, namely Jakob Hurt, who collected 27 UR from Setomaa, and pastor Matthias Johann Eisen (five UR) from Räpina. The most productive collectors were schoolteachers, who collected the work of numerous students for several years following the request of Gustav Vilbaste; J. Haring, a Värska primary school teacher in Setomaa, sent thirty-five student responses (438 UR) from the area and one student response (13 UR) from Räpina parish. There was one other productive teacher, M. Kaasikmäe from the Setomaa Košelki primary school, who sent 11 student responses in total (70 UR). Anna Vitsust, a Räpina Gymnasium teacher, also sent twenty-one student responses (259 UR) from Räpina, and Kotlep Pärg, a Pugola elementary school teacher in Vastseliina, sent seven student responses (163 UR). There were two more responses provided by two students which we excluded from the analysis. One of them had only listed 57 medicinal plants without any specification about diseases, and the other student stood out for having recipes that were too detailed, all of which were copied from the above-mentioned newspaper article [25]. In addition, Gustav Vilbaste himself (as he was a teacher in the city of Tartu) collected the response of a student from Räpina who came to study at a Tartu school (20 UR). Of the other correspondents, important information was provided by volunteers, Daniel Lepson (farmer; 29 UR) and Maria Linna (agricultural worker; 27 UR), who collected village folklore by interviewing the local inhabitants.

The use of 120 taxa belonging to 48 families was identified, of which nine were identified at the genus level and four ethnotaxa were identified as two possible botanical taxa (Table 3). In addition, 15 ethnotaxa were unidentifiable and therefore they were left out of the formal analysis; however, their uses are presented in Table 3. The most represented families were Asteraceae (sixteen taxa + two potential taxa), Rosaceae (thirteen taxa) and Ericaceae (eight taxa). The most frequently mentioned taxa were *Pinus sylvestris* (57 UR), *Matricaria discoidea* (51 UR), *Valeriana officinalis* (50 UR), *Achillea millefolium* (42 UR), *Juniperus communis* (39 UR) and *Tilia cordata* (39 UR). Thirty-four taxa had only one UR, while thirty-five taxa had ten or more UR.

Table 3. Plants and their uses from folklore collections.

Family	Taxa	Local Name	Etic Disease Category	UR
Acoraceae	*Acorus calamus* L. [L]	Jõekalmus [S], lesnagud [S], kalmus, tatersäla [V], kalmusejuur	Cardiovascular	1
			Culture-bound disease	1
			Digestive	7
			General	7
			Musculoskeletal	3
			Skin	1
			Urological	1

Table 3. *Cont.*

Family	Taxa	Local Name	Etic Disease Category	UR
Amaranthaceae	*Beta vulgaris* L. [C]	peet [S], verevä nakri [S]	Eye	1
			Skin	4
Amaryllidaceae	*Allium cepa* L. [C]	sibul, sippul [S], sipul [V]	Cardiovascular	2
			Digestive	7
			Ear	1
			General	5
			Respiratory	5
			Skin	8
			Urological	1
	Allium sativum L. [C]	kurslaga [S], kurslakk [S], küüslauk [S]	Digestive	1
			General	1
			Respiratory	1
			Skin	1
Apiaceae	*Angelica sylvestris* L. [L]	heinputk [V], pütsk [V]	Digestive	2
	Carum carvi L. [L]	köömned [V], küümned	Cardiovascular	1
			Digestive	5
			General	6
			Respiratory	2
Asparagaceae	*Convallaria majalis* L. [L]	maikelluke [S]	General	1
Asphodelaceae	*Aloe arborescens* Mill. [C]	aalo, aaloe, aalus [S], aleo [S], pakso lill [S]	Culture-bound disease	1
			General	6
			Respiratory	1
			Skin	7
Asteraceae	*Achillea millefolium* L. [L]	raudrohi, verihein, verihain, valgõ lill [S]	Culture-bound disease	2
			Digestive	11
			Female genital	2
			General	5
			Musculoskeletal	1
			Neurological	1
			Respiratory	6
			Skin	10
	Antennaria dioica (L.) Gaertn. [L]	kassikäpad [S]	Female genital	1
	Arctium tomentosum Mill. [L]	takjas, takk [S]	Blood	2
			Culture-bound disease	3
			Digestive	3
			General	3
			Respiratory	7
			Skin	11
			Urological	1
	Artemisia absinthium L. [L]	koirohi, pänül [S], pälüm	Culture-bound disease	2
			Digestive	5
			General	3
			Musculoskeletal	1
			Skin	4

Table 3. *Cont.*

Family	Taxa	Local Name	Etic Disease Category	UR
	Carduus crispus L. or *Cirsium vulgare* (Savi) Ten. [L]	karuohtja [V]	Urological	1
	Gnaphalium uliginosum L. [L]	sammaspoolehain, sammaspoolehein, sammaspoolhain [S], sammaspoolikuhain [S], sammaspoolikuhein [V], sammaspoolõhain [S], soo-kassiurb [V]	Skin	10
	Matricaria chamomilla L. [C]	kamel [V], teekummel	General	6
			Neurological	4
			Respiratory	2
	Matricaria discoidea DC. [L]	kaamel [V], kammel [V], kumelitee [V], kaamelihain [V], kummel, kummulid [S], lõhnav kummel, ubinhain, ubinhein, unõhain [S], upinhain, uppinhain [S]	Culture-bound disease	2
			Digestive	9
			Ear	1
			Eye	3
			General	15
			Musculoskeletal	1
			Neurological	2
			Pregnancy, childbearing, etc.	1
			Respiratory	11
			Skin	6
	Scorzoneroides autumnalis (L.) Moench[L], *Solidago virgaurea* L. [L] or with lower probability, many other taxa	arnikas, ärnika [S]	Culture-bound disease	6
			Digestive	3
			General	2
			Neurological	1
			Respiratory	2
	Solidago virgaurea L. [L]	ärnetsa [V]	Musculoskeletal	1
	Tanacetum vulgare L. [L]	kolladsõ lill [S], kollane lill [S], solknaheinad [S], solknarohi [V]	Digestive	4
	Taraxacum officinale F.H. Wigg. [L]	võilill	Digestive	1
			Skin	1
	Tragopogon pratensis L. [L]	piimjuur [V]	Digestive	1
	Tripleurospermum inodorum (L.) Sch.Bip. [L]	kammel [S]	Digestive	1
			Respiratory	1
	Tussilago farfara L. [L]	ämmaleht [V]	Skin	1
Betulaceae	*Alnus* spp. [L] (incl. *Alnus glutinosa* (L.) Gaertn. and *Alnus incana* (L.) Moench)	lepp, soolepp [S] imälepp [S] imälepp [S] valge lepp [V]	Culture-bound disease	2
			Skin	4
			Digestive	1
	Betula spp. [L]	kask, kõiv, kõo [S]	Culture-bound disease	2
			Digestive	2
			General	5
			Musculoskeletal	7
			Respiratory	2
			Skin	8

Table 3. *Cont.*

Family	Taxa	Local Name	Etic Disease Category	UR
Brassicaceae	*Armoracia rusticana* G. Gaertn., B. Mey. & Scherb. [C]	maarjaritska [V], mädarõigas [V]	Culture-bound disease	1
			Digestive	2
	Brassica oleracea var. *capitata* f. *Alba*[C]	kapsas [S]	Eye	1
			Neurological	2
			Skin	1
	Brassica rapa L. [C]	naar [V]	General	1
	Capsella bursa-pastoris (L.) Medik. [L]	hiirekõrv [V]	Musculoskeletal	1
	Sinapis alba L. [C]	sinep [V]	Eye	1
Cannabaceae	*Cannabis sativa* L. [C]	kanebi, kanep [V]	Culture-bound disease	3
			General	1
Caprifoliaceae	*Valeriana officinalis* L. [L]	balderjan [V], palderjaan [V], palderjan, paltõjan [S]	Cardiovascular	3
			Culture-bound disease	7
			Digestive	17
			Female genital	1
			General	2
			Neurological	20
Crassulaceae	*Hylotelephium maximum* (L.) Holub [L]	kidsihain, kidsihein [V], maapähkme [V]	Culture-bound disease	7
			General	1
	Sempervivum globiferum L. [N]	maasibul [V]	Neurological	1
Cupressaceae	*Juniperus communis* L. [L]	kadajas, kadakas, kadak [V], kattai	Blood	1
			Culture-bound disease	4
			Digestive	3
			General	19
			Musculoskeletal	1
			Neurological	2
			Respiratory	1
			Urological	8
	Linum usitatissimum L. [C]	lina	General	3
			Skin	6
Cyperaceae	*Eriophorum vaginatum* L. [L]	pikki hain [V]	Skin	1
Droseraceae	*Drosera rotundifolia* L. [L]	huulehain [S], huulhain [S], huulhein [S]	Skin	5
Equisetaceae	*Equisetum arvense* L. [L]	põldosi [V], tilkhain [V]	Skin	1
	Equisetum spp. [L]	osjad [V]	Urological	1
Ericaceae	*Andromeda polifolia* L. [L]	tsihknaõied [S]	Respiratory	1
	Arctostaphylos uva-ursi (L.) Spreng. [L]	leesikad, tsiamarja [S], lehike [S], tsiapalohka [S], tsiapalokka [S]	Culture-bound disease	1
			General	1
			Respiratory	1
			Urological	2
	Calluna vulgaris (L.) Hull [L]	kanarik [S], kanarpik [S], palokanarik [S]	Digestive	1
			General	4
			Musculoskeletal	2
			Skin	3

Table 3. *Cont.*

Family	Taxa	Local Name	Etic Disease Category	UR
	Chimaphila umbellata (L.) W.P.C.Barton [L]	obijoinihain [S], obijoinilill [S], obijoon [S], opijon [S], obijoon [S], oobium, oopiumiheinad [V], opijann [S]	Cardiovascular	2
			Culture-bound disease	3
			Digestive	4
			General	1
			Neurological	2
			Respiratory	4
	Rhododendron tomentosum Harmaja [L]	sookael [V], sookanarbik [S], sookanarik [S], suukanarik [S] sootsähknad [V], soovitsked [V], tsihk [S], tsihkna [S]	Digestive	2
			Respiratory	1
			General	4
			Musculoskeletal	2
			Skin	1
	Vaccinium myrtillus L. [L]	mustikas, mustkas [S], mustigõ [S]	Digestive	13
	Vaccinium oxycoccos L. [L]	jõhvikad [V], kuremari [S], kuremarjad	Digestive	1
			General	2
			Neurological	1
			Skin	1
	Vaccinium vitis-idaea L. [L]	palõhk [S], palohkna [S], palovka [S], palukas, pohl, pohlak [V]	Culture-bound disease	1
			General	5
			Musculoskeletal	2
			Neurological	1
			Respiratory	5
			Urological	2
Fagaceae	*Quercus robur* L. [L]	tamm, tammõ [S]	Blood	1
			Culture-bound disease	2
			Digestive	4
			General	3
			Skin	5
Gentianaceae	*Gentiana cruciata* L. [E]	vaivajarohi [V]	Musculoskeletal	1
	Gentiana spp. [E]	süäme alodsõ hain [S], südamealuse heinad [S]	Cardiovascular	3
Grossulariaceae	*Ribes nigrum* L. [C]	mustad hõrakad [V]	General	1
Hypericaceae	*Hypericum* spp. [L]	naistepuna [V]	Digestive	1
			Female genital	1
Lamiaceae	*Mentha spicata* L. [C]	rohemünt [V]	Digestive	1
	Mentha aquatica L. [L]	vesimünt [S]	Neurological	1
	Mentha spp. [C]	piparmünt, münt [V], aia-vehverments [V], pibarment [S], vehverloints [S], vehvermänts [V], vehverments	Neurological	4
			Culture-bound disease	1
			Digestive	7
			Female genital	1
			General	7
			Respiratory	5
	Thymus serpyllum L. [L]	jaanihaina [S], jaanihein, kadedushein [V], kaetiserohi [V], üheksahaiguserohi [V], kolmekordne rohi [V], liivatee, üheksatõverohi [S], maarjahein, pühamaarjahaina [S]	Digestive	2
			Eye	2
			General	12
			Musculoskeletal	3
			Neurological	1

Table 3. *Cont.*

Family	Taxa	Local Name	Etic Disease Category	UR
			Pregnancy, childbearing, etc.	1
			Respiratory	11
			General	1
	Trifolium pratense L. [L]	ristikhein	Musculoskeletal	2
			Respiratory	1
			Skin	1
	Trifolium repens L. [L]	valge ristikhein	Female genital	1
			General	1
	Trifolium sp. [L]	maarjaristikhein	General	1
Lauraceae	*Laurus nobilis* L [P]	loorber [V]	Blood	1
Lycopodiaceae	*Huperzia selago* (L.) Bernh. ex Schrank & Mart. [N]	nõiakõld [S], nõiakollad [V]	Cardiovascular	1
			Eye	1
	Lycopodium clavatum L. [N]	karukollad [V], nõiakollad [V]	Skin	1
Melanthiaceae	*Paris quadrifolia* L. [L]	–	Digestive	1
Menyanthaceae	*Menyanthes trifoliata* L. [L]	ubalehe [V], ubaleht [V]	Digestive	1
			General	2
Oleaceae	*Fraxinus excelsior* L. [L]	saar [V]	Musculoskeletal	1
Orchidaceae	*Dactylorhiza maculata* (L.) Soó [N]	jumalakäpp [V], juudakäpp [V]	Digestive	1
Papaveraceae	*Chelidonium majus* L. [L]	vererohi [V]	Urological	1
	Corydalis solida (L.) Clairv [L].	vaivaja haina [S]	Culture-bound disease	2
	Fumaria officinalis L. [L]	juuksehain [V]	Skin	1
	Papaver somniferum L. [C]	magun [S], makunna [S]	Digestive	1
			Neurological	1
Pinaceae	*Picea abies* (L.) H.Karst. [L]	kuus, kuusk	Digestive	1
			General	10
			Musculoskeletal	4
			Respiratory	2
			Skin	15
	Pinus sylvestris L. [L]	mänd, pettai [V], petäi [S], pedäjäs [S]	Digestive	2
			General	27
			Musculoskeletal	8
			Respiratory	11
			Skin	9
Piperaceae	*Piper nigrum* L. [P]	pipar	Digestive	2
			Respiratory	1
Plantaginaceae	*Plantago major* L. [L]	paiselehe, paiseleht, umbleht [V], ummelehe [S], teeleht [V]	Culture-bound disease	1
			Eye	1
			General	2
			Musculoskeletal	1
			Skin	24

Table 3. *Cont.*

Family	Taxa	Local Name	Etic Disease Category	UR
Poaceae	*Avena sativa* L. [C]	kaar, kaer	Culture-bound disease	4
			Digestive	4
			General	5
	Hordeum vulgare L. [C]	kesvad [V], oder [V]	Skin	3
	Secale cereale L. [C]	rüä [S], rüga [V], rukis	Digestive	2
			Musculoskeletal	1
			Skin	5
Polygonaceae	*Persicaria amphibia* (L.) Delarbre or *Glyceria maxima* (Hartm.) Holmb. [L]	läsnäk [S], lesnak [S], lesnäk [S]	General	3
			Respiratory	2
	Polygala amarella Crantz [L]	vahulill [S]	Skin	1
	Polygonum arenastrum Boreau [L]	morohain [S], niseldushain [V], niseldushein [V], nisõldushaina [V]	Musculoskeletal	3
			Skin	1
	Polygonum aviculare L. [L]	–	Musculoskeletal	1
	Rumex crispus L. [L]	kärnhain [S] hobuhain [S]	Skin	2
			Respiratory	1
	Rumex hydrolapathum Huds. [L]	kärnahain [S]	Skin	2
Polypodiaceae	*Dryopteris filix-mas* (L.) Schott [L] or *Pteridium aquilinum* (L.) Kuhn [L]; (Dennstaedtiaceae)	maarjasõnajalg [V], sõnajalg [V]	Culture-bound disease	2
			Digestive	7
			Musculoskeletal	2
			Neurological	1
Ranunculaceae	*Anemone nemorosa* L. [L]	haragheinad [S], haraklilled [S]	Digestive	2
Rhamnaceae	*Frangula alnus* Mill. [L]	kisõpuu [S], kitsepuu [S], kitsetoome [V], kitseuibo, vohopaadsa, soemära [S], paakspuu	Digestive	8
Rosaceae	*Alchemilla vulgaris* L. [L]	kortsleht [V]	Digestive	1
	Potentilla argentea L. [L]	verehain [S]	Skin	1
	Filipendula ulmaria (L.) Maxim. [L]	angervaks [V]	Pregnancy, childbearing, etc.	1
			Skin	1
	Fragaria vesca L. [L]	maasikas	General	3
			Respiratory	4
	Malus domestica (Suckow) Borkh. [C]	õunapuu [S], uibu [S], uip [S], uipoh [S]	Culture-bound disease	1
			General	3
			Respiratory	2
	Potentilla erecta (L.) Raeusch. [L]	kalkanajuured [S], kalgan [V], maramaar [S], maran, nabahain [V], tedremadar [V], tedremaran, tedremarja juured [S]	Blood	1
			Culture-bound disease	6
			Digestive	14
			General	1
	Prunus cerasus L. [C]	kirss, vislapuu [S]	Digestive	1
			Respiratory	2
	Prunus padus L. [L]	toome, toomingas	Cardiovascular	1
			Digestive	14
			General	7
			Neurological	3
			Respiratory	2

Table 3. *Cont.*

Family	Taxa	Local Name	Etic Disease Category	UR
			Skin	3
	Pyrus communis L. [C]	pruusa [S]	Culture-bound disease	1
	Rubus chamaemorus L. [L]	murakad [S]	Musculoskeletal	2
			Digestive	1
	Rubus idaeus L. [L]	vabarna	General	10
			Respiratory	7
	Rubus polonicus Weston [L]	mustad vabarnad	General	1
			Culture-bound disease	4
			Digestive	4
	Sorbus aucuparia L. [L]	pihlakas, pihl [S], pihlapuu [S]	General	7
			Respiratory	3
			Skin	2
Rubiaceae	*Coffea* sp. [P]	kohv [V]	Culture-bound disease	1
			Respiratory	1
	Galium boreale L. [L]	niseldushain, nikastushein	Musculoskeletal	1
Salicaceae	*Populus tremula* L. [L]	haab	General	2
			Digestive	1
	Salix spp. [L]	pai [S], paju	General	1
			Respiratory	2
			Skin	8
Sapindaceae	*Acer platanoides* L.[L]	vaher [V]	General	1
Solanaceae	*Capsicum annuum* L. (Longum Group) [C]	kõdrapipar [S], pipar söögipipõr [S], türgi pipar [V], verevä pipar [S] verikõder, veripipõr [S]	Digestive	22
			General	1
			Musculoskeletal	2
			Respiratory	6
	Hyoscyamus niger L. [N]	hambahain [S]	Digestive	2
	Nicotiana rustica L. [C]	tubaguhain [S], tubak [S], tubakas, tubakulehe [V]	Culture-bound disease	4
			Digestive	2
			Respiratory	1
			Skin	2
	Solanum dulcamara L. [L]	maavitsad [V], päris maavits	Musculoskeletal	6
			Skin	6
	Solanum tuberosum L. [C]	kartohvel [S], kartokas [S], kartul	Culture-bound disease	1
			Digestive	2
			Musculoskeletal	1
			Skin	2
Thymelaeaceae	*Daphne mezereum* L. [L]	küüvits [V]	Digestive	3
Tiliaceae	*Tilia cordata* Mill. [L]	lõhmus, pahka [V], pähn [V], pärn, pähnäpuu [V], pärnapuu [V]	Blood	2
			Digestive	2
			General	20
			Respiratory	10
			Skin	4
			Urological	1

Table 3. *Cont.*

Family	Taxa	Local Name	Etic Disease Category	UR
Urticaceae	*Urtica dioica* L. [L]	nõges	General	2
			Musculoskeletal	4
			Respiratory	2
			Skin	2
	Urtica urens L. [N]	raudnõges [S]	General	2
			Skin	2
Viburnaceae	*Viburnum opulus* L. [L]	lodjapuu	Digestive	3
			General	1
			Neurological	1
			Respiratory	2
	Adoxa moschatellina L. [L]	mättahain [S], mättalill [S]	Respiratory	1
			Cardiovascular	1
Unidentified taxon		härjapää [V], verihein [V]	Respiratory	1
		kandrohi [S]	Digestive	2
			Female genital	1
		karamarjad [S]	Urological	1
		kärnõ rohi [S]	Skin	1
		lepakukud [S]	General	1
		luuvaluheinad [V]	Musculoskeletal	1
		nätselmehein [V]	General	2
			Musculoskeletal	1
		palohain [S]	Culture-bound disease	1
		pinipussuhain [S]	General	1
		punatse lill [V]	Female genital	1
		tõrvaleht [V]	General	1
		tõrvaõied [S]	Digestive	1
		valge kassikäpp [V]	Female genital	1
		valge lill [V]	Female genital	1
		valge lill	Respiratory	2
		valgõ lill	digestive	2
		no name [S]	Skin	1

Unless recorded in both: [S]—Local plant names recorded in Setomaa, [V]—local plant names recorded in Võromaa; [C]—Cultivated, [L]—Least Concern, [N]—Near Threatened, [E]—Endangered, [P]—does not grow in Estonia. Extinction risk statuses were taken from the Estonian red list, as specified in the database at https://elurikkus.ee/ (accessed on 28 September 2022).

Of all the used taxa, 23 were cultivated, most of which were garden fruits, vegetables and crops. Only four plants can be said to have been cultivated as medicinal plants: *Mentha* spp. and *Matricaria chamomilla* in the garden, and *Aloe arborescens* and *Capsicum annuum* in flowerpots indoors. Most of the natural species on the list were common plants in Estonia. According to today's understanding, only four plants are near threatened and *Gentiana* spp. is in the endangered category. The list also includes three herbs that do not grow in Estonia, which were bought in a shop.

The most represented etic disease categories were general, digestive and skin. The proportion of culture-bound diseases was also relatively high. The most often mentioned emic disease categories were cough (129 UR), stomachache (64 UR), tuberculosis (62 UR), cold (57 UR) and stomach disease (52 UR).

The proportional division of disease categories between the different times of collection illustrates the change in the importance of some of the categories throughout the century (Figure 2).

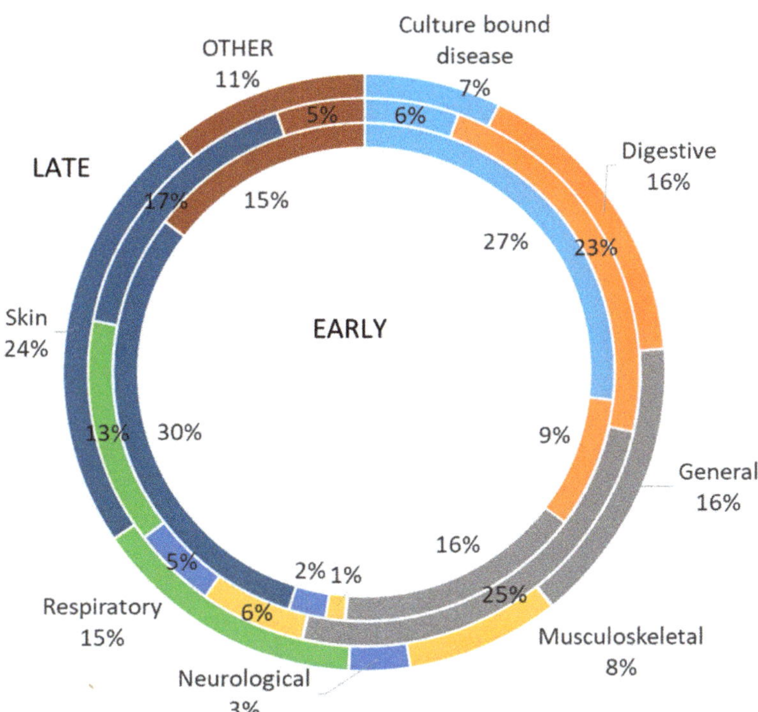

Figure 2. A proportional division of the general disease categories in early, middle and late folklore collections.

2.5. A Cross-Cultural and Diachronic Comparison

The cross-cultural comparison of the whole dataset shows high heterogeneity. Of the 119 taxa, 90 were recorded in the two Võro parishes, while 84, in Seto parish. Overall, 55 taxa overlapped (JI = 0.46), while 35 taxa overlapped for those recorded with three or more UR (JI = 0.54). Seto parish showed slightly greater consistency in the use of fewer taxa (52 out of 84 had three or more UR), while the Võro parishes exhibited more diversity (almost half of the used taxa (43 out of 90) had less than three UR). A few taxa, represented by three or more UR, were characteristic of one group; *Taraxacum officinale*, *Menyanthes trifoliata*, *Hordeum vulgare*, *Daphne mezereum*, *Angelica sylvestris* and *Dryopteris filix-mas* were used exclusively in the Võro parishes, while *Malus domestica*, *Beta vulgaris*, *Prunus ceraseus*, *Brassica oleraceae*, *Drosera rotundifolia*, *Persicaria amphibia*, *Rumex*, *Gentiana pneumonanthe* and *Urtica urens* were reported exclusively in Seto parish.

However, we need to take into consideration the fact that the folkloristic data was collected very unevenly. If we look at the data collected within the early period (before 1904), we can observe that many disease categories are represented only in one parish, which is illogical.

The comparison between all three parishes increases the diversity; however, we need to consider that the low number of UR from Vastseliina is most likely due to the lack of data, not the lack of actual uses (Figure 3).

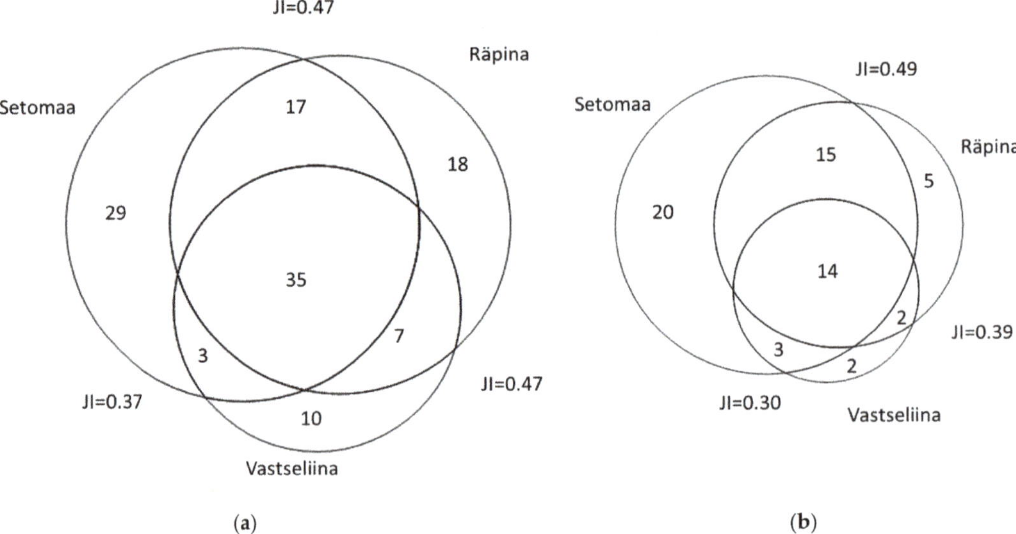

Figure 3. A comparison of (**a**) all plants and (**b**) those used in three or more UR in the three studied parishes. JI—Jaccard Index.

The 20 most used plants based on the number of UR, with a few exceptions (*Prunus padus*, *Capsicum annuum* and *Calluna vulgaris* were not mentioned in the Vastseliina parish), were present in all the parishes, but the proportion of use is not even (Figure 4).

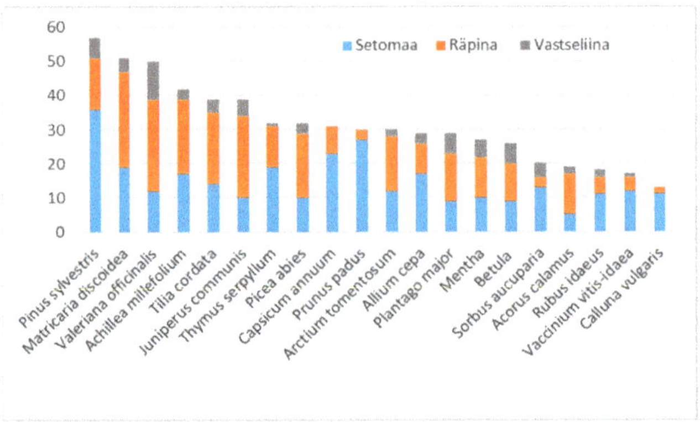

Figure 4. The 20 most used taxa and their UR distribution in the parishes.

3. Discussion

The Jaccard Indexes (JI) obtained in the cross-parish comparison are remarkably lower than those from the recently collected data [19]. It should be kept in mind that these only represent identified taxa, while unidentified ethnotaxa were not taken into account in the calculation of the JI. A high number of cultivated species was present exclusively in the Seto material, which is noteworthy as the use of cultivated species for medication is more characteristic of Slavic communities [29], indicating that the Setos had closer contact with neighboring Russians (see also [30]).

In the texts, there was a high proportion of diseases related to culture (over 6%), yet proportionally, this was higher in the early period and almost absent in the late (occupation) period. This is consistent with the tendency seen in the recently obtained data where culture-bound diseases were completely absent from the disease list [19]. The prevalence of the digestive disease category in the dataset from the 1930s may also be due to the fact that the data was collected by students, which limited to some extent the diseases covered by the data. The absence of respiratory diseases among the early data is quite indicative of historical data, guided by the understanding that minor diseases such as a runny nose or simple cough were not considered worth mentioning in peasant society and were often not even treated.

The presence of 15 unidentifiable folk taxa does not diminish the data obtained; instead, it shows the high diversity of plant names, especially in the early dataset, and the potential presence of ad hoc names. The high diversity of local plant names (mainly wild taxa reported by one person or cultivated ones that have only a single local name) demonstrates the historical diversity of plant names within the very limited geographic area. The presence of local names referring to a disease (primarily from the early and middle datasets) presents an important aspect of Estonian folk medicine, which was previously highlighted by Jakob Hurt in his identifications in 1888 [31], yet has been seldom addressed in the international literature thus far.

The use of local wild taxa growing outside the sphere of everyday human activities (such as *Eriophorum vaginatum, Menyanthes trifoliata, Dactylorhiza maculata, Daphne mezereum, Paris quadrifolia, Chimaphila umbellate, Andromeda polifolia* and *Persicaria amphibia*), which was abandoned during Soviet occupation, signals an earlier, pre-existing rich tradition of plant use and a deep relationship with nature through seasonal activities and general interaction with the surroundings. There were several reasons behind this, including the introduction of standardized medicinal plants during the Soviet era, a change in rural lifestyle and the replacement of extensive agriculture (small farm systems) with intensive agriculture (collective farm systems), causing drastic changes in rural life.

Although from today's nature conservation point of view, the majority of used local medicinal plants have a low extinction risk (Least Concern status), their general conservation status does not reflect their future sustainability. As we have shown in a recent study, even common plants in the immediate vicinity of humans may be disappearing because of changes in the management of semi-wild areas [32].

Abandoned specific cultivated species (such as *Secale cereale, Linum usitatissimum, Brassica rapa* and *Papaver somniferum*) can help to track changes in cultivation habits. For example, *Papaver somniferum* disappeared from use because it was proclaimed to be a narcotic and its cultivation in gardens was prohibited, while *Linum usitatissiumum* processing and linen cloth lost its importance with the appearance of new and more affordable textile products.

While working with pre-systematized archival datasets (Table S1), there are some important aspects to consider:

(1) Not all pre-systematized data may be suitable for the research purpose, and therefore some of the data may need to be removed due to not having sufficient information (as we did with missing information on a specific identified use).

(2) One needs to be critical of the data, especially if the collection has some competitive undertones and there is a possibility of group work or risk that the actual data was beautified.

(3) It is better to under-identify the data than over-identify it; working with historical data, especially folklore material, it is inevitable that some taxa remain unidentified.

(4) It is important to involve specialists in the local flora also having a background in historical biogeography.

(5) Working with such a dataset also requires knowledge of the purchasable material in the region.

(6) The historical epidemiology of the region also needs to be studied prior to disease identification.

Despite our efforts and previous experience in plant and disease identification in historical data, we were not able to precisely identify approximately 20% of the taxa whose uses were provided in the texts. While identification at only the genus level is inevitable when dealing with such taxa that are also not differentiated by the people themselves, the potential misidentification at the genus level (when the local name can be attributed to two or more species) can be problematic from an ethnopharmacological point of view. Although the unidentified ethnotaxa may seem like a complete loss from an ethnopharmacological perspective, they may bear important information from a cultural and ethnographic perspective, and signal the richness of the used flora.

Therefore, it should be kept in mind that there is no one-size-fits-all solution for data preparation and analysis, and therefore the utmost care needs to be taken in data interpretation in order to avoid mistakes. Working with archival data requires good preparation and the study of the historical context of the data under investigation, as well as open-mindedness and the ability to accept the fact that not all questions will be answered.

4. Materials and Methods

The basis of the analyzed texts was the HERBA database, a relational database designed by the authors [8], in which herbal folk medicine texts are searchable by local plant name and emic disease category. The HERBA database is based on numerous folklore collections housed in the Estonian Folklore Archives of the Estonian Literary Museum; most of the information has been sent in by various correspondents or collected through folklore expeditions. To construct HERBA, the texts were first identified from their original sources (Table 4) by using the registration of the collection (if present) and by carefully reading the texts within. Vilbaste's collection was worked through entirely. Plant use-related texts were transcribed (for which technical assistants were employed), checked with the original and then entered into the HERBA database by the authors. Local plant names were correlated to the botanical taxa through an additional dataset composed of different sources, mainly relying on the book of Estonian folk plant names by Gustav Vilbaste [33], in which he compiled all existing information on local plant names, and on HERBA [8] itself. We also identified folk diseases based on these two sources.

Table 4. Folklore collections and the number of pre-selected text segments from every collection. Source: https://www.folklore.ee/era/leidmine/index.html (accessed on 10 July 2022).

Abbreviation	Full Name of the Collection	Years of Collection	No. of Pages of Full Collection	No. of Texts in Setomaa	No. of Texts in Räpina	No. of Texts in Vastseliina
H	Folklore collection of J. Hurt	1860–1906	114,696	40	17	27
E	Folklore collection of M. J. Eisen	1880–1934	90,100	1	5	2
ERM	Folklore collection of Estonian National Museum	1915–1925	9398			4
ERA	Folklore collection of Estonian Folklore Archives	1927–1944	265,098	39	28	1
Vilbaste	Folklore collection of G. Vilbaste	1907–1966	20,327	520	309	196
KKI	Institute of Language and Literature folklore collection	1941–1984	35,679	6		
RKM	Folklore collection of Folklore Department of Estonian Literary Museum	1945–1996	447,231	35	37	7
	SUM of records		982,529	631	400	237

The resulting data were extracted from the database in an Excel format and the text segments (the complete narrative units corresponding to one or more plants used for one or more emic diseases) related to the selected regions were separated out for the analysis. Seven of the folklore collections housed in the Estonian Folklore Archives of the Estonian Literary Museum contained data on the medicinal use of plants for the selected regions (Table 4). As the number of resulting texts for the three parishes was uneven, some of the comparisons were made after combining texts from Vastseliina and Räpina, in order to maintain a balance and also because both parishes were historically part of Old Võromaa.

The Excel spreadsheet was further processed and divided into conditional use records (UR), referring to the plant taxon used for treating the specific health condition reported by one correspondent. Emic disease categories provided in the texts were interpreted according to current knowledge and correlated to the medicinal categories of the International Classification of Primary Care, 2nd edition (ICPC-2) [27] (hereafter etic disease categories). As there were some emic diseases described that were not univocally correlatable to the ICPC-2 classifications, we created an additional category of culture-bound diseases (CBD). Within the CBD category were included such emic diseases as the evil eye or nightmare, as well as diseases represented by specific, culturally significant disease names, such as *seest haigus*, which could correspond to either the digestive, musculoskeletal or general disease category, yet were well positioned in the culture as phenomena.

For comparative purposes, the resulting dataset was divided into three temporal categories:

1. Early: 1888–1904 (80 UR);
2. Middle: 1928–1942 (871 UR); and
3. Late: 1949–1996 (121 UR).

The Latin plant names provided in HERBA (which was formed on the basis of the Estonian Plant Name database (https://taimenimed.ut.ee/ accessed on 10 July 2022) and our resulting identifications followed those listed in the Plants of the World Online (POWO) database [34] and the European Flora [35]; family assignments followed the Angiosperm Phylogeny Group (APG) IV [36]. The correlation between the emic plant name and plant taxon was carefully checked as described above, and a number of taxa remained unidentified, e.g., at the level of ethnotaxa.

Data Comparison

The Jaccard Similarity Indices (JI) followed the methodology of González-Tejero et al. [37]: JI = (C/(A + B − C)), where A represents the number of taxa in sample A, B is the number of taxa in sample B, and C is the number of taxa common to A and B.

The proportional Venn diagrams were created using the PAST Toolkit Venn diagram plotter software program (https://omics.pnl.gov/software/venn-diagram-plotter (accessed on 10 July 2022)). The figures were visualized using RAW Graphs (RAW) [38].

5. Conclusions

Our results show a high diversity of historical medicinal plant use in three little parishes in Estonia and document the abandonment of numerous taxa growing outside the sphere of everyday human activities, which signal a deep knowledge of the wild. We also found that traditional medicinal plant foraging did not endanger local plant communities, as the majority of used plants were either very common and not endangered or cultivated, and therefore conservation should be more concerned with the reasons for the disappearance of common plants as the result of the decrease of human activities in rural areas.

We can conclude that archival data has great potential for revealing comparative data for current field studies and for understanding the historical context of medicinal plant use. However, when working with archives, the research methodology has to be carefully selected and adapted to the specific collection, while the results may not be exhaustive from the point of view of the identification of plants and diseases.

Supplementary Materials: The following supporting information can be downloaded at: https://www.mdpi.com/article/10.3390/plants11202698/s1, Table S1: Systematized texts.

Author Contributions: Conceptualization, R.S. and R.K.; methodology, R.S. and R.K.; formal analysis, R.S.; resources, R.S.; data curation, R.S. and R.K.; writing—original draft preparation, R.S.; writing—review and editing, R.S. and R.K.; visualization, R.S. All authors have read and agreed to the published version of the manuscript.

Funding: This research was supported by the European Research Council (ERC) under the European Union's Horizon 2020 Research and Innovation Programme (grant agreement No. 714874).

Institutional Review Board Statement: Not applicable.

Informed Consent Statement: Not applicable.

Data Availability Statement: Data are available in the Supplementary file.

Conflicts of Interest: The authors declare no conflict of interest. The funders had no role in the design of the study; in the collection, analyses or interpretation of data; in the writing of the manuscript; or in the decision to publish the results.

References

1. Quave, C.L.; Pieroni, A. A reservoir of ethnobotanical knowledge informs resilient food security and health strategies in the Balkans. *Nat. Plants* **2015**, *1*, 14021. [CrossRef] [PubMed]
2. Silva, T.C.D.; Medeiros, P.M.; Balcázar, A.L.; Sousa, T.A.D.A.; Pirondo, A.; Medeiros, M.F.T. Historical ethnobotany: An overview of selected studies. *Ethnobio. Conserv.* **2014**, *3*, 1–12. [CrossRef]
3. Łuczaj, Ł. Dziko rosnące rośliny jadalne w ankiecie Józefa Rostafińskiego z roku 1883. *Wiad. Bot.* **2008**, *52*, 39–50.
4. Łuczaj, Ł. Changes in the utilization of wild green vegetables in Poland since the 19th century: A comparison of four ethnobotanical surveys. *J. Ethnopharmacol.* **2010**, *128*, 395–404. [CrossRef]
5. Kalle, R.; Pieroni, A.; Svanberg, I.; Sõukand, R. Early Citizen Science Action in Ethnobotany: The Case of the Folk Medicine Collection of Dr. Mihkel Ostrov in the Territory of Present-Day Estonia, 1891–1893. *Plants* **2022**, *11*, 274. [CrossRef]
6. Kalle, R.; Sõukand, R. The name to remember: Flexibility and contextuality of preliterate folk plant categorization from the 1830s, in Pernau, Livonia, historical region on the eastern coast of the Baltic Sea. *J. Ethnopharmacol.* **2021**, *264*, 113254. [CrossRef]
7. Ningthoujam, S.S.; Talukdar, A.D.; Potsangbam, K.S.; Choudhury, M.D. Challenges in developing medicinal plant databases for sharing ethnopharmacological knowledge. *J. Ethnopharmacol.* **2012**, *141*, 9–32. [CrossRef]
8. Sõukand, R.; Kalle, R. (Eds.) *HERBA: Historistlik Eesti Rahvameditsiini Botaaniline Andmebaas*; EKM Teaduskirjastus: Tartu, Estonia, 2008. Available online: http://herba.folklore.ee (accessed on 20 August 2022).
9. Koay, A.; Shannon, F.; Sasse, A.; Heinrich, M.; Sheridan, H. Exploring the Irish National Folklore Ethnography Database (Dúchas) for Open Data Research on Traditional Medicine Use in Post-Famine Ireland: An Early Example of Citizen Science. *Front. Pharmacol.* **2020**, *11*, 584595. [CrossRef]
10. Shannon, F.; Sasse, A.; Sheridan, H.; Heinrich, M. Are identities oral? Understanding ethnobotanical knowledge after Irish independence (1937–1939). *J. Ethnobiol. Ethnomed.* **2017**, *13*, 65. [CrossRef]
11. Sõukand, R.; Kalle, R. Personal and shared: The reach of different herbal landscapes. *Est. J. Ecolog.* **2012**, *61*, 20–36. [CrossRef]
12. Pranskuniene, Z.; Bernatoniene, J.; Simaitiene, Z.; Pranskunas, A.; Mekas, T. Ethnomedicinal uses of honeybee products in lithuania: The first analysis of archival sources. *Evid.-based Complement. Altern. Med.* **2016**, *9272635*, 1–7. [CrossRef] [PubMed]
13. Sak, K.; Jürisoo, K.; Raal, A. Estonian folk traditional experiences on natural anticancer remedies: From past to the future. *Pharma. Biol.* **2014**, *52*, 855–866. [CrossRef] [PubMed]
14. Sile, I.; Romane, E.; Reinsone, S.; Maurina, B.; Tirzite, D.; Dambrova, M. Medicinal plants and their uses recorded in the Archives of Latvian Folklore from the 19th century. *J. Ethnopharmacol.* **2020**, *249*, 112378. [CrossRef] [PubMed]
15. Sile, I.; Romane, E.; Reinsone, S.; Maurina, B.; Tirzite, D.; Dambrova, M. Data on medicinal plants in the records of Latvian folk medicine from the 19th century. *Data Brief* **2020**, *28*, 105024. [CrossRef]
16. Assmuth, L. Intertwining identities: The politics of language and nationality in the Estonian-Russian borderlands. In *Crossings and Crosses: Borders, Educations, and Religions in Northern Europe*; Berglund, J., Lundén, T., Strandbrink, P., Eds.; De Gruyter: Berlin, Germany, 2015; pp. 29–46.
17. Assmuth, L. Asymmetries of gender and generation in a post-Soviet borderland. In *Border Encounters. Asymmetry and Proximity at Europe's Frontiers*; Bacas, J.L., Kavanagh, W., Eds.; Berghahn Books: New York, NY, USA, 2013; pp. 139–164.
18. Sõukand, R.; Pieroni, A. The importance of a border: Medical, veterinary, and wild food ethnobotany of the Hutsuls living on the Romanian and Ukrainian sides of Bukovina. *J. Ethnopharmacol.* **2016**, *185*, 17–40. [CrossRef]
19. Sõukand, R.; Kalle, R.; Pieroni, A. Homogenisation of Biocultural Diversity: Plant Ethnomedicine and Its Diachronic Change in Setomaa and Võromaa, Estonia, in the Last Century. *Biology* **2022**, *11*, 192. [CrossRef]
20. Kalle, R.; Sõukand, R.; Pieroni, A. Devil Is in the Details: Use of Wild Food Plants in Historical Võromaa and Setomaa, Present-Day Estonia. *Foods* **2020**, *9*, 570. [CrossRef]
21. Lüüs, A. Rahvahaigused ja rahva ravimisviisid Võrumaal 19. sajandi viimasel veerandil. II. *Mana* **1960**, *1*, 34–35.
22. Lüüs, A. Rahvahaigused ja rahva ravimisviisid Võrumaal 19. sajandi viimasel veerandil. I. *Mana* **1959**, *3*, 184–187.

23. Gustavson, H.; Pilk, F.R. Kreutzwaldi arstitegevusse. *Keel Kirjand.* **1982**, *8*, 409–416.

24. Gustavson, H. *Eesti Apteekide Ajaloost*; Eesti Apteekrite Liit: Tallinn, Estonia, 2020; 232p.

25. Anonymous. Maarohtudega haiguse vastu; Taim murrab tõve; Üks tunnike rohukauplejatega Tallinna turul. *Uudisleht* **1936**, *140*, 7. Available online: https://dea.digar.ee/page/uudisleht/1936/09/08/7 (accessed on 5 October 2022).

26. Sõukand, R.; Raal, A. How the name Arnica was borrowed into Estonian. *Trames* **2008**, *12*, 29–39. [CrossRef]

27. International Classification of Primary Care, 2nd Edition (ICPC-2). Updated March 2003 in Family Practice, OUP. Available online: https://www.who.int/standards/classifications/other-classifications/international-classification-of-primary-care (accessed on 20 August 2022).

28. Sõukand, R. "Jooksvarohud" Eesti rahvameditsiinis. *Akadeemia* **2004**, *11*, 2475–2493.

29. Sõukand, R.; Hrynevich, Y.; Prakofjewa, J.; Valodzina, T.; Vasilyeva, I.; Paciupa, J.; Shrubok, A.; Hlushko, A.; Knureva, Y.; Litvinava, Y.; et al. Use of cultivated plants and non-plant remedies for human and animal home-medication in Liubań district, Belarus. *J. Ethnobio. Ethnomed.* **2017**, *13*, 54. [CrossRef]

30. Belichenko, O.; Kolosova, V.; Kalle, R.; Sõukand, R. Green pharmacy at the tips of your toes: Medicinal plants used by Setos and Russians of Pechorsky District, Pskov Oblast (NW Russia). *J. Ethnobiol. Ethnomed.* **2022**, *18*, 1–38. [CrossRef]

31. Hurt, J. Paar palwid Eesti ärksamaile poegadele ja tütardele. *Olevik* **1888**, *12*, 1.

32. Prūse, B.; Kalle, R.; Buffa, G.; Simanova, A.; Mežaka, I.; Sõukand, R. We need to appreciate common synanthropic plants before they become rare: Case study in Latgale (Latvia). *Ethnobiol. Conserv.* **2021**, *10*, 1–26. [CrossRef]

33. Vilbaste, G. *Eesti Taimenimetused*; Emakeele Selts: Tallinn, Estonia, 1993; pp. 1–706.

34. Plants of the World Online. Facilitated by the Royal Botanic Gardens, Kew, UK. Available online: https://powo.science.kew.org/ (accessed on 20 August 2022).

35. Tutin, T.G.; Burges, N.A.; Chater, A.O.; Edmondson, J.R.; Heywood, V.H.; Moore, D.M.; Valentine, D.H.; Walters, S.M.; Webb, D.A. (Eds.) *Flora Europaea*; University Press: Cambridge, UK, 1964–1980; Volumes 1–5. Available online: http://ww2.bgbm.org/EuroPlusMed/query.asp (accessed on 20 August 2022).

36. Stevens, P.F. Angiosperm Phylogeny Website. 2001 Onwards. Version 14 July 2017. Available online: http://www.mobot.org/MOBOT/research/APweb/ (accessed on 20 August 2022).

37. González-Tejero, M.; Casares-Porcel, M.; Sánchez-Rojas, C.P.; Ramiro-Gutiérrez, J.M.; Molero-Mesa, J.; Pieroni, A.; Giusti, M.E.; Censorii, C.; de Pasquale, C.; Della, D.; et al. Medicinal plants in the Mediterranean area: Synthesis of the results of the project Rubia. *J. Ethnopharmacol.* **2008**, *116*, 341–357. [CrossRef] [PubMed]

38. Mauri, M.; Elli, T.; Caviglia, G.; Uboldi, G.; Azzi, M. RAWGraphs: A Visualisation Platform to Create Open Outputs. In Proceedings of the 12th Biannual Conference on Italian SIGCHI Chapter; Association for Computing Machinery, Cagliari, Italy, 18 September 2017; pp. 1–5.

Article

"I Climbed a Fig Tree, on an Apple Bashing Spree, Only Pears Fell Free": Economic, Symbolic and Intrinsic Values of Plants Occurring in Slovenian Folk Songs Collected by K. Štrekelj (1895–1912)

Živa Fišer

Department of Biodiversity, Faculty of Mathematics, Natural Sciences and Information Technologies, University of Primorska, Glagoljaška 8, SI-6000 Koper, Slovenia; ziva.fiser@upr.si; Tel.: +386-(5)-6117662

Abstract: In this study we examine the occurrence of plants and their symbolic, economic, and intrinsic values in Slovenian folk songs. We have analyzed songs published by the ethnologist Karel Štrekelj between 1895 and 1912. Of the 8686 songs studied, plants occur in 1246 (14%) of them. A total of 93 plant taxa were found, belonging to 48 plant families. Grapevine is the most frequently mentioned species, followed by rosemary, wheat, carnation, and lily. About half of the taxa belong to cultivated plants (52%), followed by wild plants (42%). Exotic plants (i.e., not growing in the area) are mentioned only occasionally (6%). Half of all citations (49.3%) refer to the symbolic values, such as religion, love, death, economic status, or human qualities. More than a third of the citations (36.7%) are associated with plant's usefulness, especially consumption, while only a small percentage of citations (14.0%), relate to environmental representation. Several verses show how our appreciation of some plants, especially those used as food, has changed over the centuries. Folk songs have turned out to be interesting sources of information, and although they cannot be fully trusted as historical documents, they can still be used as sources for understanding the relationship between people and plants.

Keywords: ethnobotany; folk poetry; plant symbolism; ritual plants; useful plants

Citation: Fišer, Ž. "I Climbed a Fig Tree, on an Apple Bashing Spree, Only Pears Fell Free": Economic, Symbolic and Intrinsic Values of Plants Occurring in Slovenian Folk Songs Collected by K. Štrekelj (1895–1912). *Plants* **2022**, *11*, 458. https://doi.org/10.3390/plants11030458

Academic Editors: Renata Sõukand and Raivo Kalle

Received: 17 January 2022
Accepted: 3 February 2022
Published: 7 February 2022

Publisher's Note: MDPI stays neutral with regard to jurisdictional claims in published maps and institutional affiliations.

1. Introduction

Traditional knowledge about plants, including their economic and symbolic values, is passed down from generation to generation in many different ways, such as by being written and via the oral tradition. Literary texts and poems across the globe, from the earliest times to the present, abound with plant references [1–3] and emphasize the past and present importance of plants in daily life. Although these texts cannot be fully trusted as historical documents [4], they can still be used as sources for understanding the relationship between humans and plants. Moreover, comparing plant occurrences in historical and contemporary literary texts can also reveal changes in plant symbolism and use over time and in different places.

Similarly to other literary documents, traditional folk songs also offer an interesting insight into the relationship between humans and plants. Traditional folk songs are typically studied within the domain of ethnomusicology, which focuses on how songs are performed and transmitted and the role of music in building cultural identities [5], with an emphasis on indigenous peoples and local communities. However, the study of songs from an ethnobiological perspective offers a different perspective on the same topic, focusing on the relationship between humans and nature. As several works have shown, traditional songs can serve as repositories of indigenous peoples' ethnobiological knowledge [6–8], management practices [9], linguistic expressions (e.g., archaic words; [10]), and many other cultural values.

We can speculate that folk songs contain the most important, or in some cases the most widespread, plant species in local communities that have acquired symbolic, economic, and ecological values over time. The fact that folk songs and tales were passed down orally from generation to generation allowed them to change and adapt to local characteristics over time and in different regions. In regions where different ethnic groups, cultures, or religious beliefs have met, this has led to the blending of beliefs and practices.

Since their settlement at the present site in the 8th century AD, the Slovenes (then called Carantanians) lived in an area with a vibrant political and cultural history, in a region often referred to as "where the Orient meets the West", both culturally and geographically [11]. The invaders brought with them plant and animal species, traditions, knowledge, and religious beliefs that merged with the existing cultures, forming a pagan-Christian mixture that can still be observed today in some festivals, such as the Slavic mythological figure of "Zeleni Jurij" or Green George, who is celebrated on St. George's Day in April.

In this work, we have analyzed Slovenian folk songs in order to understand how important plants were in the daily life of the predominantly peasant population in past centuries. In addition to representing a heritage on its own, folk songs may contain many more pieces of hidden information. They may reveal some long-forgotten practices in plant uses; they may also, for example, reveal what was the local people's attitude towards native plants that gradually became the basis of plant conservation.

We hypothesize that the common native plant species and the species cultivated for consumption will represent the majority of the plant references. We expect the national flower and tree—carnation and lime—to be among the most frequently mentioned taxa in folk songs.

2. Results and Discussion

Plants are common in Slovenian folk songs, which underlines their importance for the local population. Out of 8686 songs, we found 1854 plant references in 1246 songs, which corresponds to 14% of all songs. A total of 97 plant taxa were recorded: 93 taxa belonging to 48 families of seed plants and three taxa of ferns and mosses. 77 taxa were identified to species level. The list of taxa and families (or higher taxonomic categories) can be found in Table 1 and in Supplementary Materials (Supplementary Material S1 and Table S1), where all categories assigned to species or higher taxa are also listed.

A similar study was conducted by Herrero and Cardaño [8] who studied folk songs from Castile and León (Spain). The proportion of songs with plant references in Slovenian folk songs (14%) is statistically lower than that found by [8] in songs from Castile and León (18%; $p < 0.0001$), as is the number of plant taxa found in songs (96 vs. 150 for Slovenia and Castile and León, respectively; $p < 0.0001$). However, such comparisons are difficult to comment on, as they depend strongly on the amount of material studied, the size of the area studied, and its species diversity. The number of Slovenian songs analyzed was higher than that of Castile and León (8686 and 7120, respectively), but on the other hand, the surface and plant diversity of Castile and León are higher. The area of Castile and León is almost five times larger than the area of present-day Slovenia, and with about 4000 flowering plants, Castile and León has a greater diversity of plant species than Slovenia. Moreover, the Spanish Empire's trade with America [12] and other continents led to the importation of many useful and ornamental plants, which are also mentioned in the analyzed Spanish folk songs (peanuts, cocoa, cinnamon, cloves, tea, etc. [8,13]), while they are completely absent from Slovenian folk songs. However, the percentage of cultivated, wild and exotic (not growing in the area studied) taxa is surprisingly similar.

The average number of plant taxa cited per song is 1.5, which is significantly lower than in songs of Castile and León (1.7; $p < 0.0001$). The highest number of plant citations in Slovenian songs was 10, while for Castile and León it was 17 [13]. However, the most frequently cited plant taxa are surprisingly similar: rose, grapevine, carnation, wheat (*Triticum aestivum* L.), olive (*Olea europaea* L.), and laurel (*Laurus nobilis* L.) for Castile and León and grapevine, rosemary (*Rosmarinus officinalis*), wheat, carnation, lily (*Lilium candidum* L.), and lime for Slovenia. Most of these taxa are cultivated and are important features in many European cultures, especially in southern Europe. The importance of the rose is probably related to the same religion that Slovenia and Spain share, as the species is a very important symbol in Christianity [14,15]. Grapevine and wheat are also among the most important crops in Europe, so it is not surprising that they play such an important role in songs. However, a closer comparison of songs from Slovenia and Castile and León shows that Spanish songs contain many more species adapted to milder conditions, such as the Mediterranean olive or laurel, or the orange (*Citrus sinensis* (L.) Osbeck) and lemon (*Citrus limon* (L.) Osbeck), while Slovenian songs contain more temperate plants (e.g., *Fagus sylvatica, Picea abies, Abies alba)*. The higher number of exotic taxa in Spanish songs can be explained by the more intense trade with other continents in the past. Of the exotic plants in songs from Castile and León, only coffee (*Coffea arabica* L.), black pepper (*Piper nigrum* L.), frankincense (*Boswellia sacra* Flueck.), and myrrh (*Commiphora myrrha* (Nees) Engl.) also appear in Slovenian folk songs, while tamarind (*Tamarindus indica*), cinnamon (*Cinnamomum cassia* Siebold), cloves (*Syzygium aromaticum* (L.) Merr. and L.M. Perry), palms (fam. Arecaceae), bamboo (*Phyllostachys* spp.), cocoa (*Theobroma cacao* L.), and tea (*Camellia sinensis* (L.) Kuntze) are never mentioned.

Table 1. List of plants mentioned in the analyzed Slovenian folk songs.

Family Name	Scientific Name	English Name	Slovenian Name	Citations	Presence
Adoxaceae	*Sambucus nigra* L.	elder	bezeg	3	wild
Amaryllidaceae	*Allium ampeloprasum* L.	leek	por	1	cultivated
	Allium cepa L.	onion	čebula	4	cultivated
Apiaceae	*Daucus carota* L.	carrot	korenje	22	cultivated
	Foeniculum vulgare Mill.	fennel	koromač	1	cultivated *
Apocynaceae	*Vinca minor* L.	lesser periwinkle	zimzelen	1	wild
Asparagaceae	*Convallaria majalis* L.	lily-of-the-valley	šmarnica	2	wild
Berberidaceae	*Berberis vulgaris* L.	barberry	češmin	1	wild
	Alnus spp.	alder	jelša	6	wild
Betulaceae	*Betula pendula* Roth.	birch	breza	11	wild
	Carpinus spp.	hornbeam	gaber	2	wild
	Corylus avellana L.	hazel	leska	22	wild
	Brassica oleracea var. *capitata* L.	cabbage	zelje	20	cultivated
Brassicaceae	*Brassica rapa* var. *rapa* L.	turnip	repa	114	cultivated
	Erysimum cheiri (L.) Crantz	wallflower	zlati šebenik	1	cultivated
	Raphanus sativus L.	radish	redkev	3	cultivated
Burseraceae	*Boswellia sacra* Flueck.	frankincense	bozvelija	8	absent
	Commiphora myrrha (Nees) Engl.	myrrh	mira	9	absent
Buxaceae	*Buxus sempervirens* L.	box	pušpan	6	cultivated
Cannabaceae	*Cannabis sativa* L.	hemp	konoplja	17	cultivated
Caryophyllaceae	*Dianthus caryophyllus* L.	nagelj	carnation	119	cultivated
	Artemisia spp.	wormwood	pelin	11	wild **
	Helianthus tuberosus L. [2]	Jerusalem artichoke	topinambur	1	cultivated
Compositae	*Helichrysum* spp.	immortelle	smilj	2	cultivated
	Lactuca sativa L.	lettuce	vrtna solata	12	cultivated
	Santolina chamaecyparissus L.	cotton lavender	nemški rožmarin	30	cultivated
	Taraxacum officinale agg.	dandelion	regrat	1	wild
Cornaceae	*Cornus* spp.	dogwood	dren	4	wild
Cupressaceae	*Juniperus communis* L.	juniper	brin	7	wild
Ericaceae	*Erica carnea* L.[1]	heath	spomladanska resa	3	wild
Fagaceae	*Fagus sylvatica* L.	beech	bukev	35	wild
	Quercus spp.	oak	hrast	21	wild
Gentianaceae	*Gentiana verna* L.	spring gentian	spomladanski svišč	1	wild
Geraniaceae	*Pelargonium radens* H.E.Moore	rasp-leaf pelargonium	roženkravt	10	cultivated
Juglandaceae	*Juglans regia* L.	walnut	oreh	13	cultivated
	Ocimum basilicum L.	basil	bazilika	6	cultivated
Lamiaceae	*Origanum majorana* L.	marjoram	majaron	25	cultivated
	Rosmarinus officinalis L.	rosemary	rožmarin	162	cultivated
	Salvia officinalis L.	sage	žajbelj	1	cultivated
Lauraceae	*Laurus nobilis* L.	laurel	lovor	5	cultivated

Table 1. *Cont.*

Family Name	Scientific Name	English Name	Slovenian Name	Citations	Presence
	Ceratonia siliqua L.	carob	rožičevec	1	cultivated
	Lens culinaris Medikus	lentil	leča	8	cultivated
Leguminosae	*Phaseolus vulgaris* L.	common bean	fižol	12	cultivated
	Pisum sativum L.	pea	grah	4	cultivated
	Trifolium spp.	clover	detelja	25	wild
	Vicia faba L.	broad bean	bob	5	cultivated
Liliaceae	*Lilium candidum* L.	Madonna lily	lilija	103	cultivated
Linaceae	*Linum usitatissimum* L.	flax	lan	3	cultivated *
Malvaceae	*Tilia* spp.	lime	lipa	97	wild
Moraceae	*Ficus carica* L.	common fig	figa	4	cultivated *
	Morus spp.	mulberry	murva	2	cultivated
Oleaceae	*Fraxinus excelsior* L.	ash	jesen	2	wild
	Olea europaea L.	olive	oljka	2	cultivated
Papaveraceae	*Papaver rhoeas* L.,	common poppy	poljski mak	1	wild
	P. somniferum L.	opium poppy	vrtni mak	8	cultivated
	Papaver spp.	poppy	mak	6	cultivated, wild
	Abies alba Mill.	silver fir	jelka	4	wild
Pinaceae	*Picea abies* (L.) H. Karst.	Norway spruce	smreka	22	wild
	Pinus spp.	pine	bor	22	wild
Piperaceae	*Piper nigrum* L.	pepper	poper	1	absent
	Avena sativa L.	oat	oves	18	cultivated
	Hordeum vulgare L.	barley	ječmen	10	cultivated
Poaceae	*Panicum miliaceum* L.	proso millet	proso	11	cultivated
	Secale cereale L.	rye	rž	6	cultivated
	unclassified Poaceae			50	cultivated
	Zea mays L.	corn	koruza	3	cultivated
	Triticum aestivum L. em. Fiori and Paol.	wheat	pšenica	151	cultivated
Polygonaceae	*Fagopyrum esculentum* Moench	buckwheat	ajda	23	cultivated
	Rumex obtusifolius L.	bitter dock	topolistna kislica	1	wild
Ranunculaceae	*Clematis vitalba* L.	old man's beard	navadni srobot	1	wild
	Fragaria spp.	strawberry	jagoda	5	wild
	Malus domestica Borkh.	apple	jablana	72	cultivated
	Prunus avium L.	sweet cherry	češnja	11	cultivated
	Prunus cerasus L.	sour cherry	višnja	3	cultivated
Rosaceae	*Prunus domestica* L.	common plum	sliva	7	cultivated
	Prunus persica (L.) Batsch	peach	breskev	1	cultivated
	Prunus spinosa L.	blackthorn	črni trn	7	wild
	Pyrus communis L.	pear	hruška	29	cultivated
	Rosa spp.	rose	vrtnica	97	cultivated
	Rubus spp.	brambleberry	robida	3	wild
Rubiaceae	*Coffea arabica* L.	coffee	kava	13	absent
Rutaceae	*Citrus sinensis* (L.) Osbeck	sweet orange	pomaranča	6	absent
	Populus tremula L.	aspen	trepetlika	1	wild
Salicaceae	*Populus* spp.	poplar	topol	1	
	Salix spp.	willow	vrba	9	wild
Sapindaceae	*Acer campestre* L.	field maple	maklen	1	wild
	Acer spp.	maple	javor	34	wild
Solanaceae	*Nicotiana tabacum* L.	tobacco	tobak	30	cultivated
	Solanum tuberosum L.	potato	krompir	14	cultivated
Ulmaceae	*Ulmus* spp.	elm	brest	1	wild
Urticaceae	*Urtica dioica* L.	stinging nettle	kopriva	15	wild
Violaceae	*Viola* spp.	violet	vijolica	15	wild
Vitaceae	*Vitis vinifera* L.	grapevine	vinska trta	214	cultivated
Zingiberaceae	*Zingiber officinale* Roscoe	ginger	ingver	1	absent
	Unidentified seed plant			3	
	Bryophyta	moss	mah	4	wild
	Sphagnum spp.	peat moss	šota	1	wild
	Pteridophyta	fern	praprot	30	wild

* also wild, ** also cultivated. [1] could refer also (but unlikely) to *Calluna vulgaris* (L.) Hull. [2] Could refer also to *Solanum tuberosum* L.

Rosaceae is the plant family with the highest number of taxa occurring in Slovenian folk songs (10 taxa), followed by Poaceae, Leguminosae, and Compositae, all represented by 6 taxa (Table 1). Rosaceae, Poaceae, and Leguminosae are among the most economically important plant families [16], with many edible representatives. Most taxa from these families mentioned in folk songs are used for consumption, but have also acquired symbolic value because of their importance (e.g., apple, rose, wheat). Four taxa were found in Lamiaceae, Betulaceae, and Brassicaceae. Other plant families are represented by three or fewer taxa. Ferns and mosses are also mentioned, but only with the common name "praprot" (Slovenian for fern) and "mah" (moss). The only exception, where a higher taxonomic category is mentioned, is peat (*Sphagnum* sp.), which is mentioned under the Slovenian name "šota".

Grapevine is by far the most common plant species in Slovenian folk songs, appearing in 214 songs and accounting for 11.5% of all plant citations (Figure 1). This is not surprising, as viticulture and wine production have existed in this region for a long time and are of great economic and cultural importance to the Slovenian population. This led to an abundance of folk songs dedicated to wine. In the analyzed four-volume edition, an entire book section entitled Drinking Songs is dedicated to wine, containing 651 songs (Š5468–Š6119). Even the Slovenian national anthem by the poet France Prešeren, entitled "A Drinking Song" (in Slovenian "Zdravljica"), is dedicated to wine. Grapevine is followed by rosemary (162 songs, 8.7%), wheat (151 songs; 8.1%), carnation (119 songs; 6.4%), lily (103 songs; 5.6%), lime (97 songs; 5.2%), and rose (97 songs; 5.2%). These taxa account for over 50% of all plant references. Interestingly, rosemary, lilies and carnations are native to more southern parts of the Mediterranean and occur on Slovenian territory as a cultivated species. However, they have gained popularity throughout Europe and have become an important part of several European cultures. The carnation has even become the national flower of Slovenia.

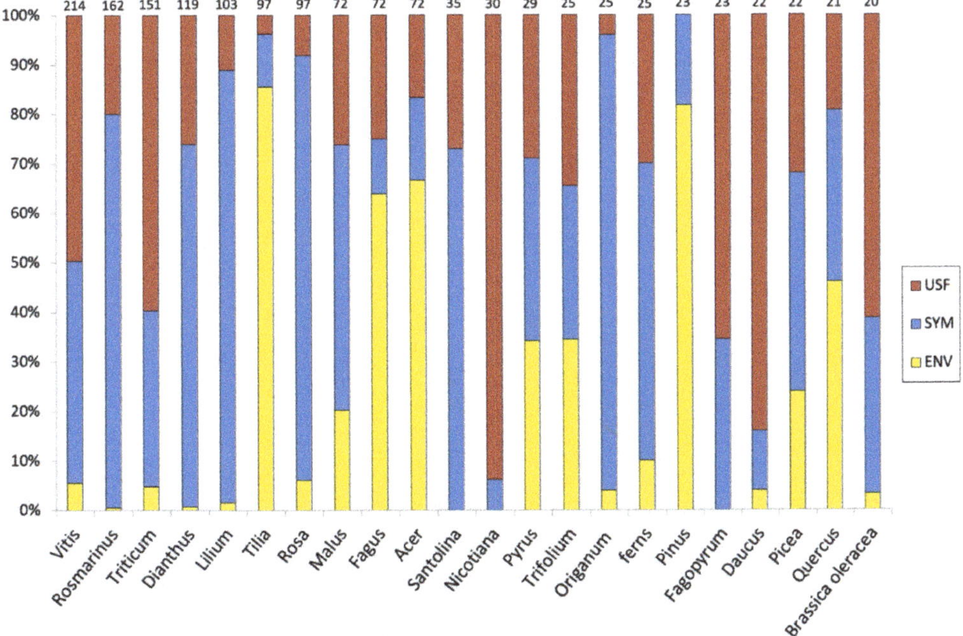

Figure 1. Plants occurring in Slovenian folk songs with 20 or more citations. The color of the bars represents the percentage of citations according to their value: USF—useful plant; SYM—plants with a symbolic value; ENV—plants as features of the environment. The number of songs containing the referenced taxon is written above each bar.

Surprisingly, some very common annuals and herbaceous perennials are rarely or never mentioned in folk songs, despite their decorative and useful qualities. One of the most common and conspicuous field flowers is the ox-eye daisy (*Leucanthemum vulgare*

(Vaill.) Lam.), while its smaller relative, the common daisy (*Bellis perennis* L.), is often found in lawns. Neither of these species are mentioned in the folk songs analyzed. The songs often mention white, yellow, red, and blue flowers without giving a specific name. For example, red flowers could be attributed to various Fabaceae, Compositae, Dipsacaceae, and Orchidaceae species; yellow ones to the species of Ranunculus, dandelions, or other representatives of the Compositae; blue ones to violets (*Viola* spp.), meadow sage (*Salvia pratensis*), forget-me-not (*Myosotis* spp.), or various Campanulaceae; and white ones to yarrow (*Achillea millefolium*), lily-of-the valley (*Convallaria majalis*), or various species of the Apiaceae family. Early spring flowering plants that indicate the end of winter, such as snowdrops (*Galanthus nivalis* L.), primrose (*Primula vulgaris*), spring crocus (*Crocus vernus* (L.) Hill), or black hellebore (*Helleborus niger* L.), are also surprisingly not represented in the songs. Cardaño and Herrero [13] also found that some common and conspicuous species (e.g., *Cistus* spp.) are rarely mentioned in songs, while those of greater importance to the habits and customs of the region (e.g., orange and lemon) are mentioned in greater numbers.

Finally, Slovenian folk songs rarely mention medicinal, poisonous, or edible wild plants. This was unexpected, as traditional medicine and wild plant gathering were widespread in the area [15]. Nowadays, several wild plants are widely collected as spices. Among the most commonly collected are the leaves of dandelions and wild garlic (*Allium ursinum* L.), the shoots of wild asparagus (*Asparagus acutifolius* L.) and black bryony (*Tamus communis* L.), the tops of Norway spruce, and the inflorescences of elder (*Sambucus nigra* L.), to name a few [17]. None of these species are mentioned in songs, at least not as food or condiments.

Over half of the mentioned taxa (52%) belong to cultivated plants (Figure 2); most of them are cultivated for consumption (grains, beans, vegetables and fruit trees, including grapevine) or for ornamental purposes (e.g., carnation, rose, lily). The remaining taxa are represented by plants that grow in the wild (42%) and to exotic species that do not grow in the area (6%). Most of the cited wild species are trees or shrubs; among them, lime, beech, and maple are the most common. Only a few native species are annuals or herbaceous perennials, and some of them are cited only once or a few times: dandelion, spring gentian (*Gentiana verna* L.), lesser periwinkle (*Vinca minor* L.), or sorrel (*Rumex obtusifolius* L.). Clover (probably *Trifolium repens* L. and/or *T. pratense* L.) is among the most often mentioned native herbaceous plants.

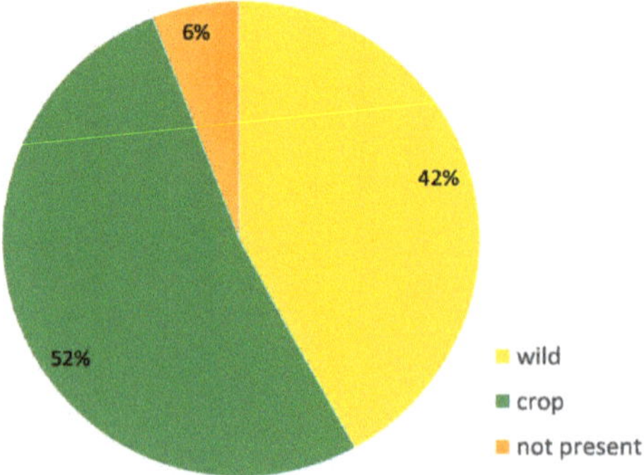

Figure 2. Percentage of wild, cultivated (crop), or exotic plant taxa cited in the analyzed folk songs.

Lastly, 6% of references belong to exotic species that do not grow in the region but are imported, or they are mentioned symbolically. Among the exotic species are the carob (*Ceratonia siliqua* L.), orange, coffee, black pepper, and ginger (*Zingiber officinale* Roscoe). The carob tree is a legume originating in the Mediterranean region and is nowadays planted elsewhere (also in Slovenia) for ornamental purposes. It is possible that some trees were also cultivated in the past, most probably in the coastal region with a milder climate. Myrrh and frankincense are mentioned in several religious songs and refer to the gifts offered to the infant Jesus by the Three Wise Men. References to exotic fruits (oranges) might also originate from the fact that in the past many Slovenian women went to work abroad, mostly as wet nurses and governesses to Egypt (the so-called Aleksandrinke): *"The bird sings, the bird sings/In a beautiful green orange tree"* (*"Vtica poje, vtica poje/V lepoj zelenoj naranči"* Š951).

3. Classification of Plants According to Their Values

Although none of the plants were assigned to one category exclusively (Supplementary Material S1, Table S1) many plants are classified predominantly to one: rose, marjoram (*Origanum majorana* L.), and lily are mostly used as symbols, while lime and pine trees are most often cited to depict the environment. Tobacco (*Nicotiana tabaccum* L.), carrot (*Daucus carota* L.), or oats (*Avena sativa* L.) are plants mentioned exclusively for consumption; the first for smoking and chewing, while the latter two as food. Only a few plants (pear (*Pyrus sylverstris* L.), clover, oak (*Quercus* spp.), Norway spruce, and apple (*Malus domestica* Borkh.) are cited almost equally in all three categories.

References with a symbolic meaning are by far the most common ones, representing almost half (49.3%) of all occurrences. Over a third of references (36.7%) are associated with their usefulness, while only a small percentage of plant references (14.0%) refer to the depiction of the environment. The number of different taxa and number of citations within each (sub)category in presented in Figure 3.

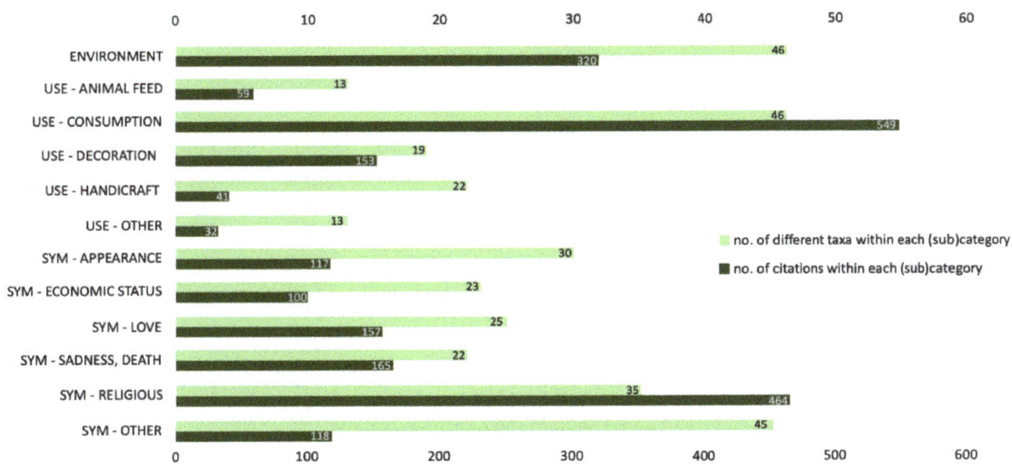

Figure 3. Number of different taxa (light green) and citations (dark green) within each (sub)category.

3.1. Plants with a Symbolic Value

3.1.1. Religious Symbolic Values

In many folk songs with religious content, plants symbolize Mary, Jesus, St. Joseph, or angels: "There is a garden by the road,/In the garden grow some flowers./The first flower is a lily,/Because Mary is merciful/The second is a rose,/Because Mary is gentle./The third is rosemary,/Because Jesus is Mary's son./The fourth is a carnation,/For the angels from heaven./The fifth is marjoram,/Because Jesus is Mary's throne./I will gather them all,/And give them in honor of Mary." (Pri cesti stoji garteljček,/V njem pa rastejo rožice./Ta prva rožca lilija,/Ker je Marija usmiljena./Ta druga roža gartroža,/Ker je MArija cartana./Ta tretka roža rožmarin,/Ker Jezus je Marijin sin. Ta šterta roža nageljnček./Kar so nebeški angeljčki./Ta šeta roža je mar'jon,/Ker Jezus je Marijin tron./Vse te rožce bom vkup pobral,/Mariji v čast jih bom dar'val." Š4919). It is clear from this song that some plant references are used exclusively to form the rhymes (in Slovenian) and, thus, some of these do not have necessarily a true symbolic value. In the traditional songs analyzed, lilies and roses are the flowers most frequently associated with Mary, which is consistent with representations in visual art in Europe since the Middle Ages [14,18–20]. Some other flowers are also used in this context, but less frequently. Among lilies, the species that symbolically represents Mary is usually the white-flowered Lilium candidum, which was also given the name Madonna lily and has been associated with the Annunciation of the Blessed Virgin Mary since the Middle Ages [19,20].

Wheat and grapevine are also very often associated with religion, symbolizing the Eucharist: *"...The first flower/Is the yellow wheat:/At Holy Mass they make/Sacramental bread of it.../The second:/Is the beloved grape:/At Holy Mass they make/The Blood of Christ from it..."* (*"Perva rožica je leta:/Oj rumena všeničica:/Per svetej maši jo nucajo/Za samo sveto hoštijo.../Druha rožca je leta:/Ljuba vinenska tertica:/Per svetej maši jo nucajo/Za samo sveto rešnjo kri..."* Š4927).

Love

In the songs analyzed, love is usually represented by ornamental plants, especially rosemary, carnation, rasp-leaf pelargonium (*Pelargonium radens* H.E. Moore), marjoram, and cotton lavender, while native plants are rarely associated with love. Several of these flowers form the traditional Slovenian love bouquet (in Slovenian, "ljubezenski pušeljc") [21–23]. This bouquet was an important part of daily life in the past. The carnation, always red in color, symbolizes life and love, not only in Slovenian folklore, but also in many other (especially Southern European) cultures [8,24]. Rosemary, which retains its fragrance both fresh and dried, represents faith, while the fragrant green rasp-leaf pelargonium represents hope. Therefore, the values represented in this bouquet are love, faith, and hope. Girls gave their boyfriends bouquets of carnations and rosemary (and sometimes other flowers) to symbolize their love and devotion.

Sadness and Death

Among the plants that symbolize negative feelings, death or sadness, or even criminal acts, many are not particularly attractive, such as clover: *"There grows green clover/in a green field./In the morning the farmer comes/and mows the clover.../Oh, sinner,/the same will happen to you.../In the evening you will lie down in your bed/all healthy and strong.../but then death will come to you/and knock you down..."* (*"Raste, raste detela,/Na zelenem travniku./Zautra kosec pride,/Jo doli pokosi.../Ravno, ravno tako/Boš grešnik ti.../Vzvečer doli ležeš,/Si frišek no si zdrav.../Kda ti smert do tebe pride,/Te doli pokosi..."* Š6120), or *"My mother asked me/"Where did you find this baby?"/"There, in those green ferns"* (*"Nas so mati prašali/" Kje ste to dete najdeli?"/"Tam le v zeleni praproti."* Š2278)), or have undesirable characteristics, such as the rush-producing nettle (*Urtica dioica* L.): *"She had three sons:/The first she threw into the sea/The second into the nettle..."*; (*"Sej je imela že sinke tri:/Enga je bla vrgla v morje,/Drugega je bla v koprivje..."* Š171), the thorny blackthorn (*Prunus spinosa* L.), or the bitter wormwood (*Artemisia* spp.). Many of these plants, for example box (*Buxus sempervirens* L.) and Norway spruce, are evergreen. De Cleene and Lejeune [14] mention that in German speaking countries, several blue flowering spring flowers, such as violets, pyramidal bugle (*Ajuga pyramidalis*), bluebell (*Hyacinthoides non-scripta*), and the spring gentian (*Gentiana verna*), were believed to bring bad luck. In some European countries (e.g., Italy, Netherlands), periwinkle (*Vinca minor*) was used in children's funeral wreaths or planted on graves (especially those of children), thus the Italian name Fiore di Morte (death's flower) [15]. Indeed, the lesser periwinkle and spring gentian are associated with death also in some songs: *"Spring gentians faded,/they were dry and faded./Mila fell asleep among them;/never again did she wake up."* (*"Spanjšice ble ocvetele,/Ble so suhe že in vele./Mina je bla v njih zaspala,/Nikdar več ni z spanjšic vstala."* Š234). Sadness or death is often represented by withering or dried flowers or plants (marjoram, spring gentian, rosemary).

Even the traditional bouquets given by girls to boys were not always associated with positive feelings. When boys went to war or when they died, girls made green bouquets that represented sadness or death: *"Last year you received/A bouquet of flowers/This year I will give you/A bush of nettle..."* (*"Vlan' si še pušelc/Iz rožic dobil,/Pa letas germušelc/Ti dam iz kopriv"* Š4461). Rosemary and cotton lavender are usually mentioned as components of green bouquets. Carnations, roses, and lilies were also often associated with death, as they were among the few cultivated ornamental plants planted on graves.

The references to ferns are interesting and show the importance of ferns both as useful and symbolic plants. Ferns grow wild in nature and their symbolic meaning relates to nature. Since dry fern leaves were used by farmers in the past to stuff mattresses, they are usually mentioned in connection with poverty and/or sleep. Some poems mention that women gave birth in the ferns and left their unwanted children there. In other poems, women who could not have children found newborns in ferns and took them home. The roots of the worm fern (*Dryopteris filix-mas* (L.) Schott) were once used to cure intestinal worms [25], but also to induce abortions [26], which might explain the connection between ferns and infanticide. On a more positive note, ferns are also mentioned as the place where lovers secretly meet and make love.

Human Appearance and Characteristics

The specific properties of plants are often compared to humans, both with a positive and negative connotation. Female beauty was for example described through the comparison with red roses, white lilies, white buckwheat flowers, white and red poppies, and black berries of blackthorn: *"She was slim as hemp,/Red as a rose,/White as a poppy flower/God himself brought her to this world!"* (*"Je b'la tenka ko konoplja,/Rudeča kakor gartroža,/Bela kakor makov cvet,/Sam Bog te zvolil na ta svet!"* Š1036). Even male beauty was sometimes compared to flowers: *"My loved one is beautiful as carnation flower"* (*"Moj ljubi je lep/ko fajdelnov cvet"* Š2900) or *"My boyfriend is handsome as laurel flower"* (*"Moj šocel je lep/Kakor lomberjev cvet"* Š2907). However, in the latter case the beauty probably does not really refer to laurel flowers as they are rather inconspicuous, but flowers serve as symbol of the plant's usefulness. Mosses, ferns, and other unattractive green plants or unappreciated vegetables are used to emphasize ugliness or stupidity: "stupid as a cabbage stem" (*"Per deli je terda/Kak zeljov kocen"* Š2926); *"Her face is green as moss/.../Everyone is scared of her"* (*"Zelena ku mah,/.../ Pa gleda ku sova,/Da je vsakig strah"* Š2887); or "If you want to be mine,/You have to buy some color,/So you won't be as green,/As sorrel and absinthe..." (*"Če češ moj bit',/Moraš barvo kupit',/Da ne boš tko zelen,/Kot ta ščavlja in pelen."* Š2654). However, the green and not particularly attractive hemp (*Cannabis sativa* L.) is usually associated with positive qualities, probably due to its many useful properties.

Economic Status

Several plants are used as symbols of economic status. Although some crops were regarded as "poor people's food", others symbolize richness or wealth. Potatoes, cabbage, beans, and turnips symbolize poverty: adjectives such as black (black potatoes-probably meaning burned potatoes), small (small beans) or stinky (stinky cabbage) are often associated with these foods: *"On Sunday there is nothing else (to eat)/Than a piece of meat/And some stinky cabbage..."* (*"V nedeljo ni druzga/Kot košček mesa/In zraven mav zelja/Usmrajenega."* Š7399). Among the grains, wheat was the most appreciated one, while rye, barley and buckwheat are regarded as food of the poor or as animal feed: *"The pilgrims are gathering/To visit the holy mother by the lake./They are preparing food for the travel:/Wheat for the rich ones,/Barley for the poor ones..."* (*"Oj romarji se zbirajo/K mater božji na jezero./Oj za brašnjo napravljajo:/Ti bogati za pšenično,/Ti ubogi za ječmenovo."* Š292). The low appreciation for barley is evident from several songs referring to eating barley porridge. Porridge was a common prison food and the saying "eating porridge" refers to being in prison [27].

3.1.2. Useful Plants

Plants Used for Human Consumption

Grapevine is by far the most frequently referenced useful plant, used exclusively for consumption. Among the other fruits, the dominating species are temperate trees which have been cultivated in the region for centuries: apples, pears, hazelnuts, cherries, walnuts, plums, figs, sour cherries, olives, and peaches. Oranges are the only fruits that are not commonly grown in the region due to the mostly unsuitable climate. Nowadays, they are occasionally cultivated in the sub-Mediterranean part of Slovenia, but folk songs indicate that they were occasionally also grown in the past outside the warmer sub-Mediterranean: *"Grow, grow, orange tree,/Orange tree, noble tree..."* (*"Rasti, rasti pomoranča,/Pomoranča, žlahtno drev'!"* Š923).

Vegetables are mentioned less frequently than fruits. This is probably because growing vegetables was not very common before the 18th century; pumpkins, cucumbers, melons, turnip, carrots, radishes, onions, and garlic are among the few vegetables grown until the end of the 17th century [28]. Vegetables also often have a negative connotation, compared to fruits or grains, and represent a symbol of poverty or contempt for foreigners, e.g., *"...You will eat white turnip,/And sit hungry by the stove . . . "* (*"Belo repo bodeš jedla,/Lačna pol' se k peči vsedla."* Š8351) or *"Vlachs-farting peas, shitting lentils!"* (*"Vlah/Prdi grah,/Seri lečo!"* Š7718). Among the vegetables present in the region since the settlement of the Slavs (between the 6th and 8th century), only cabbage, turnip, and lettuce are mentioned in more than 10 songs, while lentils, faba beans, green peas, onions, radishes, and leeks are mentioned less frequently. Introduced vegetables and grains are mentioned only rarely (except potatoes and beans, which both appear in over ten songs), but are usually despised: e.g., *"On Friday we eat/Those damn beans:/When you put them in mouth/You almost vomit./On Saturday there is nothing else (to eat)/Than black potatoes,/How can I eat them/They smell like pitch..."* (*"V petek je tisti/Prokleti fižol:/V usta ga deneš,/Bi kmalu kozlal. / V soboto ni druzga,/Kot črni krompir;/Kako ga bom jedel,/K smrdi kakor šmir!"* Š7402). None of the analyzed songs mention the now widespread bell peppers or tomatoes. Although they were introduced by the 19th century, they became more popular only in the beginning of the 20th century, after the analyzed songs had been recorded. In the last 100 years, those two vegetables became common and, together with some other new world vegetables, such as zucchini or eggplants, gained their place in traditional Slovenian cuisine.

Cultivated grains represent one of the most important food sources and the staple food of this region. Many cultivated grains are mentioned in folk songs, stressing their importance for the local population. Wheat is the most commonly cultivated and also the most appreciated grain species, which also resulted in the plentiful citations in folk songs. It was present in the region for at least 5000 years and Medieval times, taxes were paid in wheat [28]. The cheaper version of wheat is barley, one of the oldest and most cultivated crops worldwide. In Slovenia, barley is the main ingredient of the traditional porridge (in Slovenian language "ričet"). In the past, barley was even more common than today, as it was cheaper than wheat and, therefore, popular amongst the poorest: *"The pilgrims are gathering/To visit the holy mother by the lake./They are preparing food for the travel:/Wheat for the rich ones,/Barley for the poor ones..."* (*"Oj romarji se zbirajo/K mater božji na jezero./Oj za brašnjo napravljajo:/Ti bogati za pšenično,/Ti ubogi za ječmenovo."* Š292). Folk songs also mention bread and pastries made from other grains, such as a special flatbread made of common buckwheat ("ajdova prosjača") and bread made of oat or buckwheat. Buckwheat is second to wheat and is followed by oat. Other mentioned grains are proso millet, rye, and maize. Corn, now an important crop in Slovenia, is mentioned in only three songs. Corn was introduced in the 17th century and its cultivation progressed slowly in some regions as it competed with the much more popular buckwheat [11,28] and other cereals. By the end of the 19th century, corn was widely grown and its absence from songs is surprising.

Tobacco was used both for smoking and chewing, although all references except one refer exclusively to smoking: "I went to Celovec/.../I did not eat or drink anything/only smoked and chewed tobacco" (*"V Celovc sem bil/.../Nič nisem jedel, nič ne pil,/.../Le rauhtabak in čiktabak/Je bila moja špiža."* Š7345).

Plants Used as Animal Feed

Fruits and other parts of plants are also mentioned as sources of food for wild and domesticated animals. The latter—horses, pigs, chickens—were fed with wheat, oats, clover, and hazelnuts. As a rather expensive commodity, wheat was used to feed noble horses, but the more common feed for working horses was clover or oats. Pigs were fed with acorns, beechnuts, and apples. Proso millet, but also grains of wheat, were used to attract wild birds. Mice are mentioned in a few songs, where they feed on proso millet, while in one song a bear is eating oats and lentils.

Decorative Plants

Among the decorative plants, most citations refer to the previously mentioned bouquets (see the section Plants with Symbolic Values: Love and Sadness and Death). Bouquets have been used throughout Europe for several purposes and on different occasions; the composition of flowers usually reflected their purpose e.g., [29–31]. Although none of the species that composed the traditional Slovenian bouquets are native in the region, they became an essential part of the Slovenian culture. Rosemary is the second most referenced plant species in Slovenian folk songs, which gained its popularity due to its scent and evergreen character; while carnations, as mentioned previously, even became the national flower symbol, as in some other European cultures (Spain, Monaco) [14:475]. These species were grown in gardens and pots around houses, on windowsills and balconies.

Rosemary has been an important plant species in the Mediterranean since at least Ancient Greece and was used later by both pagans and Christians [32]. Rosemary has many symbolic meanings in Europe, such as love, death, eternity, fidelity, virginity, and others (see [33] and references therein). Many of these symbolic meanings are also found in the Slovenian folk songs. Although rosemary is extensively used as a condiment nowadays, the references in folk songs are mostly symbolic ones and do not relate to its useful values.

Some songs mention the "German rosemary" instead of "rosemary". The Slovenian name of *Santolina chamaecyparissus* L. is "nemški rožmarin" (translated into English as German rosemary), and in a few cases it is difficult to understand whether the reference relates to *Rosmarinus* or *Santolina*. Among the wildflowers, blue or white violets were also collected for bouquets.

Lilies and roses were (at least initially) not cultivated in gardens, therefore, they were not used in bouquets or wreaths but were often planted on graves. A special kind of religious bouquet, called "Mary's bouquet", was devoted to Virgin Mary and was composed of her symbol, the rose. Lilies used in bouquets were almost exclusively white, while roses were always red.

Plants Used for Handicrafts

Branches and trunks of various shrubs and trees were used to make objects of daily use. The songs mention the use of wood of Norway spruce for boats, silver fir for boards, apple for chests, lime and maple for cradles, hazel and dogwood for sticks, maple for musical instruments and arches, box for smoking pipes, juniper and alder for agricultural tools, such as hoes and plows, and oak for smaller decorative items such as Jesus crosses. Pliable branches of hazel and willow and stems of old man's beard (*Clematis vitalba* L.) are mentioned as sources for making reins, baskets, or for tying: "*He came to get gbanca (traditional pastry),/It's made of barely/And tied with a hazel branch*" ("*Po gbanco je prišel,/Ječmenova je,/Z 'no leskovo trtico/Zvezana je.*" Š3581). The making of homemade baskets from hazel and willow branches is well documented from some Balkan countries [34,35] and was also an important source of income in some Slovenian villages [36]. Nowadays, basketry is practiced only on a small scale, but its ethnological importance was confirmed by the inclusion of weaving in the Register of Intangible Cultural Heritage [37]. Although hazel and willow baskets are still made for commercial purposes, baskets made from old man's beard were only used for domestic purposes [36]. Textile fibers were obtained from herbaceous plants, such as hemp, flax, and stinging nettle. Similarly to the now commercially more important flax and hemp fibers, nettle fibers were used to make textiles in Central Europe before the introduction of cotton in the 19th century, but their production ceased during World War II when other, cheaper fibers became more readily available [38].

Other Miscellaneous Uses

The folk songs studied also contain sporadic references to some medicinal and magical plants or plants used in religious rituals. Dafni et al. [33] classify ritual plants as plants or their parts used in private or official ceremonies to create a tunnel with gods or supernatural forces. These plants may include sacred trees, hallucinogenic and narcotic plants, incense, and aromatic plants with other uses. Religious uses mentioned in folk songs include sprinkling the dead with evergreen plant twigs and making funeral wreaths. These consisted of green plants, usually rosemary and also cotton lavender (*Santolina chamaecyparissus*), but were sometimes combined with colorful flowers, as mentioned in one of the folk songs: "*When I die/I will have a beautiful wreath/Of rosemary/And red carnations*" ("*Če jez dekle umrla bom,/Venček lep imela bom:/Z' rožmarina blagiga,/Z' nagelna rudečiga.*" Š6250).

The sprinkling of the dead with (usually evergreen) plant shoots at funerals has its origins in antiquity and has survived as part of the Christian burial tradition. Various plants, usually evergreens, were used in burial rituals, such as myrtle, basil, olive tree, spruce, juniper, rosemary, box, and others [33,39–41]. In reports from some countries (e.g., Iraq), the evergreen habit is said to preserve the soul and is referred to as evergreen life force [42]. One of the folk songs says: "*She entered the room/And sprinkled him/With green rosemary*" ("*Prek praga je stopila,/Z zelenim rožmarinom/Ga poškropila*" Š205).

Only three plant species were mentioned as having medicinal properties, namely carrot, fennel (*Foeniculum vulgare* Mill.), and laurel, as well as the magical "koren lečen". Koren lečen is a root of a presumably magical plant that cures all diseases: "*I have an unknown taproot,/Unknown taproot, medicinal taproot (koren lečen):/I put it under my tongue,/In the evening I get very sick/In the morning I lie there dead/.../The next day the people in the castle cry/.../Only the court jester laughs,/Talks and says:/"I have a strong feeling/That the young Zora is not dead./.../He (the king) opens my mouth/and removes the root/Unknown root, healing root (koren lečen)/...*" ("*Sej imam jest neznan koren,/Neznan koren, koren lečen:/Ki ga pod jezik položim,/Precej zvečer hudo zbolim./Zjutraj pa že mertva ležim./.../Le grajski norec se smeji,/Tako-le pravi, govori:/„Močno, močno se meni zdi,/De mlada Zora mertva ni./.../Pa Zori vzame 'z ust koren,/Neznan koren, koren lečen.*" Š114) or "*...He plucks a tap root/a dear and noble one:/.../If you, my darling, are better,/this (taproot) will cure you;/But if you are dying,/you will be much worse with it./...*" ("*Vun potegne žlahten koren;/.../Če tebi k zdravji bo,/Potem ti preci bole bo;/Če pa tebi smrti bo,/Potem ti taki huje bo!*" Š129). Although there has been debate in the past as to which plant species "koren lečen" refers to, and some candidates have been mentioned, such as bryony (*Bryonia* L.) [43] or blessed thistle (*Cnicus benedictus* L.) [28], the most common opinion is that it is a fictional magical species.

Fennel is mentioned as a remedy to ward off snakebites: "*...Eat, eat the fennel,/so that the colorful snake will not bite you...*" ("*Jej, jej, jej, koromač,/Da te ne piči pisan kač!*" Š7851). This song was sung by children at Easter while eating fennel and other blessed Easter foods. In one of Pliny's works, he mentions a fable about snakes eating fennel when they shed their skin to improve their eyesight. This fable led to the development of an ointment made from viper skin, fennel, and frankincense to improve eyesight [44]. The use of fennel seeds as an antidote is reported from Hindu and Chinese cultures [45].

The absence of medicinal and magical references in Slovenian folk songs was unexpected, as traditional medicine and wild plant gathering are widespread in the area [15] and Mlakar [28] lists a large number of plants with apotropaic magical powers used in Slovenia. However, a similar observation was made by Cardaño and Herrero [13] and De Cleene and Lejeune [14]. The latter found only five references to medicinal use in folk songs of Castile and León (Spain), and, surprisingly, none of the cited species (lemon, orange, apple and rose) is primarily associated with medicine.

3.1.3. Plants as Features of the Environment

The most frequent plants used to describe the environment are native or cultivated trees: lime is by far the most cited tree, followed by maple, beech, pine, grapevine, apple, pear, and oak. This comes to no surprise, as lime is regarded as the most important tree species in the Slovenian culture. The importance of lime has roots in the Slavic times, when lime was worshiped as a ritual tree. Thus, since the early days, lime trees were planted in village centers, where all important events took place: *"There is a village named Dolina/In the middle there stands a lime tree/Underneath the gypsies gather..."* (*"Stoji, stoji Dolina vas,/Na sred vasi pa lipica./Se tam cigani zbirajo..."* Š133). This tradition is still vivid today in Slovenia and large lime trees are often protected by law as sites of cultural and natural heritage [46]. Of course, lime is not the only tree species that was planted in villages, but was sometimes replaced by oaks, pines, pears, and other trees: *"There is a green pine growing in the courtyard/a black horse is tied to it..."* (*"Na dvori vam zelen bor,/Zanj' privezan konjič vran..."* Š4746) or *"A pear is growing in front of the house/Underneath the pear is a cool shade..."* (*"Pred vrati vam hruška zrasla,/Pod njo vam je hladna senca..."* Š5083). Some tree species, such as beech, willow, or pine, are usually used to describe the natural environment outside the human settlements: *"The shepherd is herding goats/In the green pinewood . . . "* (*"Kazarič mi kazice pase/U zeljanen borawji..."* Š176).

Among the herbaceous plants used to describe the environment are both wild plants (e.g., clover, ferns, stinging nettle, violets), as well as cultivated ones (e.g., wheat, roses, cotton lavender, lilies).

4. Materials and Methods

4.1. Area of Study

Slovenia is located in Central Europe, bordering Italy to the west, Austria to the north, Hungary to the northeast, and Croatia to the south and east, including also a 43 km long stretch of the Adriatic Sea in the southwest. The country covers an area of 20,273 square kilometers and has a population of 2.06 million, with expatriate communities of Slovenian or mixed origin also living in all neighboring countries. The population density is 101 inhabitants per square kilometer, which is similar to neighboring countries, but low compared to most Western European countries.

The climate in the central and eastern parts of the country is temperate. The western and southwestern coastal parts are influenced by the Adriatic Sea and have a mild sub-Mediterranean climate, while the Alps in the north dictate an alpine climate. A large part of the country (almost 60%) is covered by dense forests, in which the beech (*Fagus sylvatica* L.) is the most common tree species. In the Alps, evergreen coniferous forests of spruce (*Picea abies* (L.) H. Karst.) and fir (*Abies alba* Mill.) are found, while the vegetation of the sub-Mediterranean climate is dominated by deciduous forests of downy oak (*Quercus pubescens* Willd.) and common (*Carpinus betulus* L.) and oriental hornbeam (*C. orientalis* Mill.). Slovenia has a rich plant diversity due to its location at the intersection of climatic and geological factors. Altogether, 3119 plant taxa (seed plants and ferns) are present [47].

In 2010, 474,432 hectares were used for agriculture. This corresponds to 23.4% of the area of Slovenia. A large part of the agricultural land is permanent grassland and pastures (58%), followed by arable land (36%) and permanent crops (6%). The trend in agricultural land use is declining, with a decrease of 2% between 2000 and 2010 [48]. In past centuries, agriculture was much more important and more land was used for it, mostly divided among small farms.

4.2. Data Collection

In this study, we have analyzed 8686 Slovenian folk songs collected by the ethnologist Karel Štrekelj (Figure 4a) and published in 14 notebooks between 1895 and 1912 (Figure 4b) [49–52]. His work was completed posthumously by his student Jože Glonar, who edited two more notebooks (Notebooks 15 and 16), published in 1923 [53]. The collection is considered the most comprehensive Slavic collection of folk literature of its time, and Štrekelj's work was distinguished by its scientific methodology, which was recognized by several foreign critics of the time [54]. The songs were transcribed in their dialects rather than converted into literary language, and all songs indicate geographical origin. Although Štrekelj encountered the problem of songs with a known author and songs "in transition" to folk songs and first resisted publishing these songs because, in his opinion, they contradicted the rules of folklore and folk poetics, in the end he decided to compromise and published a supplement of these "non-folk" songs in the 3rd and 4th editions [54]. Štrekelj divided the songs into several categories and subcategories (references to songs throughout the article are marked with a unique code composed of the letter Š and a number corresponding to the number of the song in Štrekelj's four-volume edition, namely Š1–Š8686): narrative songs (Š1–Š1006), love songs (Š1007–4729), songs for special occasions (furtherly subdivided into ritual songs (Š4730–5180), dancing songs (Š5181–5249), marriage songs (Š5250–5467), drinking songs (Š5468–6199), funeral songs (Š6120–6400)), religious songs (Š6401–6732), songs about professions (Š6733–8538), and humorous songs (Š8539–8686).

(a)   (b)

Figure 4. (**a**) Photograph of Karel Štrekelj from the period when he was teaching Slavic philology at the University of Graz; (**b**) and the title page of the second volume of the Slovenian folk songs (right).

During the analysis, each song was read and plants and products derived from plants (e.g., a juniper hoe; an alder plow; a buckwheat cake) were identified and compiled in a table. Along with the plant names, the passages were transcribed and various classifiers (see the "Data Classification" section) were assigned to each plant occurrence. When verses alluded to more than one use of a plant, we included the reference in multiple categories. In 2019, the whole collection has been published online and is now freely available in the Slovenian digital library dLib (Digitalna knjižnica Slovenije. Available online: www.dlib.si (accessed on 10 January 2022)) [55].

4.3. Plant Identification

The Slovenian language has numerous dialects, which can lead to polysemy (e.g., fajgln can be used for carnation (*Dianthus caryophyllus* L.) and wallflower (*Erysimum cheiri* (L.) Crantz)). In such cases, we selected the most plausible botanical species depending on the context. In many cases, different spellings and names were found for the same plant species (e.g., the Slovenian names 'vrtnica', 'šipek', or 'gartroža' are used for rose (*Rosa* spp.) in different regions; similarly, 'trta', 'trs', and 'grozdje' are used for grapevine (*Vitis vinifera*) or 'ajda', 'ejdica' or 'hojda' for buckwheat (*Fagopyrum esculentum* Moench)). Some plant names have changed over time and are no longer used. Species identification has, therefore, been difficult in some cases. Several written [47,56] and web-based sources [57,58] were compared to understand the plant name. In cases where identification was not possible, the plant reference was omitted (e.g., no explanation was found for 'zizer' and 'mramurka' and these references were included in the analysis as unidentified seed plants). However, for most plants, identification was straightforward. For each plant species or taxonomic category, its occurrence was recorded as growing in the wild, cultivated, or not present in the area (exotic species). The full list of plant taxa is given in Table 1, and all vernacular names as recorded in the songs and their values are explained in the Supplementary Materials. The scientific names of plants and higher taxonomic categories have been updated to the currently accepted names listed in World Flora Online [59].

4.4. Data Classification

Classification of plant occurrences was adapted from [4], who classified all plant references into three groups: 1. *Plants as features of the environment* (ENV); 2. *Useful plants* (USF); and 3. *Plants used as symbols* (SYM). According to the classification, one plant species can fit into more categories, depending on the poem. Moreover, one plant occurrence can also be classified into more than one category. Some examples of classifications of plant references into categories are mentioned below.

4.4.1. Plants as Features of the Environment (ENV)

Plants or plant formations mentioned in folk songs may be mentioned simply as elements of the landscape, without any other visible significance. Trees, especially lime (*Tilia* spp.), but also maple (*Acer* spp.) and oaks (*Quercus* spp.), represent important cultural features in Slovenian villages and are, therefore, often mentioned in songs. Trees were (and still are) planted in the centers of villages and the inhabitants met there daily; dances and important meetings of elders also took place there, e.g., *"There is a lime tree/A green lime tree/Under the lime tree a table/A stone table..."* (*"Stoji, stoji lipa,/Lipa zelena;/Pod njoj stoji miza,/Miza kamena..."* Š4929).

4.4.2. Useful Plants (USF)

In the past, plants represented one of the most important natural resources for humans. Plants and their parts have been used for: making shelters, furniture, and tools; for consumption and as medicine; and also for decoration. Because of the many different uses of plants, this category has been divided into several subcategories: (a) plants for human consumption (e.g., food, drink, and other consumption, such as smoking); (b) plants for animal feeding; (c) ornamental plants (e.g., bouquets, home decorations, and garden plants); (d) plants for handicrafts (e.g., wooden objects and other artifacts); (e) other (e.g., plants for religious rituals, medicinal plants).

4.4.3. Plants with a Symbolic Value (SYM)

Similarly to most cultures around the world, the people of Slovenia assigned symbolic meanings to various plants. Because of the many different symbolic values, this category has been divided into several subcategories: (a) plants as religious symbols (e.g., in religious folk songs, plants are often used metaphorically to represent God, Jesus, the Virgin Mary, St. Joseph, or angels); (b) plants as symbols of love; (c) plants symbolizing negative feelings, sadness, or death (e.g., flowers given as gift at the departure of a loved one to war or plants that grow out of a grave); (d) plants as symbols of economic status (e.g., some edible plants are associated with wealth, others with poverty); (e) plants representing the appearance or characteristics of a person (e.g., beauty: *"beautiful like a carnation blossom"* (*"lep ko fajdelnov cvet"* Š2900); health: *"green (sick) like absinthe and sorrel"* (*"zelen kot ta ščavlja in pelen"* Š2654); size: *"tears as (big as) grapes"* (*"debele solze kakor vinske jagode"* Š15), color: *"red like a rose"* (*"rdeča ko gatroža"* Š369), and other qualities; (f) other.

5. Conclusions

Plants play an important role in Slovenian folk songs. Half of the mentions of plants have a symbolic value, followed by the representation of the environment and, finally, by their useful aspects. The analysis revealed that indigenous plants growing in the wild were mentioned much less frequently than those that were important for survival (cereals, grapevine) or from a cultural point of view (lime, carnation, rosemary). Interestingly, the songs revealed some interesting findings not only about the plants that appear in the songs, but also about plants that do not appear in them. Wildflowers are rarely mentioned in Slovenian folk songs, suggesting that our belief in a romantic coexistence of the peasant population with nature and their appreciation for nature may not be entirely realistic. Medicinal, poisonous, or edible wild plants also rarely appear in the songs. Folk songs have proven to be interesting sources of information, and, although they cannot be fully trusted as historical documents, they can still serve as sources for understanding the relationship between people and plants.

Although this work is a comprehensive study of Slovenian folk songs, it contains only a preliminary catalog of plants and their values. The availability of digitized data combined with mathematical approaches has led to the emergence of new fields that offer a multimodal network representation of data, such as computational gastronomy [60] or, in our case, computational folkloristics [61]. Such methods allow us to explore the complexity of a folkloric corpus, such as the folk songs presented in this article, at multiple levels of resolution, allowing for a more holistic interpretation of culture.

Supplementary Materials: The following supporting information can be downloaded at: https://www.mdpi.com/article/10.3390/plants11030458/s1, Supplementary Material S1: Selected examples of plant references in Slovenian folk songs, divided according to their environmental, symbolic or useful value. All vernacular names or different spellings used in Slovenian folk songs are listed. If the most commonly used Slovenian name does not appear in the songs, it is marked with an asterisk (*). English names are in parentheses. References to songs are marked with a unique code consisting of the letter Š and a number corresponding to the song in the four-volume edition of Slovene folk songs edited by Karel Štrekelj (e.g., Š1–Š8686); Table S1: Table showing the number of plant citations for each plant taxon within each (sub)category.

Funding: This work is partially supported by the Slovenian Research Agency, infrastructure program I0—0035.

Institutional Review Board Statement: Not applicable.

Informed Consent Statement: Not applicable.

Data Availability Statement: The songs analyzed in this study are openly available in the Digital Library of Slovenia in a PDF or TXT file format at the following link: https://www.dlib.si/details/URN:NBN:SI:DOC-U1XDVBQI (accessed on 16 January 2022).

Acknowledgments: The author wishes to thank Rasto Jereb for the translation of the title song and Manica Balant, Peter Glasnović and the anonymous reviewers for the comments on the manuscript.

Conflicts of Interest: The author declares no conflict of interest.

References

1. Ellacombe, H.N. *The Plant-Lore and Garden-Craft of Shakespeare*, 2nd ed.; W. Satchell and Co: London, UK, 1884.
2. Włodarczyk, Z. Review of plant species cited in the Bible. *Folia Hortic.* **2007**, *19*, 67–85.
3. Quealy, G.; Collins, S.H.; Mirren, H. *Botanical Shakespeare: An Illustrated Compendium*; Harper Design: New York, NY, USA, 2017.
4. Pardo-de-Santayana, M.; Tardío, J.; Heinrich, M.; Touwaide, A.; Morales, R. Plants in the Works of Cervantes. *Econ. Bot.* **2006**, *60*, 159–181. [CrossRef]
5. Fernández-Llamazares, A.; Lepofsky, D. Ethnobiology through Song. *J. Ethnobiol.* **2019**, *39*, 337–353. [CrossRef]
6. Agrawal, S.R. Trees, Flowers and Fruits in Indian Folksongs, Folk proverbs and Folk Tales. In *Contribution to Indian Ethnobotany*; Jain, S.K., Ed.; Scientific Publishers: Jodhpur, India, 1997.
7. Ahmed, M.M.; Singh, P.K. Plant-lore with reference to Muslim folksong in association with human perception of plants in agricultural and horticultural practices. *Not. Bot. Horti Agrobot. Cluj-Napoca* **2008**, *36*, 42–47.

8. Herrero, B.; Cardaño, M. Ethnobotany in the Folksongs of Castilla y León (Spain). *Bot. Sci.* **2015**, *93*, 1–12. [CrossRef]
9. Turner, N.J.; Lepofsky, D.; Deur, D. Plant Management Systems of British Columbia's First Peoples. *BC Stud. Br. Columbian Q.* **2013**, *179*, 107–133.
10. Turpin, M.; Stebbins, T. The Language of Song: Some Recent Approaches in Description and Analysis. *Aust. J. Linguist.* **2010**, *30*, 1–17. [CrossRef]
11. Štih, P.; Simoniti, V.; Vodopivec, P. *A Slovene History*; Inštitut za Novejšo Zgodovino: Ljubljana, Slovenia, 2008.
12. Gänger, S. World Trade in Medicinal Plants from Spanish America, 1717–1815. *Med. Hist.* **2015**, *59*, 44–62. [CrossRef]
13. Cardaño, M.; Herrero, B. Plants in the Songbooks of Castilla y León, Spain. *Ethnobot. Res. Appl.* **2014**, *12*, 535–549. [CrossRef]
14. De Cleene, M.; Lejeune, M.C. *Compendium of Symbolic and Ritual Plants in Europe*; Man & Culture Publishers: Ghent, Belgium, 2002.
15. Mlakar, V. *Sacred Plants in Folk Medicine and Rituals. Ethnobotany of Slovenia*; The Raymond Aaron Group: Markham, ON, Canada, 2020.
16. Bennett, B. Twenty-Five Economically Important Plant Families. In *Encyclopedia of Life Support Systems, Economic Botany*; Bennett, B., Ed.; EOLSS Publishers: Oxford, UK, 2010.
17. Vulić, G. Sustainable gastronomy tourism in Slovenia in connection with wild plants of Slovenian forests and meadows. *Quaestus* **2021**, *18*, 104–117.
18. Koch, R.A. Flower Symbolism in the Portinari Altar. *Art Bull.* **1964**, *46*, 70–77. [CrossRef]
19. Kandeler, R.; Ullrich, W.R. Symbolism of plants: Examples from European-Mediterranean culture presented with biology and history of art: JUNE: Lilies. *J. Exp. Bot.* **2009**, *60*, 1893–1895. [CrossRef]
20. Kandeler, R.; Ullrich, W.R. Symbolism of plants: Examples from European-Mediterranean culture presented with biology and history of art: OCTOBER: Roses. *J. Exp. Bot.* **2009**, *60*, 3611–3613. [CrossRef]
21. Makarovič, G. *Cvetlice v Ljudski Umetnosti. Motivi v Oblikovanju za Kmetije*; Slovenski Etnografski Muzej: Ljubljana, Slovenia, 1974.
22. Šavli, J. *Slovenska Znamenja*; Založništvo Humar: Bilje, Slovenia, 1994.
23. Kumer, Z. *Pušeljc pa mora bit Rože v slovenski ljudski pesmi, Muzikološke razprave In Memoriam Danilo Pokorn*; Krstulović, N.C., Faganel, T., Kokole, M., Eds.; Založba ZRC: Ljubljana, Slovenia, 2005.
24. Belyavski-Frank, M. Flowers, Fruit, and Food: Symbolism in Bosnian and Macedonian Love Songs and Wedding Songs. In *Bosansko-hercegovački Slavistički Kongres*; Knjiga, 2, Ibrišimović-Šabić, A., Eds.; Zbornik Radova: Sarajevo, Bosnia and Herzegovina, 2019.
25. Viegi, L.; Ghedira, K. Preliminary study of plants in ethnoveterinary medicine in Tunisia and Italy. *Afr. J. Tradit. Complementary Altern. Med.* **2014**, *11*, 189–199. [CrossRef]
26. De Laszlo, H.; Henshaw, P.S. Plant Materials Used by Primitive Peoples to Affect Fertility. *Science* **1954**, *119*, 626–631. [CrossRef]
27. Keber, J. *Slovar Slovenskih Frazemov*; Založba ZRC: Ljubljana, Slovenia, 2011.
28. Mlakar, V. *Rastlina je Sveta, od Korenin do Cveta*; Samozaložba: Ljubljana, Slovenia, 2015.
29. González, J.A.; García-Barriuso, M.; Gordalizac, M.; Amich, F. Traditional plant-based remedies to control insect vectors of disease in the Arribes del Duero (western Spain): An ethnobotanical study. *J. Ethnopharmacol.* **2011**, *138*, 595–601. [CrossRef]
30. Łuczaj, J.L. A relic of medieval folklore: Corpus Christi Octave herbal wreaths in Poland and their relationship with the local pharmacopoeia. *J. Ethnopharmacol.* **2012**, *142*, 228–240. [CrossRef]
31. Łuczaj, L. Herbal Bouquets Blessed on Assumption Day in South-Eastern Poland: Freelisting versus photographic inventory. *Ethnobot. Res. Appl.* **2011**, *9*, 1–25. [CrossRef]
32. Lapucci, C.; Antoni, A.M. *La Simbologia delle Piante*; Polistampa: Florence, Italy, 2016.
33. Dafni, A.; Petanidou, T.; Vallianatou, I.; Kozhuharova, E.; Blanche, C.; Pacini, E.; Peyman, M.; Dajić Stevanovic, Z.; Franchi, G.G.; Benitez, G. Myrtle, Basil, Rosemary, and Three-Lobed Sage as Ritual Plants in the Monotheistic Religions: An Historical–Ethnobotanical Comparison. *Econ. Bot.* **2020**, *74*, 330–355. [CrossRef]
34. Andonova, M. Baskets from the Mountain: An Ethnobotanical Approach to the Balkandjii Tradition. ArchéOrient-Le Blog. CNRS/University Lyon 2: Lyon, France, 2017. Available online: https://archeorient.hypotheses.org/7769 (accessed on 30 December 2021).
35. Pieroni, A. Local plant resources in the ethnobotany of Theth, a village in the Northern Albanian Alps. *Genet. Resour. Crop Evol.* **2008**, *55*, 1197–1214. [CrossRef]
36. Bras, L. *Pletarstvo na Slovenskem. Vodnik po Razstavi*; Slovenski Etnografski Muzej: Ljubljana, Slovenia, 1973.
37. Register of Intangible Cultural Heritage. *Weaving. Republic of Slovenia*; Ministry of Culture: Ljubljana, Slovenia, 2016.
38. Vogl, C.R.; Hart, A. Production and processing of organically grown fiber nettle (*Urtica dioica* L.) and its potential use in the natural textile industry: A review. *Am. J. Altern. Agric.* **2003**, *18*, 119–128. [CrossRef]
39. Scranton, R.L. A Wreath in the Vassar Classical Museum. *Am. J. Archaeol.* **1944**, *48*, 135–142. [CrossRef]
40. Ložar-Podlogar, H. Death customs in the Slovene countryside. *Etnolog* **1999**, *9*, 101–115.
41. Kielak, O. The ethnobotanical character of the Polish Dictionary of Folk Stereotypes and Symbols. *Ethnobot. Res. Appl.* **2020**, *19*, 1–15. [CrossRef]
42. Buckley, J.J. *The Mandaeans: Ancient Texts and Modern People*; Oxford University Press: Oxford, UK, 2002.
43. Kunaver, D. *Čarodejna Moč Rastlin. Rastline v Ljudskem Izročilu*; Samozaložba: Ljubljana, Slovenia, 1996.
44. Stannard, J. Medicinal plants and folk remedies in Pliny, "Historia Naturalis". *Hist. Philos. Life Sci.* **1982**, *4*, 3–23.
45. Duke, J.A. *CRC Handbook of Medicinal Herbs*; CRC Press: Boca Raton, FL, USA, 1985.
46. Šmid Hribar, M. Kulturni vidiki drevesne dediščine. *Glas. Slov. Etnološkega Društva* **2011**, *51*, 44–54.
47. Martinčič, A.; Wraber, T.; Jogan, N.; Podobnik, A.; Turk, B.; Vreš, B. *Mala flora Slovenije*; Tehniška Založba: Ljubljana, Slovenije, 2007.

48. Kutin Slatnar, B.; Krajnc, A.; Hadžihasanović, E.L.; Stele, A. *The 2010 Agricultural Census—Every Farm Counts*; Statistical Office of the Republic of Slovenia: Ljubljana, Slovenia, 2012.
49. Štrekelj, K. (1895–1898) Slovenske Narodne Pesmi. Zvezek I. Available online: https://www.dlib.si/stream/URN:NBN:SI:DOC-U1XDVBQI/210eacec-ae9d-4f38-a238-361583041248/PDF (accessed on 10 January 2022).
50. Štrekelj, K. (1900–1903) Slovenske Narodne Pesmi. Zvezek II. Available online: https://www.dlib.si/stream/URN:NBN:SI:DOC-U1XDVBQI/4e48d9b6-9656-4155-b43d-5ff604d27ce5/PDF (accessed on 10 January 2022).
51. Štrekelj, K. (1904–1907) Slovenske Narodne Pesmi. Zvezek III. Available online: https://www.dlib.si/stream/URN:NBN:SI:DOC-U1XDVBQI/56f88125-275e-4f24-9d7b-9f263c0905b5/PDF (accessed on 10 January 2022).
52. Štrekelj, K. (1908–1923) Slovenske Narodne Pesmi. Zvezek IV. Available online: https://www.dlib.si/stream/URN:NBN:SI:DOC-U1XDVBQI/64d8601d-3f14-4aba-888b-42da232e7ebb/PDF (accessed on 10 January 2022).
53. Kropej, M. Karel Štrekelj: Prerez skozi življenje in delo. *Traditiones* **1995**, *24*, 25–48.
54. Terseglav, M. Štrekljev sindrom. *Traditiones* **2001**, *30*, 201–220.
55. Slovenian Digital Library dLib. Available online: www.dlib.si (accessed on 2 January 2022).
56. Mastnak, H. *Skrivnostne Moči Narave*; Založništvo Pozoj: Velenje, Slovenia, 1998.
57. Fran. Dictionaries of the Institute of Slovenian Language Fran Ramovš ZRC SAZU. Version 8.0. 2014. Available online: www.fran.si (accessed on 10 January 2022).
58. Nikolič, T. Flora Croatica Database. 2021. Available online: http://hirc.botanic.hr/fcd (accessed on 10 January 2022).
59. World Flora Online. Available online: http://www.worldfloraonline.org/ (accessed on 8 December 2021).
60. Ahnert, S.E. Network analysis and data mining in food science: The emergence of computational gastronomy. *Flavour* **2013**, *2*, 1–3. [CrossRef]
61. Abello, J.; Broadwell, P.; Tangherlini, T.R. Computational Folkloristics. *Commun. ACM* **2012**, *55*, 60–70. [CrossRef]

Article

Temporal Changes in the Use of Wild Medicinal Plants in Trentino–South Tyrol, Northern Italy

Giulia Mattalia [1,2,*], Felina Graetz [3], Matthes Harms [3], Anna Segor [3], Alessio Tomarelli [3], Victoria Kieser [3], Stefan Zerbe [4] and Andrea Pieroni [3,5,*]

[1] Institut de Ciència i Tecnología Ambientals (ICTA-UAB), Universitat Autònoma de Barcelona, 08193 Barcelona, Spain

[2] New York Botanical Garden, New York, NY 14058, USA

[3] University of Gastronomic Sciences, 12042 Pollenzo, Italy; felina.graetz@gmail.com (F.G.); matthes.harms@gmail.com (M.H.); segoranna@gmail.com (A.S.); alessio.tomarelli@gmail.com (A.T.); victoria.kieser@gmail.com (V.K.)

[4] Faculty of Agricultural, Environmental and Food Sciences, Free University of Bozen-Bolzano, 39100 Bolzano-Bozen, Italy; stefan.zerbe@unibz.it

[5] Department of Medical Analysis, Faculty of Applied Science, Tishk International University, Erbil 44001, Iraq

[*] Correspondence: giulia.mattalia@uab.cat (G.M.); a.pieroni@unisg.it (A.P.)

Abstract: Mountain regions are fragile ecosystems and often host remarkably rich biodiversity, and thus they are especially under threat from ongoing global changes. Located in the Eastern Alps, Trentino–South Tyrol is bioculturally diverse but an understudied region from an ethnobotanical perspective. We explored the ethnomedicinal knowledge of the area from a cross-cultural and diachronic perspective by conducting semi-structured interviews with 22 local inhabitants from Val di Sole (Trentino) and 30 from Überetsch–Unterland (South Tyrol). Additionally, we compared the results with ethnobotanical studies conducted in Trentino and South Tyrol over 25 years ago. The historical comparison revealed that about 75% of the plants currently in use were also used in the past in each study region. We argue that the adoption of "new" medicinal species could have occurred through printed and social media and other bibliographical sources but may also be due to limitations in conducting the comparison (i.e., different taxonomic levels and different methodologies). The inhabitants of Val di Sole and Überetsch–Unterland have shared most medicinal plants over the past few decades, yet the most used species diverge (perhaps due to differences in local landscapes), and in South Tyrol, people appear to use a higher number of medicinal plants, possibly because of the borderland nature of the area.

Keywords: Alps; biocultural diversity; borders; ethnomedicine; historical ethnobotany; local ecological knowledge; mountain regions

Citation: Mattalia, G.; Graetz, F.; Harms, M.; Segor, A.; Tomarelli, A.; Kieser, V.; Zerbe, S.; Pieroni, A. Temporal Changes in the Use of Wild Medicinal Plants in Trentino–South Tyrol, Northern Italy. *Plants* **2023**, *12*, 2372. https://doi.org/10.3390/plants12122372

Academic Editor: Daniel Sánchez-Mata

Received: 24 May 2023
Revised: 14 June 2023
Accepted: 16 June 2023
Published: 19 June 2023

1. Introduction

Mountain regions often host rich biodiversity, and thus they are especially under threat from the ongoing global changes (e.g., [1]), particularly those related to climate and habitat shifts resulting from land-use changes during the last century [2]. European mountain ranges are no exception to such a trend (e.g., Gurung et al. [3] for the Carpathians, Miranda Cebrian et al. [2] for the Pyrenees, Rew et al. [4] for the Alps).

The Alps are one of the largest mountain ranges in Europe, and the Alpine region features complex historical, political, and ecological trajectories. For instance, the region of Trentino–South Tyrol in the Southern Eastern Alps was one of the last regions to be added to the country of Italy [5]. Until 1919, Trentino–South Tyrol was part of the Habsburg Empire, a political entity that was characterized by considerable linguistic and ethnic heterogeneity [6]. Despite attempts at "Italianization", particularly under the Mussolini government (1922–1943), the region remained highly diverse in terms of culture and language. Accordingly, German, Italian, and Ladin coexist in this small mountainous territory [7,8]. In 1972, such diversity led to autonomy status for the provinces of Trento (mainly Italian-speaking) and Bozen–South Tyrol (mainly German-speaking).

Political and linguistic contexts often play a key role in the way people relate to plants (e.g., [9,10] for example from Bukovina, the easternmost region of the former Habsburg Empire). Visible (e.g., institutional/political) and invisible (e.g., cultural/linguistic) borders could offer valuable tools for detecting the effect of political, cultural, or linguistic heterogeneity on local ecological knowledge (LEK) (e.g., [9,11,12]).

The Alps have been often perceived as borders, but as well noted by Salsa [13], this mountain range has also served as zippers, as spaces for exchange among different sociocultural groups. For instance, in the last ten years, five publications have addressed the ethnomedicine of cultural groups, three linguistic minorities and one religious minority of the Western Alps [14–18], and four addressed the folk medicinal uses of the Central Alps [19–22], and two of the Eastern Alps [23,24].

In addition to a spatial and socio-political perspective, a diachronic perspective could further contribute to our understanding of the complex dynamics of local ecological knowledge in the Alpine environment. Historical ethnobotanical studies in Europe have often focused on the eastern portion of the continent, where ethnobotanical archival data from the 19th and 20th centuries [25–33] are frequently available. These studies have generally shown a decrease in wild food plant-centered LEK, while local medicinal plant reports are not only fading but often re-arranged via influences of printed media.

In recent years, public interest in the use of traditional ethnobotanical knowledge and medicinal plants has steadily increased [34]. The cultivation and use of medicinal plants have become a growing market niche in the past few decades, which underlines the importance of folk medical knowledge for the promotion of ecotourism, eco-gastronomy, and organic farming [35]. This development could play an important role in conserving biodiversity and cultural heritage in South Tyrol and the Alps in general while revitalizing the relationship between humans and nature [36].

Despite the growing interest, the Trentino–South Tyrol region has been little studied from an ethnobotanical perspective compared to other Italian regions (see Cavalloro et al. [37] for an overview). Indeed, we found only four publications on the local ecological knowledge of medicinal plants in this region. Two studies listing the medicinal plants used by local populations were conducted over 25 years ago, in Val di Sole, Trentino [38], and the Puster Valley, South Tyrol [39]. Two more recent studies conducted, in several parishes of Trentino [37] in South Tyrol based on published material [36] have warned about the ongoing plant-related LEK erosion. Additionally, the concept of Cultural Keystone Species with regard to traditionally used medicinal plants was studied by Petelka et al. [40] and interviews were conducted in South Tyrol by Scherrer et al. [41] to gain insight into, and reflections on, the cultural value of traditional medicinal plants and their interplay within the local landscape, nature conservation and their role in environmental education and knowledge transfer across generations.

Although under-researched, Trentino–South Tyrol is rich in cultural, linguistic, and ecological diversity. Additionally, considering the peculiar vulnerability of the Alpine context and the current threats from climate change and social changes, documenting the LEK held by culturally and linguistically distinct groups could be crucial to fostering future biocultural conservation strategies. Therefore, the objective of this study is to explore LEK related to medicinal plants from a cross-cultural and diachronic perspective and, specifically:

1. to document current LEK related to wild medicinal plants and its transmission strategy in Val di Sole (Trentino) and South Tyrol,
2. to identify similarities and differences with historical ethnomedicinal data regarding Trentino and South Tyrol published over 25 years ago, and
3. to identify similarities and differences among current LEK held by inhabitants of Val di Sole and South Tyrol.

2. Results

2.1. Current Use of Wild Medicinal Plants in Überetsch–Unterland (South Tyrol)

We documented the use of 73 plant species, belonging to 30 families (including 29 gymnosperms and an angiosperm), and one lichen (Table 1). The dominant plant families were Asteraceae with 13 species, followed by Lamiaceae with 11 species, and Rosaceae with eight species. Among the mentioned medicinal plants, six were cultivated (*Ocimum basilicum, Salvia rosmarinus, Tropaeolum minus*, Mentha × *piperita, Levisticum officinale, Melissa officinalis*) while eight were partially cultivated (*Calendula officinalis, Lavandula angustifolia, Salvia officinalis, Borago officinalis, Alchemilla vulgaris, Ribes nigrum, Thymus vulgaris, Phyllanthus niruri*); all other plants were collected in the wild.

Table 1. List of the quoted botanical taxa, their folk names, used parts, the local preparations, and medicinal uses/treated illnesses (as reported by the emic perceptions that study participants mentioned), and the frequencies of quotation in the two study areas. [C: cultivated; W: wild; * also in the corresponding historical source].

Botanical Taxon and Family	Local Vernacular Names	Status	Part(s) Used	Trentino Frequency of Quotation	Südtirol Frequency of Quotation	Preparation	Local Perceived Medicinal Uses or Treated Illnesses
Abies alba Mill. (Pinaceae)	Abete bianco, Avez (T)	W	Buds	2 *		Bath, tincture	Chilblains, bones, and tooth support
Achillea erba-rotta subsp. *moschata* (Wulfen) Vacc. (Asteraceae)	Medico gentile, Erba iva (T)	W	Flowers	7 *		Schnaps, tea	Digestive, cramps
Achillea millefolium L. (Asteraceae)	Achillea millefoglie, Milifoi (T), Schafgarbe, Jungfrauenkraut, Garbenkraut (S)	W	Flowers	3 *	12 *	Oil infusion, tea, fresh juice, decoction	Cicatrizing, disinfectant, soothing, digestive, depurative, menstrual pain, power source, mineral salt source, antibacterial, antibiotic, anti-inflammatory, arthritis, fever, gastrointestinal cramps, wounds
Acorus calamus L. (Acoraceae)	Kalmus (S)	W	Roots		1 *	Tea, tincture, oil	Digestive disorders, respiratory diseases
Aegopodium podagraria L. (Apiaceae)	Gewöhnlicher Geißfuß, Giersch (S)	W	Aerial parts of the flowering plant		4 *	Tea	Rheumatism, gout
Agrimonia eupatoria L. (Rosaceae)	Odermennig (S)	W	Aerial parts of the flowering plant		1 *	Tea, tincture	Inflammation in the oral cavity, diarrhea
Alchemilla vulgaris L. (Rosaceae)	Frauenmantel (S)	W, C	Aerial parts		2 *	Tea	Menstrual and hormonal disorders
Alchemilla xanthoclora Rothm. (Rosaceae)	Alchemilla (T)	W, C	Aerial parts	2 *		Tea	Stomach cramps
Allium ursinum L. (Amaryllidaceae)	Bärlauch (S)	W	Flowers, leaves, bulbs		2 *	Tea, pesto, salt	Arteriosclerosis, hypertension, digestive disorders
Althaea officinalis L. (Malvaceae)	Eibisch, Weiswurzel (S)	W	Flowers, leaves, roots		1 *	Tea, tincture, syrup	Bronchitis, stomach, and intestinal inflammation
Angelica archangelica L. (Apiaceae)	Engelwurz (S)	W	Roots, leaves		1 *	Tea	Digestive disorders
Angelica sylvestris L. (Apiaceae)	Engelwurz (S)	W	Roots, leaves		1 *	Tea	Digestive disorders
Armoracia rusticana G.Gaertn., B.Mey. & Scherb. (Brassicaceae)	Meerrettich, Kren (S)	C	Roots		2 *	Tincture, honey, syrup, salve	Antibiotic, bronchitis, dry cough, bladder and kidney infections
Arnica montana L. (Asteraceae)	Arnica (T), Arnika, Bergwohlverleih (S)	W	Flowers, leaves, roots	2 *	9 *	Tincture, ointment oil infusion, salve	Rheumatism, musculoskeletal pain, muscle and bone injuries
Artemisia absinthium L. (Asteraceae)	Wermut, Bitterer Beifuß (S)	W	All parts of the plant		3 *	Tea, tincture	Digestive
Artemisia umbelliformis subsp. *umbelliformis* (Asteraceae)	Genepy, Genepi, Erba blanc (T)	W	Leaves, flowers	2		Schnaps	Digestive
Artemisia vulgaris L. (Asteraceae)	Beifuss, Besenkraut, Sonnenwendgürtel (S)	W	All parts of the plant		4 *	Tea, tincture, powder	Antiviral, antibacterial
Betula pendula Roth. (Betulaceae)	Betulla (T), Birke, Maibaum, Bedól (S)	W	Buds, sap, leaves, seeds	1 *	4 *	Birch sap, tea, tincture	Cholesterol, good for the kidneys, immune system booster, gives power, rheumatism

Table 1. *Cont.*

Botanical Taxon and Family	Local Vernacular Names	Status	Part(s) Used	Trentino Frequency of Quotation	Südtirol Frequency of Quotation	Preparation	Local Perceived Medicinal Uses or Treated Illnesses
Borago officinalis L. (Boraginaceae)	Borretsch (S)	W, C	Flowers, leaves, seeds		1	Salad	Lifts the mood
Calendula officinalis L. (Asteraceae)	Ringelblume, Totenblume, Goldblume (S)	W, C	Flowers		10 *	Tea, salve, syrup, oil infusion	Wounds, scars, stomach and intestinal disorders, anti-inflammatory, skin-regenerating properties
Carlina acaulis L. (Asteraceae)	Silberdistel (S)	W	Roots		1 *	Tea, tincture, powder	Cold, fever, diuretic
Centaurea benedicta (L.) L. (Asteraceae)	Benediktenkraut (S)	W	Flowers, leaves		1 *	Tea, tincture	Stomach disorders
Centaurea cyanus L. (Asteraceae)	Fiordaliso (T), Kornblume (S)	W	Flowers	1	1	Tea, oil	Liver stimulator, good for digestion, skin disorders
Cetraria islandica (L.) Ach. (Parmeliaceae)	Lichen, Lichene islandico, Alga di montagna (T)	W	Aerial parts	7 *		Tincture, oil, ointment, boiled	Cough, bronchitis, COVID-19 prevention
Corylus avellana L. (Betulaceae)	Nocciolo (T)	W, C	Buds	1 *		Oil	Hemorrhoids
Crataegus laevigata (Poir.) DC. (Rosaceae)	Biancospino (T), Weißdorn, Mehlbeere (S)	W	Flowers, leaves, fruit, buds	1 *	1 *	Tea, tincture, powder, oil	Blood pressure balancer, calmant, diuretic, antispasmodic
Crataegus monogyna L. (Rosaceae)	Biancospino (T), Weißdorn, Mehlbeere (S)	W	Flowers, leaves, fruit, buds	1	1 *	Tea, tincture, powder, oil	Blood pressure balancer, calmant, diuretic, antispasmodic
Dipsacus fullonum L. (Caprifoliaceae)	Karde, Igelkopf, Weberdistel (S)	W	Roots		2	Tea, tincture, powder	Antibacterial, blood cleansing, detox
Dryas octopetala L. (Rosaceae)	Weiße Silberwurz, Stängellose Eberwurz (S)	W	Roots		1 *	Tea, tincture	Indigestion, appetite stimulator, laxative
Elymus repens (L.) Gould (Poaceae)	Quecke (S)	W	Roots		1 *	Tea	Anti-inflammatory, respiratory diseases
Equisetum arvense L. (Equisetaceae)	Equiseto, Erba cavallina, Coda cavallina (T), Ackerschachtelhalm, Zinnkraut (S)	W	Summer sprouts, leaves, stems	1 *	6 *	Tea, tincture, salve, powder, decoction	Osteoporosis, diuretic, kidney stones, vertebral and connective tissue strengthener
Euphrasia officinalis subsp. *pratensis Fr.* (Orobanchaceae)	Augentrost (S)	W	All parts of the herb		3 *	Tea	Eye disorders
Ficus carica L. (Moraceae)	Fico (T)	W, C	Buds	1		Oil	Digestion support supplement, stomachache
Filipendula ulmaria (L.) Maxim. (Rosaceae)	Spirea, Regina dei prati (T), Echtes Mädesüß, Geißbart (ST)	W	Aerial parts	1 *	3 *	Tea, powder, salve	As aspirin, cellulitis, cramps, diarrhea
Gentiana lutea L. (Gentianaceae)	Genziana, Enzian, Radis de genziana (T), Enzian (ST)	W	Roots, flowers	8	1 *	Tea, tincture	Depurative, stomach and intestinal disorders
Geranium robertianum L. (Geraniaceae)	Storchschnabel, Ruprechtskraut (S)	W	All parts of the herb		6 *	Tea, tincture, salve	Infertility, hormonal balance support supplement, heavy metal removal from the body
Hedera helix L. (Araliaceae)	Efeu (S)	W	Leaves		1	Tea, tincture	Cough
Humulus lupulus L. (Cannabaceae)	Ligaboschi, Luppol, Luppolo (T)	W	Fruit	8 *		External application	Sedative
Hypericum perforatum L. (Hypericaceae)	Iperico perforato, Erba di San Giovanni, Iperico (T), Johanniskraut, Blutkraut (ST)	W	Flowers	7 *	7 *	Tea, tincture, salve, oil	Skin burns, back pain, hemorrhoids, wounds, antidepressant, anti-wrinkle, antioxidant, anti-inflammatory, antiviral
Juniperus communis L. (Cupressaceae)	Wacholder, Kranewitt, Brusin (S)	W	Fruit, sprouts, roots,		1 *	Tea, tincture, syrup	Diuretic, kidney antiseptic, bladder inflammation, flatulence
Lamium album L. (Lamiaceae)	Weiße Taubnessel, Kuckucksnessel (S)	W	Flowers, leaves, roots		1 *	Tea, tincture	Menstrual cramps
Lamium galeobdolon (L.) L. (Lamiaceae)	Goldnessel, Gelbe Taubennessel (S)	W	Flowers, leaves, roots		1	Tea, tincture	Kidney and bladder infections
Larix decidua Mill. (Pinaceae)	Larice, Làres dalla resina (T)	W	Cones, buds, resin, flowers	11 *		Syrup, ointment, fomentation, topically applied	Cough, thorn removal, furuncles, burns
Laurus nobilis L. (Lauraceae)	Alloro (T)	C	Leaves, fruit	2 *		Oil	Dermatitis

Table 1. *Cont.*

Botanical Taxon and Family	Local Vernacular Names	Status	Part(s) Used	Trentino Frequency of Quotation	Südtirol Frequency of Quotation	Preparation	Local Perceived Medicinal Uses or Treated Illnesses
Lavandula angustifolia Mill. (Lamiaceae)	Lavendel, Zöpfe (S)	W, C	Flowers		2 *	Tea, oil	Calmant, somniferum, headache, indigestion, acne, sunburn
Leontopodium nivale (Ten.) A.Huet ex Hand.-Mazz. (Asteraceae)	Edelweiß, Alpenedelweiß (S)	W	Flowers		2	Tincture	Good for memory, skin support supplement against premature ageing, sunscreen
Leonurus cardiaca L. (Lamiaceae)	Echtes Herzgespann, Bärenschweif (S)	W	Aerial parts		2	Tea, tincture, powder	Heart calmant
Levisticum officinale W.D.J.Koch (Apiaceae)	Liebstöckel, Maggikraut (S)	C	All parts of the plant		1 *	Powder, salt	Flatulence, digestive disorders
Malva sylvestris L. (Malvaceae)	Malva selvatica, Male va (T)	W	Flowers	10		Tea, topical application, bath, oil	Cystitis, constipation, skin rashes, vaginal lavages, anti-inflammatory, emollients, cough
Matricaria chamomilla L. (Asteraceae)	Camomilla (T), Kamille, Kummerblume, Mutterkraut (S)	W	Flowers	2 *	4 *	Oil infusion, salve, tea, ointment	Skin care, cramps, calmant, wounds, acne, bladder infections, stomach and intestinal disorders
Melissa officinalis L. (Lamiaceae)	Melisse, Bienenkraut (S)	C	Leaves		3 *	Tea, syrup, salve, oil	Nervous system calmant, heart support supplement; bladder disorders, herpes
Mentha × *piperita* L. (Lamiaceae)	Minze, Pfefferminze (S)	C	Leaves		2 *	Tea, syrup, oil	Headache, stomach and intestinal disorders
Ocimum basilicum L. (Lamiaceae)	Basilikum, Basilienkraut (S)	C	Leaves		1 *	Tea, tincture, pesto	Digestive disorders, flatulence
Peucedanum ostruthium (L.) W.D.J.Koch (Apiaceae)	Meisterwurz (S)	W	Roots		1 *	Tea	Gout, rheumatism, fever
Phyllanthus niruri L. (Phyllanthaceae)	Spacca pietra, Spacca muri (T)	C	Leaves, flowers	1		Decoction	Kidney stones
Picea abies (L.) H. Karst. (Pinaceae)	Aghi, Abete (T)	W	Leaves, buds	1 *		Fomentation, syrup	Respiratory system support supplement, cough
Pimpinella major (L.) Huds. (Apiaceae)	Bibernelle, Bockwurz (S)	W	Roots		1 *	Tincture, drying the root	Anti-inflammatory, antibiotic
Pimpinella saxifraga L. (Apiaceae)	Bibernelle, Bockwurz (S)	W	Roots		1 *	Tincture, drying the root	Anti-inflammatory, Antibiotic
Pinus cembra L. (Pinaceae)	Cirmolo, Cimbro (T)	W	Cones	3		Syrup	cough
Pinus mugo Turra (Pinaceae)	Pino mugo, Mughi (T)	W	Buds	8 *		Syrup, oil	cough
Plantago lanceolata L. (Plantaginaceae)	Piantaggine lanceolata (T), Spitzwegerich, Heilwegerich, Foie dei tai (S)	W	Leaves, roots, seeds	3	8 *	Tea, tincture, salve, bath additive, oil, topical application	Cough, insect bites, bronchial disorders, nervous system strengthener, eye support supplement, respiratory support supplement, hormone regulation
Plantago major L. (Plantaginaceae)	Breitwegerich, Wegtritt (S)	W	Leaves		2 *	Tea, salve	Cough, earache, sore throat foot blisters
Polypodium vulgare L. (Polypodiaceae)	Tüpfelfarn (S)	W	Roots		1 *	Tea, tincture	Mild laxative, diuretic
Potentilla erecta (L.) Raeusch. (Rosaceae)	Blutwurz, Rotwurz (S)	W	Roots		3 *	Tea, tincture, powder, salve	Throat disorders, stomach disorders, anti-inflammatory, astringent, wounds
Quercus petraea (Matt.) Liebl. (Fagaceae)	Eiche, Mosteiche (S)	W	Bark, acorns		1 *	Tea, tincture	Diarrhea, strengthener,
Quercus robur L. (Fagaceae)	Eiche, Mosteiche (S)	W	Bark, acorns		1 *	Tea, tincture	Diarrhea, strengthener,
Rhodiola rosea L. (Crassulaceae)	Rodiola rosea (T)	W	Roots	1		Tincture	Tonic
Ribes nigrum L. (Grossulariaceae)	Schwarze Johannisbeere, Ribisel (S)	W, C	Fruit, buds, leaves		1	Tea, tincture	Adrenal gland stimulator, anti-inflammatory, diarrhea, stomach pain
Rosa canina L. (Rosaceae)	Rosa canina (T), Hagebutte, Heckenrose (S)	W	Flowers, leaves, fruit, buds	2 *	2 *	Tea, powder, jam, oil	Immune system booster, allergies, vitamin C source, chronic bladder diseases, bronchitis, rheumatic complaints relief, arthrosis
Rumex acetosa L. (Polygonaceae)	Sauerampfer, Erba brusca (S)	W	Leaves		1 *	Soup, salad	Digestion stimulator, iron source
Salix spp. (Salicaceae)	Salice (T), Silberweide, Palmkätzchen, Salgar (S)	W	Bark, leaves, branches	2	1	Tea, tincture, salve, tincture	As aspirin

Table 1. *Cont.*

Botanical Taxon and Family	Local Vernacular Names	Status	Part(s) Used	Trentino Frequency of Quotation	Südtirol Frequency of Quotation	Preparation	Local Perceived Medicinal Uses or Treated Illnesses
Salvia officinalis L. (Lamiaceae)	Salbei, Königssalbei, Zahnblätter (S)	W, C	Leaves		5 *	Tea, oil infusion, salt, toothpaste	Lymphatic system cleanser, excessive sweating, anti-inflammatory properties
Salvia rosmarinus Spenn. (Lamiaceae)	Rosmarin, Antonkraut, Weihrauchkraut (S)	C	Flowers, leaves		4 *	Oil infusion, salve, tincture	Calmant, heart and circulation strengthener, digestive system strengthener, flatulence
Sambucus nigra L. (Viburnaceae)	Sambuco (T), Holunder, Hollerbusch (S)	W	Flowers, fruit, roots	8 *	5 *	Tea, tincture, salve, syrup, Hollermulla (jam), deep-fried flowers	Cold, cough, fever, antioxidant, antiviral, Immune system strengthener
Solidago virgaurea L. (Asteraceae)	Goldrute, Pferdskraut (S)	W	All parts of the flowering plant		3 *	Tea, tincture, salve	Kidney stones, bladder disorders, fungal infections
Sorbus aucuparia L. (Rosaceae)	Eberesche, Vogelbeere (S)	W	Fruit		1 *	Jam, liqueur, syrup, powder, tea	Diarrhea, wounds, vitamin C source
Symphytum officinale L. (Boraginaceae)	Consolida (T), Beinwell, Wundallheil (S)	W	Roots, leaves	2	6 *	Tea, tincture, salve, ointment, oil	Swelling relief, muscular pain, wounds
Taraxacum sect. *Taraxacum* F.H.Wigg. (Asteraceae)	Denti de can, Denti de leone, Zicoria, Cicoria (T), Löwenzahn, Pustblume (S)	W	Flowers, leaves, roots, buds	5 *	12 *	Tea, tincture, powder, oil	Metabolism activator, dissolves deposits in the joints, cholesterol, secretory gland activator, depurative, anemia, diuretic, skin rashes
Thymus serpyllum L. (Lamiaceae)	Timo selvatico (T)	W	Leaves	6 *		Oil	Cough
Thymus vulgaris L. (Lamiaceae)	Thymian, Marienbettstroh (S)	W, C	Flowers, leaves, roots		8 *	Tea, tincture, salt, oil infusion	Bronchitis and dry cough, metabolism stimulator, menstrual cramp relief
Tilia platyphyllos Scop. (Malvaceae)	Tiglio (T), Linde, Steinlinde, Tiar (S)	W	Flowers, leaves, bark, buds	2	4 *	Tea, oil infusion	Fever, sore throat, diaphoretic properties, antistress, relaxant
Tilia cordata Mill. (Malvaceae)	Tiglio (T), Linde, Steinlinde, Tiar (S)	W	Flowers, leaves, bark, buds	2	4 *	Tea, oil infusion	Fever, sore throat, diaphoretic properties, antistress; relaxant
Trifolium pratense L. (Fabaceae)	Wiesenklee, Futterklee, Rotklee (S)	W	Aerial parts		1	Tea, tincture, salve, syrup	Eye infections, blood cleansing, mind calmant
Tropaeolum minus L. (Tropaeolaceae)	Kapuzinerkresse (S)	C	All parts of the herb		1	Decoration, salad	Good for the kidneys, support supplement for the immune system, gives power, rheumatism
Tussilago farfara L. (Asteraceae)	Huflattich, Bachblümlein, Fohlenfuß (S)	W	Flowers, leaves		2 *	Tea, tincture, fresh leaves	Mucolytic, cough, skin rashes and burns
Urtica dioica L. (Urticaceae)	Ortica (T), Brennnessel, Donnernessel, Piola (S)	W	Aerial parts, roots, seeds	6 *	12 *	Aerial part: tea, fresh juice, to make Knödel, Spätzle; Seed: to eat on bread or in salad; Root: to make a powder	Depurative, anemia, diuretic, anti-inflammatory, hair loss
Valeriana officinalis L. (Caprifoliaceae)	Baldrian (S)	W	Roots		1 *	Tea, tincture, bath additive	Sleep disorders, calmant
Verbascum densiflorum Bertol. (Scrophulariaceae)	Könikskerze, Fackelblume (S)	W	Flowers, leaves		3 *	Tea	Painful cough, mucolytic
Verbascum thapsus L. (Scrophulariaceae)	Tasso barbasso, Tasso verbasco (T)	W	Flowers	1 *		N.D.	Cold
Veronica officinalis L. (Plantaginaceae)	Echter Ehrenpreis (S)	W	All parts of the flowering plant		1 *	Tea, tincture	Digestive disorders, respiratory diseases, rheumatism
Viola odorata L. (Violaceae)	Veilchen, Wohlriechendes (S)	W	Flowers		1 *	Tea, tincture, syrup, vinegar	Bronchitis
Viola tricolor L. (Violaceae)	Stiefmütterchen (S)	W	All parts of the flowering plant		1 *	Tea, tincture	Cough, acne

Interviewees reported that medicinal plants can cure numerous illnesses.

Different parts of medicinal plants, especially the aerial parts, leaves, and flowers, were used for the preparation of a variety of remedies.

The interviewees indicated that the collected medicinal plants are mainly used to prepare tinctures, oil, infusions, salves, teas, and powders. Herbs are also sometimes simply mixed and used in smoothies or juices (i.e., *Urtica dioica* and *Plantago lanceolata*). Jams and syrups are made from a variety of fruit, berries, and herbs. Examples include jam made from *Sambucus nigra* berries, which have antiviral and antimicrobial effects, and jam made from *Rosa canina* fruit to supplement vitamin C. Syrups are made, for example, from *Salvia rosmarinus* and *Lavandula angustifolia*. Many respondents replace table salt with aromatic herb-based salts as a healthier seasoning. Additionally, bath salts or bath additives are prepared with herbs such as *Thymus vulgaris*, *Plantago major*, *Urtica dioica*, and *Valeriana officinalis*.

The three medicinal plants most frequently mentioned by our respondents were *Urtica dioica* (13 interviewees), *Achillea millefolium* (12), and *Calendula officinalis* (10). The most commonly mentioned plant, *Urtica dioica*, is considered a good source of iron for the body. It detoxifies, refreshes the brain, strengthens the immune system, is a diuretic, induces labor during pregnancy, and supports women with menstrual and hormonal disorders. The plant can be dried and used as a tea or bath additive, freshly squeezed or blended as a juice or smoothie, and used, as is common, as a substitute for spinach in traditional dishes such as "Spätzle" (special kind of egg noodle) or "Knödel" (dumplings). Participants also mentioned some knowledge about nettle, possibly arising from popular media, i.e., that the root, which can be eaten dried or fresh, has good properties for the liver, and that the seeds are rich in nutraceuticals. *Achillea millefolium* was described by the interviewees as having various positive effects on the body. It not only has a calming and antidepressant effect but also relieves stomach pain, detoxifies the body, stimulates the appetite, helps women with menstruation and menopause, and contains all the necessary trace elements. According to our interviewees, the whole plant can be cut approximately a hand's width from the ground during the flowering period, tied into a bouquet, and dried upside down. *Achillea millefolium* can also be used as a tea, as a bath infusion for menstrual cramps, as a culinary mineral salt, for example in soups, and as a seasoning powder. The third most frequently mentioned plant was *Calendula officinalis*, which is said to be good for digestion and to help with diarrhea. It can be applied to closed wounds to promote healing and improve scarring. *Calendula officinalis* also helps to refresh the skin. As another external treatment, it can be applied as a cream or oil on bruises. The herbs are infused in oil for at least three weeks, which can then be further processed into a cream. The infused oil can also be consumed to alleviate stomach pain. Finally, it is often prepared as a tea.

Our interviewees gave several pieces of advice on the proper use of medicinal herbs. For instance, several interviewees advised that, when infusing medicinal herbs, it is important to steep them only briefly, usually about two to three minutes, so as not to extract the wrong active ingredients and burn the herbs. Infused oils and creams should not be stored for too long to avoid rancidity. It was also recommended that attention be paid to the correct dosage of the active ingredients, as interviewees know that overdoses may have harmful or undesirable side effects.

In South Tyrol, several interviewees stated that gathering medicinal herbs was part of their childhood, a family activity associated with hiking in the region's mountains. Most often, mothers and grandmothers possessed the traditional knowledge of local medicinal plants. Many participants mentioned a well-known saying of their grandmothers: "For every ailment, there is an herb". From the recollections of our interviewees, we learned that many of these collected plants were subsequently incorporated into a variety of dishes cooked by their mother or grandmother and that they were also used to treat numerous diseases. Thus, female family members were often the primary source of their knowledge and interest in traditional medicinal herbs. Knowledge of local medicinal herbs was acquired not only from female family members, but also through a long-standing personal interest in wild plants that led to self-study via books, literature, or participation in formal courses, lectures, workshops, and seminars. The young people we interviewed liked to use books to acquire knowledge about medicinal plants, as did elderly individuals who used many different "Kräuterbüchlen" (literally: books about herbs). When reviewing the utilized literature, it is evident that it consisted mostly of writings by the authors Maria Treben, Hildegard von Bingen, and Gottfried Hochgruber. These authors enjoy a great reputation, mainly because of the success respondents have experienced healing themselves using their books.

2.2. Current Use of Wild Medicinal Plants in Val di Sole, Trentino

We documented the use of 36 species belonging to 21 families (20 angiosperms and a gymnosperm). The dominant families were Asteraceae with seven species, followed by Pinaceae with five species, and Rosaceae with four species. Among the mentioned medicinal plants, one was cultivated (*Laurus nobilis*) and three were semi-cultivated (*Alchemilla* spp., *Corylus avellana*, *Ficus carica*).

According to our interviewees, cough and other respiratory ailments were the most commonly cured with medicinal plants, along with problems of the digestive and integumentary systems. The top-mentioned medicinal plants included *Larix decidua* (12 interviewees) and *Malva sylvestris* (11), followed by *Sambucus nigra*, *Humulus lupulus*, *Gentiana lutea*, and *Pinus mugo* (8). The recipes of the mentioned preparations do not include exact measurements, as the preparations are never written down but rather passed on orally through generations; the required amount is often simply "a handful", "a bit", or "a basket".

One example of current medicinal use is the collection of the green buds of mountain pine (*Pinus mugo*) for the preparation of a cough syrup to be used during the winter months. Additionally, Icelandic moss (*Cetraria islandica*), a species that grows close to the ground in only a few high-altitude areas, is used to alleviate coughing and bronchitis when boiled with honey-sweetened milk. The most used and important root is that of gentian (*Gentiana lutea*). Interviewees reported the use of this root to flavor digestive liqueurs and aquavits. After the collection of the root between September and October, the cut-up roots are put into white aquavit for about three weeks. The end product is still frequently consumed and loved by the valley population. The root was once also used to prepare a depurative decoction to be drunk, especially by women, during the seasonal change from winter to spring.

Another fundamental ingredient was extracted from the responses of the interviewees: larch resin, "l'Argà" or "Resina del Lares" in the local dialect. This ingredient, mainly used in creams and unguents, is collected by a single man in the entire valley, and according to him, he is the last man in the world known to do this job. Larch resin collector, or "Largaiòl", is a laborious profession: each year larch trees are hand drilled at the base; the hole, around 40 cm in length, is then closed with a larch wood cork. After three to four years, the "Largaiòl" will then collect the resin with a special tool called a "sgorbia" and proceed to filter the resin. One larch tree produces, on average, 100 g of resin every four years, making the job extremely strenuous and time-consuming. The most widespread preparation with larch resin is "l'Ont dei Tai", an unguent used to remove splinters or to help the healing of infected cuts. Larch is also used to help alleviate coughs and colds by warming, in a double boiler, the resin and breathing in the steam while covering the head with a cloth.

Another well-known and used plant is dandelion (*Taraxacum* sect. *Taraxacum*). Depending on the area, the dialectal name changes from "zicoria" or "cicoria" to "dente de cagn" or "dente di leone". Young dandelion stems, which are less bitter, are eaten raw in salads. More mature and larger leaves are used to prepare a blood and liver depurative decoction. This plant was once picked everywhere, even along roadsides and within villages; however, today, because of pollution and agricultural expansion, it must be picked from other locations, far from roads and settlements to avoid pesticides and smog pollution. In the kitchen, dandelion is used to prepare gnocchi, frittata, and canederli (a regional dish of boiled dumplings). The consumption of this plant is so rooted in valley tradition that every year in April there is a festival called "Zicoria in Val di Rabbi".

In traditional phytotherapy, buds and young pinecones of mountain pine, larch, and stone pine, believed to have the highest concentration of medicinal properties, are used to prepare bud extracts. Bud extract is a product obtained through a 15- to 20-day maceration in alcohol and glycerine. After the maceration period, the liquid is filtered and bottled. The bud extracts are mainly used to strengthen immunity or as depuratives. As with all phytotherapy practices, the bud extracts need prolonged consumption, even two to three months, to have their full effect. Bud extracts are to be taken up to three times per day in variable dosages, in drops (from four to ten) under the tongue. Another method of extraction is tincture, in which the fresh plant is macerated in a solution of water and alcohol for up to 30 days. After this period, the liquid is then filtered and bottled. A more homemade and easier preparation is sugar syrups: pinecones and buds are put in jars with sugar and exposed to sunlight for 30 days to obtain a thick and aromatic syrup used to treat coughs and sore throats.

Even though the number and variety of plants and their curative properties are numerous, most of the interviewees use phytotherapy preparations in conjunction with conventional medicine. For smaller, less severe problems they seem to prefer phytotherapeutic methods, often reducing the severity of the illness via the consumption of a tincture, decoction, or bud extract. Most of the interviewees only gather plants near their houses, while a few individuals travel to specific areas and locations to pick only the best quality ingredients for traditional preparations.

In Val di Sole, the interviewees showed a strong interest in the gathering and usage of wild plants. Much of the information was learned from their parents and grandparents. Some of them reported gaining a renewed interest in wild plants following classes held by Eulalia Panizza, a local naturopath and important teacher in the valley.

2.3. Diachronic Comparison with Historical Sources

The comparison between medicinal plants mentioned in South Tyrol in summer 2022 (in Überetsch–Unterland) and those reported by Pick-Herk in 1995 [39] (in the Puster Valley) revealed that 67 species (out of a total of 74, corresponding to about 90%) were also reported in the past for medicinal use (Figure 1). Thus, only seven plants (*Leontopodium nivale, Dipsacus fullonum, Lamium galeobdolon, Tropaeolum minus, Leonurus cardiaca, Centaurea cyanus, Borago officinalis*) were not reported by Pick-Herk [39]. The Jaccard Index is 25.

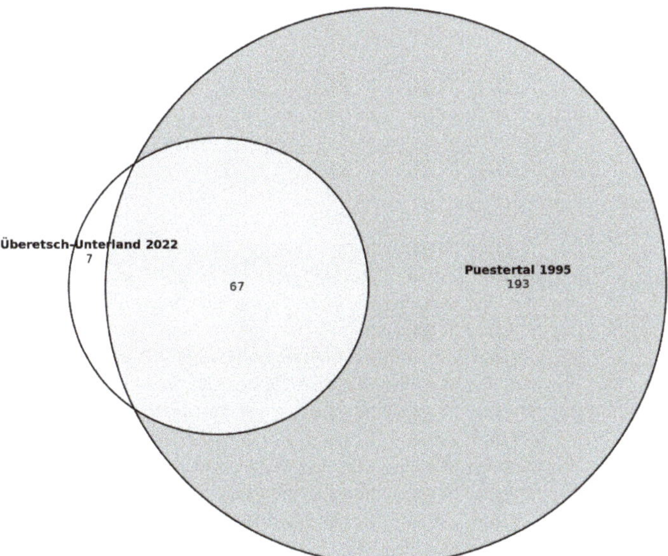

Figure 1. Venn diagram of the number of medicinal plant species in the Puster Valley (presumably collected in 1994) and Überetsch–Unterland (collected in 2022).

The comparison between plants mentioned in the study of Cappelletti and Fanzago [38] and our research in Val di Sole revealed that 23 species (out of a total of 37, corresponding to about 60%) were also reported in the past (Figure 2). Thus, 14 species currently used were not reported as used in the past in Val di Sole. The most common species mentioned now but not in the past are *Alchemilla alpina*, *Plantago lanceolata*, *Crataegus monogyna*, *Tilia platyphyllos*, *Ficus carica*, *Phyllanthus niruri*, *Symphytum officinale*, *Pinus cembra*, *Rhodiola rosea*, *Artemisia umbelliformis* subsp. *umbelliformis*, *Malva sylvestris*, *Salix* spp., *Gentiana lutea* and *Tilia cordata*. The Jaccard Index is 20.

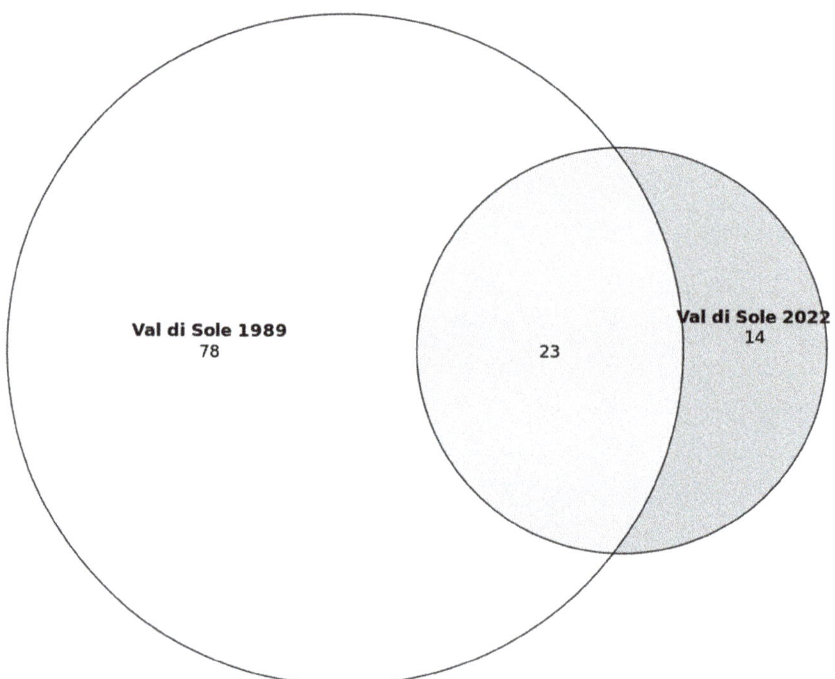

Figure 2. Venn diagram of the number of medicinal plant species in Val di Sole (collected in 1978–1979 and 2022).

2.4. Cross-Cultural Comparison

Nineteen medicinal plants were mentioned in both Val di Sole (Trentino) and Überetsch–Unterland (South Tyrol) (Figure 3). Twenty-one percent (*n* = 19) of the mentioned plant species were currently common to the two groups, while 20% (*n* = 18) were mentioned only in Val di Sole and 59% (*n* = 55) only in Überetsch–Unterland. The Jaccard Index is 21. The most common taxa include *Urtica dioica*, *Taraxacum* sect. *Taraxacum*, *Achillea millefolium*, *Hypericum perforatum* and *Sambucus nigra*. However, none of the top three mentioned plants are common to the two communities (Figure 4). Indeed, the top-cited species in Val di Sole was *Larix decidua* (11 interviewees), followed by *Malva sylvestris* (10) and *Pinus mugo*, *Humulus lupulus* L., *Sambucus nigra*, and *Gentiana lutea* (8 each). In Überetsch–Unterland, *Urtica dioica*, *Taraxacum* sect. *Taraxacum* and *Achillea millefolium*. were the most used plants (12 interviewees mentioned each plant), followed by *Calendula officinalis* (10) and *Arnica montana* (9).

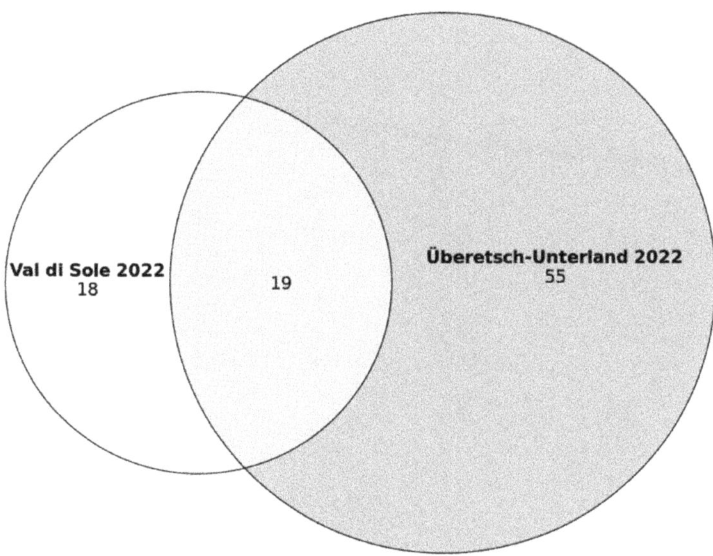

Figure 3. Venn diagram of the number of medicinal plant species in Val di Sole and Überetsch–Unterland (collected in 2022).

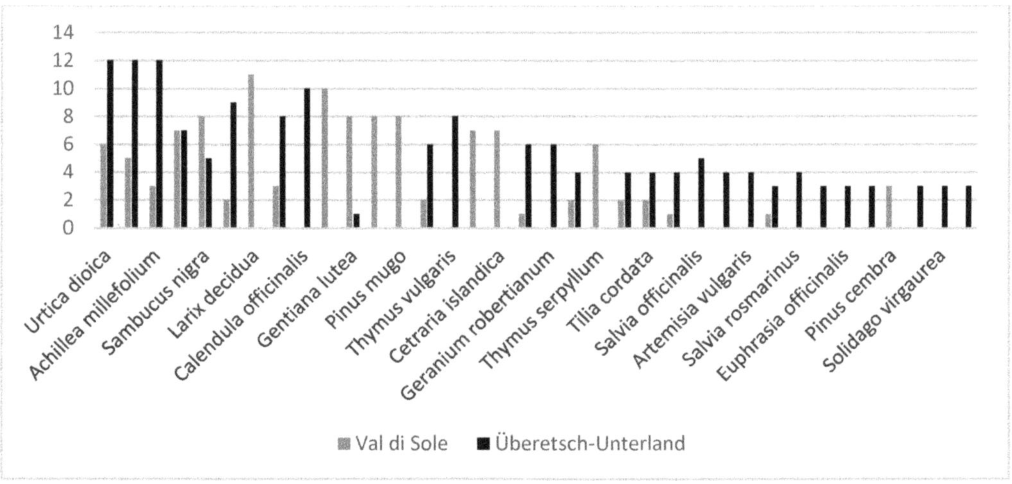

Figure 4. Comparison of plant taxa used by at least three interviewees in Val di Sole and Überetsch–Unterland.

The most frequently reported medicinal uses were for the digestive system (e.g., healing stomach and intestine) and the respiratory system (bronchitis, sore throat, cough, cold) (Figure 5). Moreover, supplements play an important role and several plants are used to this scope. In both Val di Sole and South Tyrol medicinal plants are also used to treat integumentary and genitourinary systems (although this was less relevant in Überetsch–Unterland).

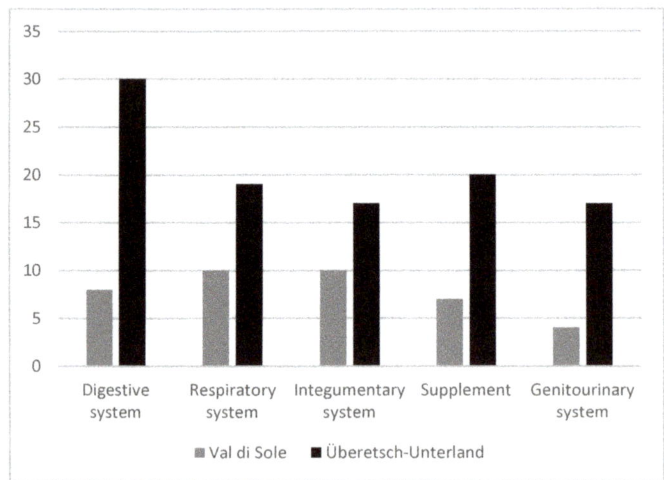

Figure 5. Top five apparatuses per number of mentioned ailments. Supplement refers to substances used as antiviral, antibacterial, vitamin and mineral supplements, energizers, etc. which do not refer to any specific apparatus.

The cross-cultural comparison of the historical sources reveals that 27% ($n = 77$) of the mentioned plant species were common to the two groups, while 8% ($n = 24$) were mentioned only in Val di Sole and 65% ($n = 55$) only in the Puster Valley (Figure 6). The Jaccard Index is 27.

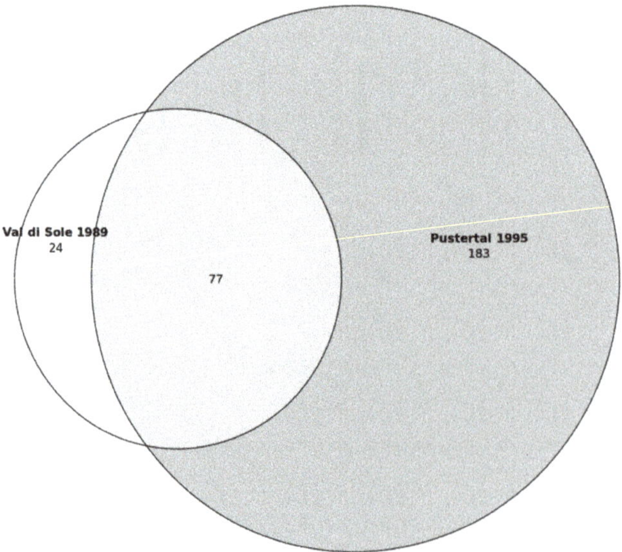

Figure 6. Venn diagram of the number of medicinal plant species in Val di Sole (collected in 1978–1979) and the Puster Valley (presumably collected in 1994).

3. Materials and Methods

3.1. Study Area

Val di Sole in the Autonomous Province of Trento is the largest branch of Val di Non in northwest Trentino (Figures 7 and 8). The valley floor extends from 700 m a.s.l. in Malè to over 3500 m a.s.l. on the mountain peaks that surround the valley. The population of Val di Sole, called Solander in the local dialect, totals approx. 15,000 inhabitants. Each of the 13 municipalities is characterized by numerous hamlets that were historically all independent municipalities. Even if the size of the population has not changed much, Val di Sole has changed drastically since the end of the Second World War and the economic boom that followed. Land-use changes, including the expansion of cultivated areas and the intensification of agriculture (particularly for apple, wine grape, and legume production), the opening of large ski resorts and the construction of new infrastructure (e.g., roads, hotels, mountain bike paths), have shaped the valley considerably.

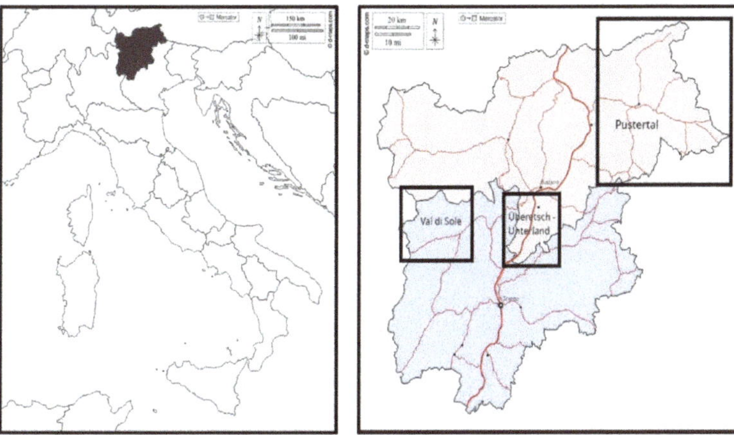

Figure 7. On the left, the Autonomous Provinces of Trento and Bozen–South Tyrol in northeastern Italy and, on the right, the location of the three case study areas of Val di Sole, Überetsch–Unterland, and Puster Valley (the latter for the historical comparison).

Figure 8. On the left, a typical panorama of Val di Sole (credit: Anna Segor); on the right, a typical view of Überetsch–Unterland (credit: Lena Seebacher).

Überetsch–Unterland (*Oltradige-Bassa Atesina* in Italian) is a district in the southern part of the Autonomous Province of Bozen–South Tyrol (Figures 7 and 8). It consists of 18 small municipalities (ranging from 393 to 18,000 inhabitants each). Among the 76,000 total inhabitants, about 67% belong to the German language group and 32% to the Italian language group [42]. The area is well known for its intense viticulture and fruticulture. With almost two million overnight stays per year, tourism makes an important contribution to income generation [42].

The main characteristics of the two studied valleys are summarized in Table 2.

Table 2. Main characteristics of our sample and the study areas.

Groups	Val di Sole	Überetsch–Unterland
Ecological environment of the area	Forested area with mainly spruce (*Picea abies* (L.) H.Karst.), fir (*Abies alba* Mill.) and stone pine (*Pinus cembra* L.) at higher altitudes; birch (*Betula* spp.), elder (*Alnus* spp.) and mixed shrubs at lower altitudes near pastures; large mountain pastures and cultivated areas, with mainly fruit trees	Area equally distributed between forests and agricultural surfaces (fruticulture and viticulture)
Altitudinal range (Highest village and altitude)	700–3757 m a.s.l. (Peio, 1200 m a.s.l.)	217–2439 m a.s.l. (Aldino, 1225 m a.s.l.)
Climate type (Koppen-Geiger)	Warm-summer humid continental climate	
Number of inhabitants of the studied area	Approximately 15,000	Approximately 76,000
Main economic sectors	Tourism, followed by animal husbandry and, in the lower valley, fruticulture	Mainly viticulture, fruit cultivation (mostly apple production), and tourism
Language and dialect	Italian, Solandro (local dialect)	German, South Tyrolean (local dialect), Italian
Historical background	Until 1815: Prince-Bishopric of Trent 1815–1918: part of the Habsburg Empire 1919: Paris Peace Conference annexed Trentino and South Tyrol to the Kingdom of Italy	Until 1919: Habsburg Empire 1919: Paris Peace Conference annexed Trentino and South Tyrol to the Kingdom of Italy 1946: Being mainly a German-speaking area, it was moved from Trentino to South Tyrol

3.2. Data Collection

Primary data were collected in July 2022. Interviews were carried out in both "Bezirks-gemeinschaft Überetsch–Unterland" ("Comunità Comprensioriale Oltradige-Bassa Atesina") in South Tyrol, consisting of 18 municipalities, as well as Val di Sole, and more specifically the villages of Folgarida, Malga di Dimaro, Dimaro, Mastellina, Costa Rotian, Castello, Pejo, Malè, Presson, Monclassico, San Bernardo in Val di Rabbi, and Pracorno. Ethnobotanical information was collected through qualitative semi-structured interviews with a total of 52 interviewees (30 in Bezirksgemeinschaft Überetsch–Unterland and 22 in Val di Sole) chosen via convenient sampling. The main recruitment criterion was expertise in local medicinal plants. No special consideration was given to socio-demographics, such as age, income, or education level. Personal contacts were used to select the first individuals for interviews, and then we used the snowball method [43]. We interviewed the local population using ten open-ended questions about their current and past use of medicinal plants by mentioning one part of the body at a time. We strictly followed the ethical guide-lines of the International Society of Ethnobiology [44]. Prior and Free Informed Consent was obtained before starting the interviews, which were conducted in German in South Tyrol and Italian in Val di Sole, with respondents using dialect and vernacular names. Verbal informed consent was always obtained before each interview, with the purpose of the study clearly stated; interviewees' data were later anonymized. The interviews were mostly conducted in person, although the location varied according to the respondents' wishes. In some cases, when informants invited us to their homes, they showed us dried parts of medicinal plants or gave us "Schnaps" (liqueur) or tinctures to taste. Because of the COVID-19 pandemic which started in 2020, some interviews had to be conducted by telephone. Depending on the expertise of the interviewees, the interviews lasted between

15 and 60 min. Upon agreement, the interviews were recorded. Botanical identification was carried out linking the local plant names to those recorded in previous field ethnobotanical studies (see following paragraph) that were conducted in the same area a few decades ago and in which proper taxonomic identifications were performed [38,39]. For the very few new taxa quoted in the current study only the identification was presumed upon their quoted common Italian and German plant names.

Secondary data for conducting the diachronic analysis were collected through a literature search of studies conducted through semi-structured interviews over 25 years ago in the area. This resulted in the discovery of an article pertaining to the plant-based ethnomedicine of Val di Sole, published in Italian in 1989 [38]. Unfortunately, no ethnobotanical data were available for the Überetsch–Unterland area, so we used the closest historical ethnobotanical research, which corresponded to a master thesis about medicinal plants of the Puster Valley completed at the University of Wien in 1995 [39]. The Puster Valley is in the northeastern region of South Tyrol (Figure 1), and it is a German-speaking valley, with a minority of Ladin-speaking people. Further details of the methodology of these sources are available in Table 3.

Table 3. Main characteristics of the four studies including the original data we collected and the published data for comparison [* SSI = semi-structured interviews].

	Trentino		South Tyrol	
Data	Cappelletti and Fanzago, 1989 [38]	Our data (2022)	Pick-Herk, 1995 [39]	Our data (2022)
Collection period	1978–1979	2022	Presumably, 1994	2022
Method	SSI *	SSI *	SSI *	SSI *
Number of interviews	N.D.	22 (average age 59 years old)	83	30 (average age 48 years old)
Interviewees	Local elderly individuals (only knowledge orally obtained was documented)	Local people	Local plant experts, including midwives, monks, and farmers	Local people
Included wild and cultivated plants	yes	only wild	yes	yes

3.3. Data Analysis

After conducting the interviews, we organized the data into two Excel tables, one for each of our case studies. For each plant taxon, we provided the vernacular names, the botanical family to which it belongs, how often our respondents mentioned it, whether it is a wild or cultivated medicinal plant, where it is most commonly found, what parts are used, and what medicinal uses it has. Plant species and families were verified against Plants of the Word Online and APG IV [45,46]. Medicinal uses were checked against the categories of ICD-11 [47] whenever possible. The tables were then merged, and two more columns were added to compare the obtained data with the historical data. In this case, we considered that use was also reported in historical data if the same species was reported with a medicinal purpose. We then used an online freely available tool for drawing Venn diagrams [48] and we calculated the Jaccard Index according to González-Tejero et al. [49].

4. Discussion

The results reveal two main findings. First, 75% of the plants currently in use were also used in the past. Second, the inhabitants of Val di Sole and Überetsch–Unterland have shared most of the local ecological knowledge related to medicinal plants over the past few decades, yet the most used species diverge, and in South Tyrol people appear to use a higher number of medicinal plants.

Before discussing our results, we would like to critically elaborate on the diachronic methodology. Although temporal comparison with previous studies is a valuable tool for better understanding the evolution of LEK, the methodology used (especially sample selection, the number of interviewees, and methods of data collection) is often vaguely described or altogether absent, which hampers precise comparability of the studies. For instance, in our comparison, we are missing information regarding the number of informants for the research conducted in Val di Sole in 1978–1979 [38] and the year of data collection for Pick-Herk [39]. Moreover, even when mentioned, data collection is rarely precisely described, thus undermining replicability. For instance, Pick-Herk [39] reported obtaining data from expert individuals, including farmers, monks, and midwives, without clearly describing their percentages and whether they reported knowledge acquired in the area. Finally, the evolving socio-economic context presents the interviewees and researchers with a different relationship given the different access to information and especially social media now and in the past.

The first finding relates to diachronic continuity in the use of medicinal plants in the studied areas, especially as most of the medicinal species reported by our interviewees in the summer of 2022 were reported in previous studies. This is not surprising considering that knowledge is mainly vertically transmitted. In South Tyrol, only 9% of the mentioned species were not reported in the past, while in Trentino 38% were not reported. We argue that various factors could have led to the adoption of "new" medicinal species. First, they could have been introduced through social media and other bibliographical sources; this could be the case for the use of *Tropaeolum minus*, *Leonurus cardiaca*, *Centaurea cyanus*, and *Borago officinalis*. In Trentino, the use of medicinal plants could have been influenced by the long-term activity of local herbalist Eulalia Panizza and her writings (which could be the case for *Alchemilla alpina*, *Phyllanthus niruri*, *Symphytum officinale*, *Rhodiola rosea*, and *Gentiana lutea*). Second, different taxonomic levels or species belonging to the same genus were reported for the same ailments. For example, in South Tyrol, *Leontopodium nivale* (Ten.) A. Huet ex Hand.-Mazz. was mentioned in our research, while *Leontopodium nivale* subsp. *Alpinum* (Cass.) Greuter was reported in historical sources. In Trentino, *Crataegus monogyna* Jacq. was not included in historical sources where only *Crataegus laevigata* (Poir.) DC. was mentioned. Similarly, *Tilia platyphyllos* Scop. and *Tilia cordata* Mill. were observed in 2022, while *Tilia × europaea* L. was reported by Cappelletti and Fanzago [38]. Third, the discrepancy between past and current medicinal species could be due to the methodology applied in the studies (our data collection in Val di Sole did not include cultivated plants, while Cappelletti and Fanzago [38] did include them). Finally, some species which are very important now, such as *Pinus cembra* L. in Trentino, were not mentioned in the past, possibly because of the recent expansion of forests as a consequence of the abandonment of mountain grasslands.

The second result indicates a similarity between the two groups as they show a fair amount of overlapping of medicinal plant use (19 taxa, corresponding to a fifth of the total). These include easily recognizable species with wide distribution, availability, and versatility (e.g., *Urtica dioica*, *Taraxacum* spp., *Hypericum* spp., *Equisetum arvense*, *Matricaria chamomilla*, *Rosa canina*) and other species specific to the Alpine region (*Arnica montana*, *Alchemilla vulgaris*, *Gentiana acaulis*). Despite some species being listed in the South Tyrolean Red Book [50], these seem to be popular medicinal plants in different areas of the Alps (e.g., [15,20,23,51]). The findings also reveal divergences in the most used plants. This may be due to differences in the landscape, as Val di Sole is mainly forested and thus conifers such as *Larix decidua* and *Pinus mugo* play an important role. These taxa, however, were not among the species mentioned in Überetsch–Unterland, likely because most of the interviewees live in lower altitude areas where agricultural land use dominates. Overall, our study found a richer corpus of local ecological knowledge in South Tyrol, which is consistent with the comparison of historical sources. This may be due to the intrinsic cultural nature of the area lying between the Italian and Mediterranean sphere and the German and Continental sphere from a geographical, linguistic, and cultural perspective. Such blending may have led to a richer corpus of knowledge that draws from these cultural, historical, geographical, and linguistic areas.

Our results are only partially in line with the findings of two recent studies conducted in the two autonomous provinces. In 2020 and 2021, Cavalloro et al. [37] interviewed

200 people across Trentino about their herbal ethnomedicinal knowledge. Two of the top five used plants mentioned in that study [37] (*Pinus mugo* and *Malva sylvestris*) were also reported by our participants in Val di Sole, while the other mentioned species were common ethnomedicinal remedies (*Achillea millefolium, Arnica montana, Hypericum perforatum, Larix decidua, Sambucus nigra, Gentiana lutea*). One top species in each study group was not mentioned in the other (*Satureja montana* was not mentioned in Val di Sole, while *Humulus lupulus* was mentioned in Val di Sole but not reported in the study by Cavalloro et al.).

Petelka et al. [36] conducted a literature review of the medicinal plants used in South Tyrol. The number of medicinal plants reported by Petelka and colleagues is much larger than the number of ethnomedicinal plants we found in Überetsch–Unterland (255 versus 74), possibly due to differences in methodology. We refrain from quantitative analysis because it is particularly difficult to compare data obtained from regional reviews with those obtained through semi-structured interviews conducted in more restricted areas. Among the most frequently cited species, one—*Urtica dioica*—was also among the most used species in the review by Petelka et al. [36], while the others were mentioned in both studies but were not among the most cited species in the other study (*Hypericum perforatum, Plantago lanceolata, Taraxacum* sect. *Taraxacum* and *Achillea millefolium*).

The study has, however, two main limitations: as in the majority of historical ethnobotanical investigations, the set of interviews and the area in which we conducted them was not large, and the actual adopted field methods may have not been the same as those in the historical research, as the earlier works used in the comparative analysis were published at a time when it was not yet common to precisely describe how the study participants were recruited (sampling) or how the interviews were conducted. This study sheds light on the importance of local ecological knowledge as an intangible heritage which nevertheless is undergoing a process of hybridization with different written sources of knowledge including social media and books published in Italian and/or German.

5. Conclusions

This diachronic and cross-cultural comparison of the ethnomedicine of Trentino and South Tyrol reveals that about 75% of the plants currently in use were also used in the past and that the inhabitants of Val di Sole and Überetsch–Unterland have shared the majority of medicinal plants over the past few decades, yet the most used species diverge and, in South Tyrol, people appear to use a higher number of medicinal plants. This highlights the dappled pattern of local medicinal plant use in a small but culturally diverse Alpine region. Indeed, mountain regions not only host rich biodiversity but are also often reservoirs of local ecological knowledge, which could serve as a basis for the sustainable use of local natural resources and local small-scale circular economies. For instance, in Trentino and South Tyrol, some smallholders run local businesses cultivating and foraging medicinal plants (e.g., Bergila in the Tures Valley or Schmiedthof in the Isarco Valley), or such local ecological knowledge is taught in seminars and activities for children and adults (e.g., the mountain school of Eulalia Panizza in Vermiglio, Val di Sole). Moreover, the revitalization of local wild plant-centered LEK should be further fostered by similar bottom-up initiatives.

Considering the complex biological and cultural diversity of this mountain area, more research is needed to further and better document the diachronic evolution of local medicinal plant knowledge as this could be pivotal in valorizing (and preserving) its embedded biocultural diversity.

Author Contributions: Conceptualization and methodology, G.M., A.P. and S.Z.; field investigations, F.G., M.H., A.S., A.T. and V.K.; writing—original draft preparation, G.M.; writing—review and editing, A.P., G.M. and S.Z.; supervision and funding acquisition, A.P. All authors have read and agreed to the published version of the manuscript.

Funding: The editing of the paper was funded by the University of Gastronomic Sciences, Pollenzo, Italy.

Data Availability Statement: All data are provided in the main article text.

Acknowledgments: We sincerely thank all the interviewees in both Val di Sole and Überetsch–Unterland for sharing their time and knowledge with us. We are also grateful to Giorgia Morselli for contributing to data collection.

Conflicts of Interest: The authors declare no conflict of interest.

References

1. Rahbek, C.; Borregaard, M.K.; Antonelli, A.; Colwell, R.K.; Holt, B.G.; Nogues-Bravo, D.; Rasmussen, C.M.Ø.; Richardson, K.; Rosing, M.T.; Whittaker, R.J.; et al. Building Mountain Biodiversity: Geological and Evolutionary Processes. *Science* **2019**, *365*, 1114–1119. [CrossRef] [PubMed]
2. Miranda Cebrián, H.; Font, X.; Roquet, C.; Pizarro Gavilán, M.; García, M.B. Assessing the Vulnerability of Habitats through Plant Rarity Patterns in the Pyrenean Range. *Conserv. Sci. Pract.* **2022**, *4*, e12649. [CrossRef]
3. Gurung, A.B.; Bokwa, A.; Chełmicki, W.; Elbakidze, M.; Hirschmugl, M.; Hostert, P.; Ibisch, P.; Kozak, J.; Kuemmerle, T.; Matei, E.; et al. Global Change Research in the Carpathian Mountain Region. *Mt. Res. Dev.* **2009**, *29*, 282–288. [CrossRef]
4. Rew, L.J.; McDougall, K.L.; Alexander, J.M.; Daehler, C.C.; Essl, F.; Haider, S.; Kueffer, C.; Lenoir, J.; Milbau, A.; Nuñez, M.A.; et al. Moving up and over: Redistribution of Plants in Alpine, Arctic, and Antarctic Ecosystems under Global Change. *Arct. Antarct. Alp. Res.* **2020**, *52*, 651–665. [CrossRef]
5. Cattaruzza, M. *L'Italia e il Confine Orientale. 1866–2006*; Il Mulino: Bologna, Italy, 2008.
6. Feichtinger, J.; Cohen, G.B. *Understanding Multiculturalism: The Habsburg Central European Experience*; Berghahn Books Incorporated: New York, NY, USA, 2014; ISBN 978-1-78238-265-2.
7. Di Michele, A.D. *L'italianizzazione Imperfetta: L'amministrazione Pubblica dell'Alto Adige tra Italia Liberale e Fascismo*; Edizioni dell'Orso: Alessandria, Italy, 2003; ISBN 978-88-7694-637-0.
8. Leonardi, M.M.V. Famiglie Plurilingui in Alto Adige. Pratiche Linguistiche e Appartenenza Linguistica. In *Studi AItLA 11: Lingue Minoritarie tra Localismi e Globalizzazione*; Marra, A., Dal Negro, S., Eds.; Studi AitLA: Milano, Italy, 2020; pp. 167–181.
9. Mattalia, G.; Stryamets, N.; Pieroni, A.; Sõukand, R. Knowledge Transmission Patterns at the Border: Ethnobotany of Hutsuls Living in the Carpathian Mountains of Bukovina (SW Ukraine and NE Romania). *J. Ethnobiol. Ethnomed.* **2020**, *16*, 41. [CrossRef]
10. Mattalia, G.; Stryamets, N.; Grygorovych, A.; Pieroni, A.; Sõukand, R. Borders as Crossroads: The Diverging Routes of Herbal Knowledge of Romanians Living on the Romanian and Ukrainian Sides of Bukovina. *Front. Pharmacol.* **2021**, *11*, 598390. [CrossRef] [PubMed]
11. Kazancı, C.; Oruç, S.; Mosulishvili, M. Medicinal Ethnobotany of Wild Plants: A Cross-Cultural Comparison around Georgia-Turkey Border, the Western Lesser Caucasus. *J. Ethnobiol. Ethnomed.* **2020**, *16*, 71. [CrossRef] [PubMed]
12. Sõukand, R.; Pieroni, A. The Importance of a Border: Medical, Veterinary, and Wild Food Ethnobotany of the Hutsuls Living on the Romanian and Ukrainian Sides of Bukovina. *J. Ethnopharmacol.* **2016**, *185*, 17–40. [CrossRef] [PubMed]
13. Salsa, A. *Il Tramonto delle Identità Tradizionali. Spaesamento e Disagio Esistenziale Nelle Alpi*; Priuli & Verlucca: Scarmagno, Italy, 2007; ISBN 978-88-8068-378-0.
14. Bellia, G.; Pieroni, A. Isolated, but Transnational: The Glocal Nature of Waldensian Ethnobotany, Western Alps, NW Italy. *J. Ethnobiol. Ethnomed.* **2015**, *11*, 37. [CrossRef] [PubMed]
15. Mattalia, G.; Quave, C.L.; Pieroni, A. Traditional Uses of Wild Food and Medicinal Plants among Brigasc, Kyé, and Provençal Communities on the Western Italian Alps. *Genet. Resour. Crop Evol.* **2013**, *60*, 587–603. [CrossRef]
16. Cornara, L.; La Rocca, A.; Terrizzano, L.; Dente, F.; Mariotti, M.G. Ethnobotanical and Phytomedical Knowledge in the North-Western Ligurian Alps. *J. Ethnopharmacol.* **2014**, *155*, 463–484. [CrossRef] [PubMed]
17. Fontefrancesco, M.F.; Pieroni, A. Renegotiating Situativity: Transformations of Local Herbal Knowledge in a Western Alpine Valley during the Past 40 Years. *J. Ethnobiol. Ethnomed.* **2020**, *16*, 58. [CrossRef] [PubMed]
18. Danna, C.; Poggio, L.; Smeriglio, A.; Mariotti, M.; Cornara, L. Ethnomedicinal and Ethnobotanical Survey in the Aosta Valley Side of the Gran Paradiso National Park (Western Alps, Italy). *Plants* **2022**, *11*, 170. [CrossRef] [PubMed]
19. Bottoni, M.; Milani, F.; Colombo, L.; Nallio, K.; Colombo, P.S.; Giuliani, C.; Bruschi, P.; Fico, G. Using Medicinal Plants in Valmalenco (Italian Alps): From Tradition to Scientific Approaches. *Molecules* **2020**, *25*, 4144. [CrossRef]
20. Vitalini, S.; Puricelli, C.; Mikerezi, I.; Iriti, M. Plants, People and Traditions: Ethnobotanical Survey in the Lombard Stelvio National Park and Neighbouring Areas (Central Alps, Italy). *J. Ethnopharmacol.* **2015**, *173*, 435–458. [CrossRef]
21. Dei Cas, L.; Pugni, F.; Fico, G. Tradition of Use on Medicinal Species in Valfurva (Sondrio, Italy). *J. Ethnopharmacol.* **2015**, *163*, 113–134. [CrossRef]
22. Vitalini, S.; Iriti, M.; Puricelli, C.; Ciuchi, D.; Segale, A.; Fico, G. Traditional Knowledge on Medicinal and Food Plants Used in Val San Giacomo (Sondrio, Italy)—An Alpine Ethnobotanical Study. *J. Ethnopharmacol.* **2013**, *145*, 517–529. [CrossRef]
23. Mattalia, G.; Sõukand, R.; Corvo, P.; Pieroni, A. Dissymmetry at the Border: Wild Food and Medicinal Ethnobotany of Slovenes and Friulians in NE Italy. *Econ. Bot.* **2020**, *74*, 1–14. [CrossRef]
24. Poncet, A.; Schunko, C.; Vogl, C.R.; Weckerle, C.S. Local Plant Knowledge and Its Variation among Farmer's Families in the Napf Region, Switzerland. *J. Ethnobiol. Ethnomed.* **2021**, *17*, 53. [CrossRef]

25. Kalle, R.; Sõukand, R. Historical Ethnobotanical Review of Wild Edible Plants of Estonia (1770s–1960s). *Acta Soc. Bot. Pol.* **2012**, *81*, 271–281. [CrossRef]
26. Sõukand, R.; Kalle, R. The Appeal of Ethnobotanical Folklore Records: Medicinal Plant Use in Setomaa, Räpina and Vastseliina Parishes, Estonia (1888–1996). *Plants* **2022**, *11*, 2698. [CrossRef]
27. Prakofjewa, J.; Anegg, M.; Kalle, R.; Simanova, A.; Prūse, B.; Pieroni, A.; Sõukand, R. Diverse in Local, Overlapping in Official Medical Botany: Critical Analysis of Medicinal Plant Records from the Historic Regions of Livonia and Courland in Northeast Europe, 1829–1895. *Plants* **2022**, *11*, 1065. [CrossRef]
28. Svanberg, I. The Use of Wild Plants as Food in Pre-Industrial Sweden. *Acta Soc. Bot. Pol.* **2012**, *81*, 317–327. [CrossRef]
29. Kalle, R.; Sõukand, R. The Name to Remember: Flexibility and Contextuality of Preliterate Folk Plant Categorization from the 1830s, in Pernau, Livonia, Historical Region on the Eastern Coast of the Baltic Sea. *J. Ethnopharmacol.* **2021**, *264*, 113254. [CrossRef] [PubMed]
30. Kalle, R.; Pieroni, A.; Svanberg, I.; Sõukand, R. Early Citizen Science Action in Ethnobotany: The Case of the Folk Medicine Collection of Dr. Mihkel Ostrov in the Territory of Present-Day Estonia, 1891–1893. *Plants* **2022**, *11*, 274. [CrossRef]
31. Kujawska, M.; Klepacki, P.; Łuczaj, Ł. Fischer's Plants in Folk Beliefs and Customs: A Previously Unknown Contribution to the Ethnobotany of the Polish-Lithuanian-Belarusian Borderland. *J. Ethnobiol. Ethnomed.* **2017**, *13*, 20. [CrossRef] [PubMed]
32. Berisha, R.; Sõukand, R.; Nedelcheva, A.; Pieroni, A. The Importance of Being Diverse: The Idiosyncratic Ethnobotany of the Reka Albanian Diaspora in North Macedonia. *Diversity* **2022**, *14*, 936. [CrossRef]
33. Köhler, P.; Bystry, A.; Łuczaj, Ł. Plants and Other Materials Used for Dyeing in the Present Territory of Poland, Belarus and Ukraine According to Rostafiński's Questionnaire from 1883. *Plants* **2023**, *12*, 1482. [CrossRef]
34. Schippmann, U.; Leaman, D.; Cunningham, A. Impact of Cultivation and Gathering of Medicinal Plants on Biodiversity: Global Trends and Issues. In *Biodiversity and the Ecosystem Approach in Agriculture, Forestry and Fisheries*; FAO: Rome, Italy, 2002; pp. 142–167.
35. Schunko, C.; Lechthaler, S.; Vogl, C.R. Conceptualising the Factors That Influence the Commercialisation of Non-Timber Forest Products: The Case of Wild Plant Gathering by Organic Herb Farmers in South Tyrol (Italy). *Sustainability* **2019**, *11*, 2028. [CrossRef]
36. Petelka, J.; Plagg, B.; Säumel, I.; Zerbe, S. Traditional Medicinal Plants in South Tyrol (Northern Italy, Southern Alps): Biodiversity and Use. *J. Ethnobiol. Ethnomed.* **2020**, *16*, 74. [CrossRef]
37. Cavalloro, V.; Robustelli Della Cuna, F.S.; Quai, E.; Preda, S.; Bracco, F.; Martino, E.; Collina, S. Walking around the Autonomous Province of Trento (Italy): An Ethnobotanical Investigation. *Plants* **2022**, *11*, 2246. [CrossRef] [PubMed]
38. Cappelletti, E.M.; Fanzago, N. Fitoterapia Tradizionale Della Val Di Sole (Trentino Occidentale): Phytotherapy in the Traditional Medicine of Val Di Sole (Trento, Italy). *Studi Trentini Sci. Nat. Acta Biol.* **1989**, *66*, 17–68.
39. Pickl-Herk, W. Volksmedizinische Anwendung von Arzneipflanzen Im Norden Südtirols. Unpublished. Master's Thesis, University of Vienna, Vienna, Austria, 1995.
40. Petelka, J.; Bonari, G.; Säumel, I.; Plagg, B.; Zerbe, S. Conservation with Local People: Medicinal Plants as Cultural Keystone Species in the Southern Alps. *Ecol. Soc.* **2022**, *27*, 14. [CrossRef]
41. Scherrer, M.M.; Zerbe, S.; Petelka, J.; Säumel, I. Understanding Old Herbal Secrets: The Renaissance of Traditional Medicinal Plants beyond the Twenty Classic Species? *Front. Pharmacol.* **2023**, *14*, 1141044. [CrossRef]
42. Provincia Autonoma di Bolzano; Istituto Provinciale di Statistica (ASTAT). Banche dati e dati comunali. Available online: https://astat.provincia.bz.it/it/banche-dati-comunali.asp (accessed on 15 June 2023).
43. Zocchi, D.M.; Mattalia, G.; Aziz, M.A.; Kalle, R.; Fontefrancesco, M.F.; Sõukand, R.; Pieroni, A. Searching for Germane Questions in the Ethnobiology of Food Scouting. *J. Ethnobiol.* **2023**, *43*, 19–30. [CrossRef]
44. International Society of Ethnobiology. ISE Code of Ethics. Available online: https://www.ethnobiology.net/what-we-do/core-programs/ise-ethics-program/code-of-ethics/ (accessed on 15 June 2023).
45. POWO Plants of the World Online | Kew Science. Available online: https://powo.science.kew.org/ (accessed on 24 April 2023).
46. The Angiosperm Phylogeny Group Update of the Angiosperm Phylogeny Group Classification for the Orders and Families of Flowering Plants: APG IV | Botanical Journal of the Linnean Society | Oxford Academic. Available online: https://academic.oup.com/botlinnean/article/181/1/1/2416499 (accessed on 2 June 2023).
47. WHO. ICD-11 for Mortality and Morbidity Statistics. Available online: https://icd.who.int/browse11/l-m/en#/http://id.who.int/icd/entity/224336967/mms/unspecified (accessed on 14 June 2023).
48. Draw Venn Diagram. Available online: http://bioinformatics.psb.ugent.be/webtools/Venn/ (accessed on 24 April 2023).
49. González-Tejero, M.R.; Casares-Porcel, M.; Sánchez-Rojas, C.P.; Ramiro-Gutiérrez, J.M.; Molero-Mesa, J.; Pieroni, A.; Giusti, M.E.; Censorii, E.; de Pasquale, C.; Della, A.; et al. Medicinal Plants in the Mediterranean Area: Synthesis of the Results of the Project Rubia. *J. Ethnopharmacol.* **2008**, *116*, 341–357. [CrossRef]
50. Wilhalm, T.; Hilpold, A. Rote Liste der Gefährdeten Gefäßpflanzen Südtirols. *Glederiana* **2006**, *6*, 115–198.
51. Abbet, C.; Mayor, R.; Roguet, D.; Spichiger, R.; Hamburger, M.; Potterat, O. Ethnobotanical Survey on Wild Alpine Food Plants in Lower and Central Valais (Switzerland). *J. Ethnopharmacol.* **2014**, *151*, 624–634. [CrossRef]

Article

From Şxex to *Chorta*: The Adaptation of Maronite Foraging Customs to the Greek Ones in Kormakitis, Northern Cyprus

Andrea Pieroni [1,2,*], Naji Sulaiman [3], Zbynek Polesny [3] and Renata Sõukand [4]

[1] University of Gastronomic Sciences, 12042 Bra, Italy
[2] Department of Medical Analysis, Tishk International University, Erbil 44001, Iraq
[3] Department of Crop Sciences and Agroforestry, Faculty of Tropical AgriSciences, Czech University of Life Sciences Prague, Kamýcká 129, 16500 Praha-Suchdol, Czech Republic
[4] Department of Environmental Sciences, Informatics and Statistics, Ca' Foscari University of Venice, 30172 Venezia, Italy
* Correspondence: a.pieroni@unisg.it

Abstract: The traditional foraging of wild vegetables (WVs) has played an important role in the post-Neolithic development of rural local food systems of the Near East and the Mediterranean. This study assessed the WVs gathered by the ancient Maronite Arabic diaspora of Kurmajit/Kormakitis village in Northern Cyprus and compared them with those gathered by their Cypriot and Arab Levantine neighbors. An ethnobotanical field survey focusing on WVs was conducted via twenty-two semi-structured interviews among the few remaining Maronite elderly inhabitants (approximately 200); and the resulting data were compared with those described in a few field studies previously conducted in Cyprus, Lebanon, and coastal Syria. Wild vegetables in Kormakitis are grouped into a folk category expressed by the emic lexeme Şxex, which roughly corresponds to the Greek concept of *Chorta* (wild greens). The large majority of Şxex have Greek folk phytonyms and they overlap for the most part with the WVs previously reported to be gathered by Greek Cypriots, although a remarkable number of WVs are also shared with that of the other groups. The findings address a possible adaptation of Maronite WV foraging to the Greek one, which may be explained by the fact that the Maronite minority and the majority Greek communities lived side by side for many centuries. Additionally, after Turkish occupation in 1974, a remarkable migration/urbanization of Maronites to the main Greek centers on the southern side of the isle took place, and Kurmajit became part of Cypriot trans-border family networks.

Keywords: Cyprus; ethnobotany; Maronites; Mediterranean diet; wild greens

Citation: Pieroni, A.; Sulaiman, N.; Polesny, Z.; Sõukand, R. From Şxex to *Chorta*: The Adaptation of Maronite Foraging Customs to the Greek Ones in Kormakitis, Northern Cyprus. *Plants* **2022**, *11*, 2693. https://doi.org/10.3390/plants11202693

Academic Editor: Laura Cornara

Received: 24 September 2022
Accepted: 9 October 2022
Published: 12 October 2022

Publisher's Note: MDPI stays neutral with regard to jurisdictional claims in published maps and institutional affiliations.

1. Introduction

The Mediterranean diet (MD) was proposed for the first time in the cross-cultural epidemiological "Seven Countries Study" by the American nutritionist Ancel Benjamin Keys [1] and it concerns "food patterns typical of Crete, much of the rest of Greece, and southern Italy in the early 1960s, where adult life expectancy was among the highest in the world and rates of coronary heart disease, certain cancers, and other diet-related chronic diseases were among the lowest" [2]. This dietary system was recognized one decade ago as a UNESCO Intangible Cultural Heritage of Humanity [3] and has been described by an important spectrum of bioscientific literature, which have recently not only underlined its major health and nutrition benefits, but also its low environmental impacts, high sociocultural value, and positive local economic returns [4,5].

However, systematic detailed studies on the foraging and culinary use of wild plant ingredients in the Eastern Mediterranean and the Near East, and how these change cross-temporally and cross-culturally, are rare, despite the fact that investigating these regions is essential for better understanding the possible development and evolution of the MD. In fact, it has been proposed that the spread of the use of the wild vegetables (WVs) present in the MD may have originated in Neolithic settlements of the Fertile Crescent, where these plant resources often represented local foods and medicines [6,7].

The ethnobotany of WV foraging is, however, more generally relevant to two main environmental and social sustainability pillars: (a) community-centered sustainable management of local biodiversity resources; and (b) wild food plant cuisine heritagization; i.e., the revitalization of knowledge concerning wild plant ingredients into new educational platforms for increasing overall community botanical literacy, as well as in possible new gastronomic arenas and local-food-centered rural eco-tourism economic initiatives.

In the field of wild food ethnobotany, over the past 20 years, Cyprus has been the object of two main field studies: one conducted among the Greek population in the south [8] and one focusing on the Turkish population in the NE of the isle [9].

On the other hand, our research groups have devoted considerable effort in the past two decades to understanding how minority linguistic, ethic, and religious groups in Eastern Europe and the Near East strengthen or adapt/homogenize their identities to the majority system through wild food and medicinal plants, given the fact that these processes could be relevant to the revitalization of food or herbal heritage [7,10–17], as well as educational platforms devoted to the celebration of local bio-cultural diversity [18].

We therefore decided to conduct ethnobotanical research on WVs among the Maronites of Cyprus, who moved into the area between the 8th and 13th Centuries and still represent the totality of the population of Kurmajit/Kormakitis village [19].

The specific objectives of this research were:

(a)   to record local phytonyms and traditional uses of WVs in Kormakitis; and
(b)   to compare the gathered data with that collected and reported for Greek and Turkish Cyprus, Lebanon, and coastal Syria.

2. Materials and Methods

2.1. Study Village and Brief Historical Background

The field study was carried out in the small village of Kormakitis, Northern Cyprus, in May 2022 (Figures 1 and 2). Most of the Maronites in Kormakitis came from a Lebanese village called Kour located in the Batroun district of Northern Lebanon [20]. The name of the studied village can be transcribed in the spoken Lebanese dialect as "Kor ma giti", which literally means "Kour did not come", referring to the fact that Maronites emigrated from the village of Kour, but were not able to bring the village with them; this may somehow reflect the Maronite migrants' nostalgia for their homeland and the original Arabic identity of the village. Although Maronites represent one of the main ethno-religious groups in the Eastern Mediterranean and preserve an essential part of the region's identity, they are still underdocumented from an ethnobotanical perspective. The uniqueness of the group is rooted in their merger of eastern Assyrian culture and rituals on one side, and the Church of Rome on the other [21]. Historically, the group originated in the Orontes Valley of present-day Syria. Between the 8th and 10th centuries, a mass migration of the group led to the transfer of the Maronite patriarchal residence to Mount Lebanon [19]. Currently, Maronites live mainly in Lebanon, where they form a quarter of the country's population, as well as in Syria, France, USA, Mexico, and South America. Similar to other ethnic-religious groups of the region, Maronites have suffered from instability in the region over the past millennia due to various empires frequently clashing over the eastern Mediterranean (e.g., Byzantine, Persian, Roman, Umayyad, Abbasi, and Ottoman), political riots, and intricate and delicate religious affairs. Therefore, many members of the group have migrated out of the region, with Cyprus being one of the destinations [21].

Figure 1. Location of Kormakitis in Cyprus.

Figure 2. View of Kormakitis.

Kormakitis (170 m.a.s.l.) is home to approximately 300, mainly elderly, Maronite Christian inhabitants. Its surrounding rural landscape is characterized by olive orchards and the traditional local economy is based on small-scale farming. The village has a bar that serves as a gathering place for local senior citizens where they chat and play backgammon, and a family-run small restaurant famous all over the region, which serves lamb baked in the traditional Cypriot oven (*Kleftiko*) (Figure 3).

Figure 3. View of the village center, with private courtyard, grape pergola, and traditional external baking oven.

The village remained an enclave of Turkish Cyprus after division of the island in the 1970s. Those who were born there and continue living in the village receive bi-weekly food provisions by the Republic of Cyprus (Greek Cypriot) authorities via the UN troops. However, tension in the village is still high and can be sensed in the way people behave, especially in stressful situations. We witnessed a fire in a nearby village and the inhabitants were very worried because of recent experiences of help arriving late, and also possibly by signs on the streets (such as vandalized EU-support project signs), since the general feeling of the people is that they have been abandoned by Western Europeans with the exception of the Catholic Pope. The inhabitants, especially the elderly, speak a specific dialect of Arabic (Cypriot Maronite Arabic) as their mother tongue while being fluent in Greek and Turkish as well. An elderly woman recalled that Arabic was their cultural savior after the Turkish occupation, and Kormakits is the only active and entirely Christian village in present-day Northern Cyprus.

2.2. Ethnobotanical Field Study

The ethnobotanical field study was carried out in Kormakitis in May 2022. The main purpose of the survey was to record local knowledge of wild vegetables currently gathered and consumed by locals. Twenty-two elderly residents (range: 64 to 84 years old), especially rural farmers and elderly women who were considered potential WV local knowledge holders in the area, were recruited through snowball techniques to participate in semi-structured interviews. Prior to each interview, verbal consent was obtained from each of the participants and the Code of Ethics adopted by the International Society of Ethnobiology was followed [22]. Full anonymity was observed during the interviews. For each of the WVs free listed during the study, the local name and local food uses were documented.

We deliberately excluded from the survey other wild food plants, such as wild fruits. The quoted wild food taxa were collected from the study area, when available, and identified by the authors using standard reference works and checklists [23,24].

Voucher specimens (bearing the accession code UVVETBOT) were deposited at the Herbarium of the Biocultural Diversity Lab of the Department of Environmental Sciences, Informatics and Statistics, Ca' Foscari University of Venice, Italy. The identification of wild plants that were not available during the field study was conducted on the basis of folk names and detailed plant descriptions; in this case, pictures of the presumed plants were shown to the study participants after a preliminary evaluation of the quoted folk name and description. Nomenclature always followed The World Flora Online database [25], while plant family assignments were consistent with the Angiosperm Phylogeny website [26]. Recorded local Arabic plant names were reported in Romanized transliterations following standard reference works [27].

2.3. Data Analysis

A historical comparison was conducted in May–August 2022, analyzing the data gathered in the current study together with those reported by three food ethnobotanical studies conducted in the past two decades in Cyprus among Greeks [8] and Turks [9], Lebanon [28], and coastal Syria [29]. In addition, we used both our observations and unpublished data from the field work conducted in coastal Syria for qualitative interpretation of our present data from Kormakitis.

3. Results

We recorded the use of twenty-nine folk plant taxa used as leafy vegetables, corresponding to thirty-six botanical species (Table 1). The most represented family was Asteraceae (twelve species) and the use of eight genera (*Asparagus*, *Capparis*, *Crithmum*, *Cynara*, *Foeniculum*, *Glebionis*, *Portulaca*, and *Sinapis*) was named by more than 40% of interviewees.

Figure 4 shows the overlap between the WVs gathered and consumed in Kurmajit, and those of their Greek and Turkish neighbors and communities in Lebanon and Syria. The highest number of overlaps (twenty-eight out of thirty-two genera) are with the records from Greek Cyprus, four of them exclusively (*Amaranthus*, *Echinops*, *Glebionis*, *Onopordum*). Of the eighteen genera shared with Turkish Cyprus, only one taxon (*Rumex*) was shared exclusively, while twenty-one taxa were also used in Lebanon and coastal Syria. The only taxon found exclusively in the Maronite community was *Tordylium*, which was used as a seasoning.

Figure 5 shows the overlap of Cypriot Maronite folk plant names and those referring to WVs documented among Cypriot Greeks and Turks, as well as among Arabs in Lebanon and coastal Syria. The overlap with Greek Cypriot phytonyms is remarkable and includes approximately two thirds of the recorded folk plant names.

Table 1. Recorded Cypriot Maronite wild vegetables in Kormakitis, including their folk names, local culinary uses, frequency of quotation, and comparison to reports from Lebanon, coastal Syria, and Greek and Turkish Cyprus.

Botanical Taxon or Taxa, Botanical Family; Voucher Specimen Code (UVVETBOT)	Local Name(s)	Used Parts	Local Food Uses	Frequency of Quotation	Folk Names and Frequency of Quotation in Lebanon	Folk Names and Frequency of Quotation in Coastal Syria	Folk Names and Frequency of Quotation in Greek Cyprus	Folk Names in Turkish Cyprus
Allium ampeloprasum L., Amaryllidaceae; Cr01	**Agrioskordo, Skordo**	Whole plant	Salads, cooked, seasoning	+	Kurrat +++	Kerrat +++	Agriopraso, **Agrioskordo** +	Yabani pırasa
Amarantus album L., Amaranthaceae; Cr23	**Ghlindo**	Leaves	Cooked, especially with beans	++			**Glindos**, Vlito +	
Ammi majus L., Apiaceae	**Agriosellino**	Aerial parts	Salads	+		Khelleh +	**Arkoseleno** +	
Asparagus acutifolius L., Asparagaceae; CY09	**Agrelia, Agrenli**	Young shoots	Boiled or cooked with eggs	+++	Halyoun ++	Halyoun +++	**Agrelia** +	
Beta vulgaris L., Amaranthaceae; CY10	**Agriolahano, Lahanutkia**	Leaves	Boiled	+		Selq barri +	**Agriolachano**, Agrioteflo +	Yabani ıspinak, Pazı
Capparis spinosa L., Capparaceae; CY02	**Kappari**	Flower buds, unripe fruits, and young aerial parts	Pickled or in mixed salads	+++			**Kappari** +++	**Gabbar**
Centaurea calcitrapa L., Asteraceae; Cr78	**Trisagia**	Young rosettes	Cooked, especially with beans	++	Dardar, Dardrieh ++	Dardar, Qellaibeh ++	Agratsia, Atrachia, Atrachouna, **Trisatsia** +++	
Cichorium intybus L., Asteraceae; Cr97	**Radica**, Vallişxex	Rosettes	In mixed salads or boiled	+	Aalet, Barrieh, Hindbeh +++	Hendbeh +++	**Agrioradikia** ++	
Crithmum maritimum L., Apiaceae, CY74	**Kirdama, Kirtam**	Young aerial parts	Boiled, then pickled	+++			**Kirtamo** +	Deniz otu, **Girtama**
Cynara cornigea Lindl. and *C. cardunculus* L., Asteraceae; CY54 and CY58	**Agriaginara, Anginares, Hostos, Karkarua**	Flower receptacles and stems	Raw (consumed with lemon and salt), pickled, cooked	+++			**Agrioagkinara, Arzotsinara, Chosti, Kafkarua** +	**Hostes**, Gafgarit
Echinops spinossimum Turra, Asteraceae; CY01	Şerabit, Şrabir, Şrabit	Young stems	Raw as snacks or in salads	+			Saratzinos +	
Eruca vesicaria (L.) Cav., Brassicaceae; CY38	**Roxa**	Leaves	Salads	+	Jarjeer ++		**Roca** +	
Eryngium campestre L., Asteraceae; CY25	**Bangallo, Pangaros**	Tender upper roots and shoots	Pickled	++	Aarakbeen, Jibaaneh, Quraset el aaneh, Qursaneeh +++			Kazayağı, **Mangallo**
Foeniculum vulgare Mill., Apiaceae; Cr09, Cr19	**Şummar**	Aerial parts	Raw, cooked, seasoning (especially snails)	+++	**Shumar, Shumra** +++	Shamra ++	Marathos ++	Dere otu, Maraho, Rezene

Table 1. *Cont.*

Botanical Taxon or Taxa, Botanical Family; Voucher Specimen Code (UVVETBOT)	Local Name(s)	Used Parts	Local Food Uses	Frequency of Quotation	Folk Names and Frequency of Quotation in Lebanon	Folk Names and Frequency of Quotation in Coastal Syria	Folk Names and Frequency of Quotation in Greek Cyprus	Folk Names in Turkish Cyprus
Glebionis segetum Fourr., Asteraceae; CY07	Lazaros, **Similia**	Young shoots	Raw, as a snack and in salads	+++			**Similia** +	
Leopoldia comosa (L.) Parl., Asparagaceae; Cr75	Arkoskorto	Bulbs	Cooked in various ways, pickled	++	Agriohyacinthos +			
Malva sylvestris L., Malvaceae; CY08	**Moloxa**	Leaves	Boiled	++	Khubbaizeh +++	Khebbaizeh +++	**Molocha, Molochoua** ++	Gömeç
Mentha spicata L., Lamiaceae; CY45	**Nana**	Young leaves	Seasoning	++	**Naana barri** +++		Dyosmos +	**Dere nanesi, Yabani nana**
Nasturtium officinale W. T. Aiton, Brassicaceae; Cr34	Gardamo	Aerial parts	Salads	+		Jarjeer +++		
Notobasis syriaca (L.), Cass., Asteraceae; CY29	Şerabit, Şrabir, Şrabit	Young stems	Raw as snacks or in salads	+		Kherfesh, Qailouh, Shok aljamal ++	Gauroukavlos, Nerokavlos, Patsalokavlos +	Sütleğen, Tuzlu gavulya
Onopordum cyprium Eig, Asteraceae; CY19	Şerabit, Şrabir, Şrabit	Young stems	Raw as snacks or in salads	+			AsGaedouragkatho +	
Origanum syriacum (Boiss.) Kuntze and *O. majorana* L., Lamiaceae	**Rigani**	Flowering aerial parts	Seasoning	+	Zaatar +++	Zauba' +++	**Rigani**, Sapsissia +	Dağ kekiği
Pistacia terebinthus L. Anacardiaceae; CY04	**Tremiθia**	Fruits	Seasoning sausages	+	Shaashoub ++	Betem ++	**Tremithia** ++	Çitlembic, Menengiç
Portulaca oleracea L. aggr., Portulacaceae; CY11	**Glistirìa, Glistriδa,** Nistriδa	Aerial parts	Salads	+++	Baqleh, Farfahin +++	Beqail barriah ++	**Glystirida** +	Semiz otu
Rumex crispus L., Polygonaceae; CY8	**Laxana**, Xamedui, Zamedui	Leaves	Cooked, dolmades	+				Labada, Yabani ıspanak
Scolymus hispanicus L., Asteraceae; CY25	**Zalatuna**	Young shoots, tender peduncles, and rachis of leaves (sometimes with parts of the stem); underground part of stems and external coat of the roots	Cooked	++			Alatouna, Aspragkatho, Atrachounes, Christagkatho, Galaktites, **Galatouna**, Plotarka +++	Kara diken, Kara ot, Otluk, Sahura, Saracino
Silene vulgaris (Moench) Garcke, Caryophyllaceae; CY75	**Struθkia**	Young shoots	Cooked	+	Dwaylineh +		**Stroufouthkia, Strouthi**, Tsakridia +	Gıcır, Kuş otu, Serçe otu, Yumurta tu
Sinapis alba L. and *S. arvensis* L., Brassicaceae; CY25 and CY26	**Lapsana**	Young aerial parts	Boiled	+++			**Lapsana** +	**Lapsana**, Hardal

Table 1. *Cont.*

Botanical Taxon or Taxa, Botanical Family; Voucher Specimen Code (UVVETBOT)	Local Name(s)	Used Parts	Local Food Uses	Frequency of Quotation	Folk Names and Frequency of Quotation in Lebanon	Folk Names and Frequency of Quotation in Coastal Syria	Folk Names and Frequency of Quotation in Greek Cyprus	Folk Names in Turkish Cyprus
Sonchus oleraceus L., Asteraceae; CY12	Radiča, Şxex	Young aerial parts	In mixed salads or boiled	+		Khesaiseh, Asat alraa'i, Elk alghazal ++	**Sonchos**, Tsiofos, **Tsionchos** ++	
Taraxacum cyprium H.Lindb. and *T. hellenicum* Dahlst., Asteraceae; CY23 and Cr07	**Pikraradiča, Radiča** morren, Şxex	Young aerial parts	In mixed salads or boiled	+	Chissaisseh ++		**Agriodrakia, Agrioraditzia** +	Karahindiba
Thymus capitatus (L.) Hoffmanns. & Link, Lamiaceae; Cr13	**Zaatar**	Flowering tops and leaves	Seasoning	+	**Zaatar** +		Throuembi, Thymari +++	
Tordylium apulum L., Apiaceae; Cr02	Miriaθeo	Young aerial parts	Seasoning	++				

Frequency of quotation: +++: quoted by 40–100% of the study participants; ++: quoted by 10–39% of the study participants; +: quoted by less than 10% of the study participants. Folk linguistic cognates are reported in bold.

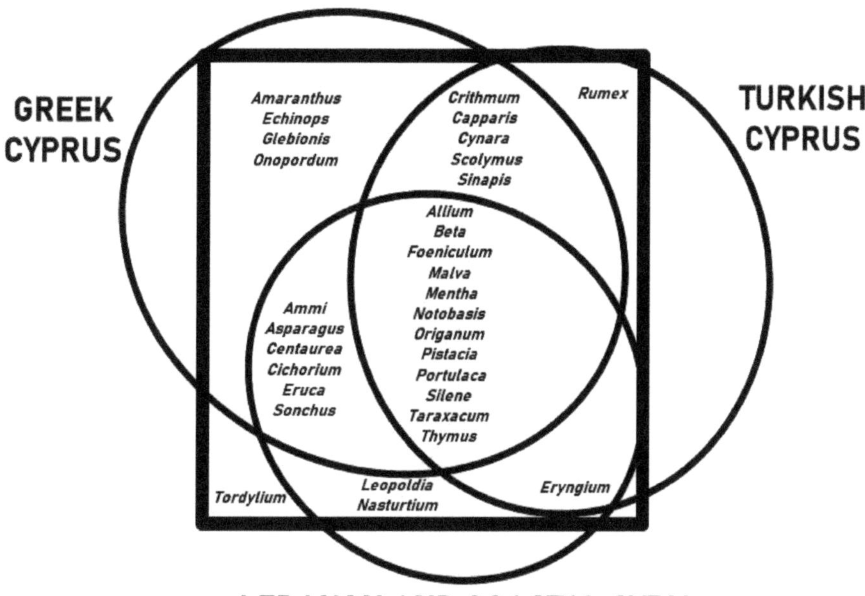

Figure 4. Venn diagram showing the overlap between the wild vegetables gathered and consumed in Kormakitis, and those recorded among Greek and Turkish Cypriots, and among Arabs in Lebanon and coastal Syria.

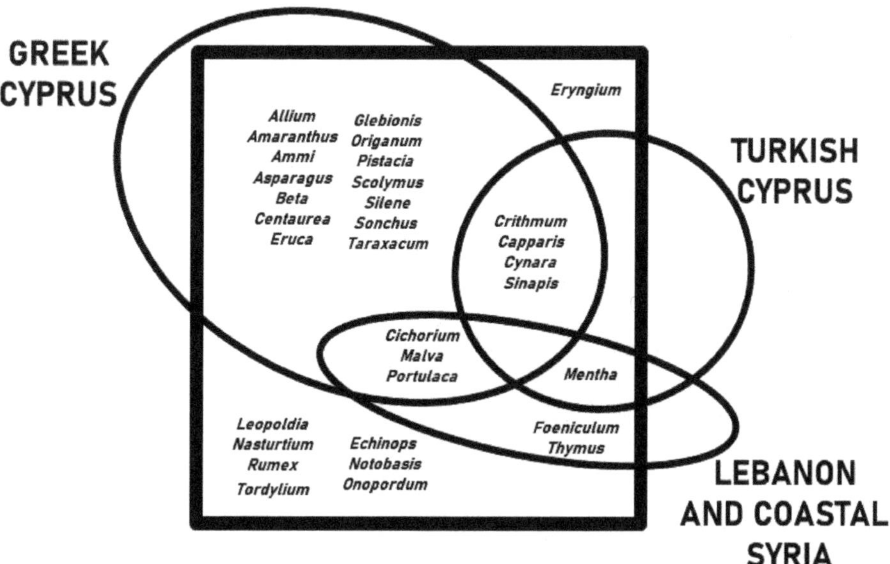

Figure 5. Venn diagram showing the overlaps between the folk names of wild vegetables in Kormakitis and those recorded among Greek and Turkish Cypriots, and among Arabs in Lebanon and coastal Syria.

4. Discussion

Wild vegetables in Kormakitis are grouped into a folk category expressed by the lexeme *Şxex*, which roughly corresponds to the Greek concept of *Chorta*; i.e., wild greens. Locals regard the bitter, weedy Asteraceae, i.e., *Cichorium*, *Taraxacum*, and *Sonchus* spp., as prototypical for *Şxex*. The local name *Şxex* could possess some Arabic roots as the word *Skhakh* (سخاخ) means "the free soft land", which somehow refers to the wild. The Maronites in coastal Syria refer to wild greens as *Mhabbleh*, which literally means "the steamed food" [29]; however, another local name for wild leafy vegetables in coastal Syria is *Sleeq*, which overlaps in some phonemes with *Şxex*. Nonetheless, the large majority of the WVs traditionally foraged by Maronites have Greek folk names and they overlap for the most part with the WVs previously reported to be gathered by Greek Cypriots. Only minor overlaps with Lebanese and Turkish Cypriot WV ethnobotanies were found; for instance, overlap with Arabic names was found in the species *Foeniculum vulgare*, *Thymus capitatus*, and *Mentha spicata*. On the other hand, Table 1 shows similarities in the local names of a few species (*Portulaca oleracea*, *Malva sylvestris*, *Cichorium intybus*) between our study participants in Kormakitis and the Greek minority in coastal Syria. The findings from Kormakitis, however, show a remarkable adaptation of possible original Maronite WV foraging customs to Greek Cypriot ones. This finding can be explained by the fact that the Maronites of Kormakitis lived together with their Greek neighbors for several centuries, although in the past intermarriages between the two communities were nearly impossible (Maronites are Catholic Christians, while Greeks are Orthodox Christians). This means that before the Turkish invasion of Northern Cyprus in 1974, original Maronite customs and thus, plant foraging, may have been "hellenicized".

In contrast, minimal influences from Turkish Cypriot WV ethnobotany may have played a role only recently, after Kormakitis and Northern Cyprus were occupied approximately five decades ago. While the use of *Rumex* was borrowed from Turkish Cyprus, they still maintained a distinct name for it, which may indicate they had some other need for differentiating this plant.

However, the custom of foraging seems to have disappeared in Kormakitis among the middle-aged generation; this, apart from global trends, may be due to the fact that after Turkish occupation, a remarkable migration/urbanization of Maronites from Kormakitis and other Northern Cypriot villages to the main Greek centers on the southern side of the isle took place, with extended families becoming transnational and commuting more or less regularly. Even now, individuals are moving back to the village; thus, further enhancing "urbanization" and the dilution of the village's foraging heritage.

Finally, the fact that the village has been, and continues to be, supported by the Republic of Cyprus (Greek Cypriot) authorities, who, via UN troops, deliver food (along with water, fuel, and medical supplies) across the border every two weeks, may have diluted the local need for foraged foods during the past few decades.

5. Conclusions

Plant biodiversity and the traditional/local ecological knowledge attached to it are essential for the holistic sustainability of "socio-ecological systems". This study suggests that the recorded Maronite wild vegetables could provide crucial baseline data for revitalizing local wild plant knowledge and practices in educational arenas, as well as their sustainable foraging, ideally aimed at promoting local food heritage and the community-centered management of plant resources.

Author Contributions: A.P. and R.S. planned and designed the research, collected data in the field, and identified the plant specimens; A.P. and N.S. performed the data analysis; A.P. drafted the first version of the manuscript, whose discussion was later improved and approved by Z.P., N.S. and R.S. All authors have read and agreed to the published version of the manuscript.

Funding: A.P. and R.S. fieldwork was supported by the University of Gastronomic Sciences, Pollenzo, Italy and of Ca' Foscari University of Venice; APC was funded by Ca' Foscari University of Venice;

editing was supported by the Internal Grant Agency of the Faculty of Tropical AgriSciences, Czech University of Life Sciences, Prague, Czech Republic (IGA FTZ, Project No. 20223104).

Institutional Review Board Statement: Not applicable.

Informed Consent Statement: Informed consent was obtained from all involved study participants, according to the ISE ethical guidelines [22].

Data Availability Statement: Not applicable.

Acknowledgments: Special thanks are due to all the marvelous Maronite inhabitants of Kormakitis; their endurance and resilience is inspiring.

Conflicts of Interest: The authors declare no conflict of interest.

References

1. Keys, A. *Seven Countries: A Multivariate Analysis of Death and Coronary Heart Disease*; Harvard University Press: Cambridge, MA, USA, 1980.
2. Willett, W.C.; Sacks, F.; Trichopoulou, A.; Drescher, G.; Ferro-Luzzi, A.; Helsing, E.; Trichopoulos, D. Mediterranean diet pyramid: A cultural model for healthy eating. *Am. J. Clin. Nutr.* **1995**, *61* (Suppl. S6), 1402S–1406S. [CrossRef]
3. UNESCO. Mediterranean Diet. Cyprus, Croatia, Spain, Greece, Italy, Morocco and Portugal. Inscribed in 2013 (8.COM) on the Representative List of the Intangible Cultural Heritage of Humanity. 2013. Available online: https://ich.unesco.org/en/RL/mediterranean-diet-00884 (accessed on 18 July 2022).
4. Dernini, S.; Berry, M.E. Mediterranean diet: From a healthy diet to a sustainable dietary pattern. *Front. Nutr.* **2015**, *7*, 2–15. [CrossRef]
5. Dernini, S.; Berry, E.M.; Serra-Majem, L.; La Vecchia, C.; Capone, R.; Medina, F.X.; Aranceta-Bartrina, J.; Belahsen, R.; Burlingame, B.; Calabrese, G.; et al. Med Diet 4.0: The Mediterranean diet with four sustainable benefits. *Public Health Nutr.* **2017**, *20*, 1322–13301. [CrossRef]
6. Leonti, M. The co-evolutionary perspective of the food-medicine continuum and wild gathered and cultivated vegetables. *Genet. Resour. Crop Evol.* **2012**, *59*, 1295–1302. [CrossRef]
7. Pieroni, A.; Sõukand, R.; Amin, H.I.M.; Zahir, H.; Kukk, T. Celebrating multi-religious co-existence in central Kurdistan: The bio-culturally diverse traditional gathering of wild greens among Yazidis, Assyrians, and Muslim Kurds. *Hum. Ecol.* **2018**, *46*, 217–227. [CrossRef]
8. Della, A.; Paraskeva-Hadjichambi, D.; Hadjichambis, A. An ethnobotanical survey of wild edible plants of Paphos and Larnaca countryside of Cyprus. *J. Ethnobiol. Ethnomedicine* **2006**, *2*, 34. [CrossRef] [PubMed]
9. Ciftcioglu, G.C. Sustainable wild-collection of medicinal and edible plants in Lefke region of North Cyprus. *Agrofor. Syst.* **2015**, *89*, 917–931. [CrossRef]
10. Pieroni, A.; Quave, C.L. Traditional pharmacopoeias and medicines among Albanians and Italians in southern Italy: A comparison. *J. Ethnopharmacol.* **2005**, *101*, 258–270. [CrossRef]
11. Pieroni, A.; Sõukand, R.; Bussmann, R.W. The Inextricable Link Between Food and Linguistic Diversity: Wild Food Plants among Diverse Minorities in Northeast Georgia, Caucasus. *Econ. Bot.* **2020**, *74*, 379–397. [CrossRef]
12. Vlková, M.; Kubátová, E.; Šlechta, P.; Polesný, Z. Traditional Use of Plants by the Disappearing Czech Diaspora in Romanian Banat. *Sci. Agric. Bohem.* **2018**, *46*, 49–56. [CrossRef]
13. Pieroni, A.; Cattero, V. Wild vegetables do not lie: Comparative gastronomic ethnobotany and ethnolinguistics on the Greek traces of the Mediterranean Diet of southeastern Italy. *Acta Bot. Bras.* **2019**, *33*, 198–211. [CrossRef]
14. Kalle, R.; Sõukand, R.; Pieroni, A. Devil Is in the Details: Use of Wild Food Plants in Historical Võromaa and Setomaa, Present-Day Estonia. *Foods* **2020**, *9*, 570. [CrossRef]
15. Mattalia, G.; Sõukand, R.; Corvo, P.; Pieroni, A. Dissymmetry at the Border: Wild Food and Medicinal Ethnobotany of Slovenes and Friulians in NE Italy. *Econ. Bot.* **2020**, *74*, 1–14. [CrossRef]
16. Aziz, M.A.; Abbasi, A.M.; Saeed, S.; Ahmed, A.; Pieroni, A. The Inextricable Link between Ecology and Taste: Traditional Plant Foraging in NW Balochistan, Pakistan. *Econ. Bot.* **2022**, *76*, 34–59. [CrossRef]
17. Sõukand, R.; Kalle, R.; Pieroni, A. Homogenisation of Biocultural Diversity: Plant Ethnomedicine and Its Diachronic Change in Setomaa and Võromaa, Estonia, in the Last Century. *Plants* **2022**, *11*, 192. [CrossRef]
18. Aziz, M.A.; Volpato, G.; Fontefrancesco, M.F.; Pieroni, A. (forthcoming). Perceptions and Revitalization of Local Ecological Knowledge in Four Schools in Yasin Valley, North Pakistan. *Mt. Res. Dev.* **2022**, *in press.*
19. Hourani, G. A Reading in the history of the Maronites of Cyprus from the eighth century to the beginning of British rule. *J. Maronite Stud.* **1998**, *2*, 1–16.
20. Tsoutsouki, C. Continuity of National Identity and Changing Politics: The Maronite Cypriots. In *The Minorities of Cyprus: Development Patterns and the Identity of the Internal-Exclusion*; Cambridge Scholars Publishing: Newcastle upon Tyne, UK, 2009; pp. 192–240.
21. Abouzayd, S. The Maronite Church. In *The Syriac World*; Taylor & Francis: London, UK, 2022; pp. 731–750.

22. International Society of Ethnobiology/ISE. Code of Ethics. 2006. Available online: https://www.ethnobiology.net/what-we-do/core-programs/ise-ethics-program/code-of-ethics/ (accessed on 19 July 2022).

23. Hand, R.; Hadjikyriakou, G.N.; Christodoulou, C.S. Flora of Cyprus—A Dynamic Checklist. 2011. Available online: http://www.flora-of-cyprus.eu/ (accessed on 2 June 2022).

24. Meikle, M.D. *Flora of Cyprus*; Bentham-Moxon Trustees: Kew, UK, 1977.

25. The World Flora Online/WFO. 2022. Available online: http://www.worldfloraonline.org/ (accessed on 20 July 2022).

26. Stevens, P.F. Angiosperm Phylogeny Website. Version 14. 2017. Available online: http://www.mobot.org/MOBOT/research/APweb/ (accessed on 23 July 2022).

27. Borg, A. *Cypriot Arabic: A Historical and Comparative Investigation into the Phonology and Morphology of the Arabic Vernacular Spoken by the Maronites of Kormakiti Village in the Kyrenia District of North-Western Cyprus*; Deutsche Morgenländische Gesellschaft: Stuttgart, Germany, 1985.

28. Marouf, M.; Batal, M.; Moledor, B.; Talhouk, S.N. Exploring the Practice of Traditional Wild Plant Collection in Lebanon. *Food Cult. Soc.* **2015**, *18*, 355–378. [CrossRef]

29. Sulaiman, N.; Pieroni, A.; Sõukand, R.; Polesny, Z. Food behavior in emergency time: Wild plant use for human nutrition during the conflict in Syria. *Foods* **2022**, *11*, 177. [CrossRef]

MDPI

St. Alban-Anlage 66

4052 Basel

Switzerland

www.mdpi.com

Plants Editorial Office

E-mail: plants@mdpi.com

www.mdpi.com/journal/plants